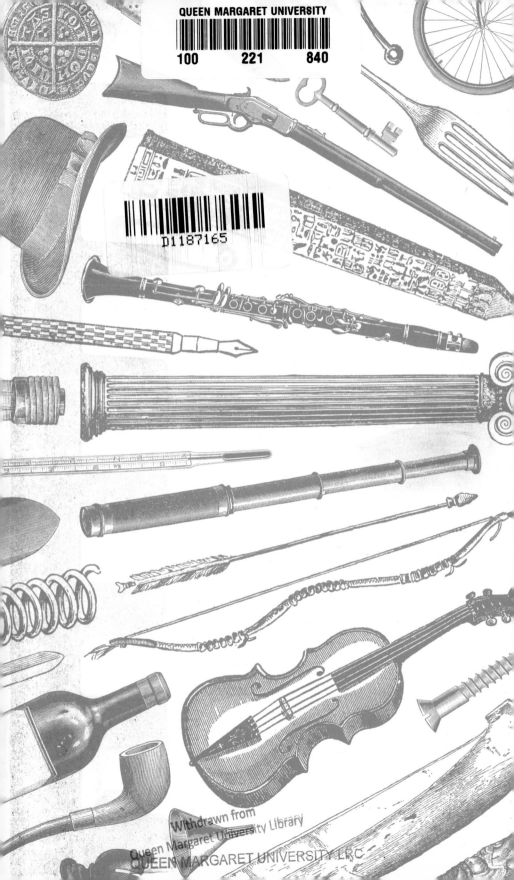

QUEEN MARGARET UNIVERSITY

100   221   840

D1187165

Withdrawn from
Queen Margaret University Library
QUEEN MARGARET UNIVERSITY LRC

More praise for *The Secret of Our Success*

"*The Secret of Our Success* provides a valuable new perspective on major issues in human evolution and behavior. Bringing together topics from such diverse areas as economics, psychology, neuroscience, and archaeology, this book will provoke vigorous debates and will be widely read."

—Alex Mesoudi, author of *Cultural Evolution*

"Is the ability to acquire highly evolved culture systems like languages and technologies the secret of humans' success as a species? This book convinces us that the answer is emphatically 'yes.' Moving beyond the sterile nature-nurture debates of the past, Joseph Henrich demonstrates that culture—as much a part of our biology as our legs—is an evolutionary system that works by tinkering with our innate capacities over time."

—Peter J. Richerson, University of California, Davis

"In the last decade, in the interstices between biology, anthropology, economics, and psychology, a remarkable new approach to explaining the development of human societies has emerged. It's the most important intellectual innovation on this topic since Douglass North's work on institutions in the 1970s and it will fundamentally shape research in social science in the next generation. This extraordinary book is the first comprehensive statement of this paradigm. You'll be overwhelmed by the breadth of evidence and the creativity of ideas. I was."

—James Robinson, coauthor of *Why Nations Fail*

"With compelling chapter and verse and a very readable style, Joseph Henrich's book makes a powerful argument—in the course of the gene-culture coevolution that has made us different from other primates, culture, far from being the junior partner, has been the driving force. A terrific book that shifts the terms of the debate."

—Stephen Shennan, University College London

"A delightful and engaging expedition into and all around the many different processes of genetic and cultural evolution that have made humans such 'a puzzling primate.'"

—Michael Tomasello, codirector of the Max Planck Institute for Evolutionary Anthropology

THE
SECRET
OF OUR
SUCCESS

How Culture Is Driving Human Evolution,
Domesticating Our Species,
and Making Us Smarter

JOSEPH HENRICH

PRINCETON UNIVERSITY PRESS
Princeton & Oxford

Copyright © 2016 by Joseph Henrich

Requests for permission to reproduce material from this work
should be sent to Permissions, Princeton University Press

Published by Princeton University Press
41 William Street, Princeton, New Jersey 08540

In the United Kingdom: Princeton University Press
6 Oxford Street, Woodstock, Oxfordshire OX20 1TW

press.princeton.edu

Jacket art courtesy of ClipArt ETC at the Florida Center for Instructional
Technology and *Victorian Goods and Merchandise: 2,300 Illustrations*, selected and
arranged by Carol Belanger Grafton, © 1997 by Dover Publications, Inc.

All Rights Reserved

ISBN 978-0-691-16685-8

Library of Congress Control Number: 2015934779

British Library Cataloging-in-Publication Data is available

This book has been composed in Sabon Next LT Pro and DIN Pro

Printed on acid-free paper. ∞

Printed in the United States of America

10 9 8 7 6 5 4 3 2 1

For Jessica, Joshua, and Zoey

CONTENTS

PREFACE

We humans are not like other animals. Sure, we are obviously similar to monkeys and other apes in many ways, but we also variously play chess, read books, build missiles, enjoy spicy dishes, donate blood, cook food, obey taboos, pray to gods, and make fun of people who dress or speak differently. And though all societies make fancy technologies, follow rules, cooperate on large scales, and communicate in complex languages, different societies do all this in very different ways and to significantly different degrees. How could evolution have produced such a creature, and how does answering this question help us understand human psychology and behavior? How can we explain both cultural diversity and human nature?

My journey to addressing these questions, and writing this book, began in 1993 when I quit my engineering position at Martin Marietta, near Washington DC, and drove to California, where I enrolled as a graduate student in UCLA's Department of Anthropology. I had two interests at the time, which I'd developed while pursuing undergraduate degrees in both anthropology and aerospace engineering at the University of Notre Dame. One interest focused on understanding economic behavior and decision-making in the developing world, with the idea that new insights might help improve people's lives around the globe. In part, I was attracted to anthropology because the research involved in-depth and long-term fieldwork, which I felt had to be crucial to understanding people's decisions and behavior, and the challenges they faced. This was my "applied" focus. Intellectually, I was also keenly interested in the evolution of human societies, particularly in the basic question of how humans went from living in relatively small-scale societies to complex nation-states over the last ten millennia.

The plan was to study with two well-known anthropologists, one a sociocultural anthropologist and ethnographer named Allen Johnson, and the other an archaeologist named Tim Earle.

After a summer of research in Peru, traveling by dugout canoe among indigenous Matsigenka communities in Amazonia, I wrote my master's thesis on the effects of market integration on farming decisions and deforestation. Things were going fine, my advisors were happy (though Tim had departed for another university), and my thesis was accepted.

Nevertheless, I was dissatisfied with what anthropology had to offer for explaining why the Matsigenka were doing what they were doing. For starters, why were Matsigenka communities so different from the nearby indigenous Piro communities, and why did they seem to have subtly adaptive practices that they themselves couldn't explain?

I considered bailing out of anthropology at this point and heading back to my old engineering job, which I had quite liked. However, during the previous few years I'd gotten excited about human evolution. I had also enjoyed studying human evolution at Notre Dame, but I hadn't seen how it could help me with explaining either economic decision-making or the evolution of complex societies, so I'd thought of it more as a hobby. At the beginning of graduate school, to narrowly focus my energies on my main interests, I tried to get out of taking the required graduate course on human evolution. To do this, I had to appeal to the instructor of the graduate course in biological anthropology, Robert Boyd, and argue to him that my undergraduate work met the course requirements. I'd already successfully done this for the required sociocultural course. Rob was very friendly, looked carefully over the classes I had taken, and then denied the request. If Rob hadn't denied my request, I suspect I'd be back doing engineering right now.

It turned out that the field of human evolution and biological anthropology was full of ideas one could use to explain important aspects of human behavior and decision-making. Moreover, I learned that Rob and his long-time collaborator, the ecologist Pete Richerson, had been working on ways to model culture using mathematical tools from population genetics. Their approach also allowed one to think systematically about how natural selection might have shaped human learning abilities and psychology. I didn't know any population genetics, but because I knew about state variables, differential equations, and stable equilibria (I was an aerospace engineer), I could more or less read and understand their papers. By the end of my first year, working on a side project under Rob's guidance, I'd written a MATLAB program to study the evolution of conformist transmission (you'll hear more about this in chapter 4).

Entering my third year, with a master's under my belt, I decided to go back to the drawing board—to start over, in a sense. I consciously took a "reading year," though I knew it would extend my time to the PhD by one year. You could probably get away with this only in a department of anthropology. I had no classes to take, no advisors to work for, and no one really seemed to care what I was doing. I started by going to the library to take out a stack of books. I read books on cognitive psychology, decision-making, experimental economics, biology, and evolutionary psychology. Then I moved to journal articles. I read every article ever written on an economics experiment called the Ultimatum Game, which I'd used during my second and third summers with the Matsigenka. I also read a lot by the psychologists Daniel Kahneman and Amos Tversky, as well as by a political scientist named Elinor Ostrom. Kahneman and Ostrom would, years later, both receive Nobel Prizes in economics. Of course, along the way, I never stopped reading anthropological ethnographies (this was my "fun" reading). In many ways, that year was the first year of research on this book, and by the end of it, I had developed a murky vision for what I wanted to do. The goal was to integrate insights from across the social and biological sciences to build an evolutionary approach to studying human psychology and behavior that takes seriously the cultural nature of our species. We needed to harness the full arsenal of available methods, including experiments, interviews, systematic observation, historical data, physiological measures, and rich ethnography. We had to study people, not in university laboratories, but in their communities and over their life course (from babies to the elderly). From this vantage point, disciplines like anthropology, and especially subdisciplines like economic anthropology, began to look small and insular.

Of course, Boyd and Richerson, building on work by Marc Feldman and Luca Cavalli-Sforza, had already laid down some of the key theoretical foundations in their 1985 book, *Culture and the Evolutionary Process.* However, in the mid-1990s there was still essentially no program of empirical research, no toolbox of methods, and no established ways of testing the theories generated by the evolutionary models. Moreover, the existing ideas about psychological processes hadn't been developed very far or in ways that easily connected with the rising intellectual tides in cultural or evolutionary psychology, neuroscience, or even with the scientific wings of cultural anthropology.

During this time, two new graduate students had arrived to work with Rob Boyd: Francisco Gil-White and Richard McElreath (now a director at the Max Planck Institute for Evolutionary Anthropology). A little later, Natalie Smith (now Natalie Henrich) moved over from archeology to work with Rob on cooperation. Suddenly, I was no longer alone; I had like-minded friends and collaborators with shared interests. This was an exciting and fast moving time, as new ideas and fresh intellectual avenues seemed to be bubbling up all over the place. It felt like someone had suddenly taken the brakes off and removed the stops. Rob and I were assembling a team of field ethnographers and economists to conduct behavioral experiments around the globe and study human sociality. This was virtually unheard of, since ethnographers do not work in teams and they certainly don't (or, didn't) use economic games. Based on my first experiments in Peru, I sent off a paper entitled "Does Culture Matter in Economic Behavior?" to a journal I'd found in the library called the *American Economic Review*. As an anthropology graduate student, I had no idea this was the top journal in economics, or how skeptical economists were at that time about culture. Meanwhile, Francisco was importing methods from developmental psychology to test his ideas about folk sociology and ethnicity (see chapter 11) among herders in Mongolia. Natalie and I invented the Common Pool Resources (CPR) Game to study conservation behavior in Peru. (To our dismay, we later learned that it had already been invented.) Richard was writing computer programs to create and study "cultural phylogenies," something no one had done, and was discussing with a Caltech economist named Colin Camerer how to use computer-based experimental techniques to test theories of social learning. Francisco and I came up with a new theory of human status over coffee one morning (see chapter 8). And, inspired by reading the diffusion of innovations literature in sociology, I started wondering if it was possible to detect a "signature" of cultural learning from data on the diffusion of new ideas and technologies over time. Several of these early efforts later became substantial research endeavors in various disciplines.

It has now been twenty years since this began for me in 1995, so I'm laying down this book as a waymark, a work in progress. I'm more convinced than ever that to understand our species and to build a science of human behavior and psychology, we need to begin with an evolutionary theory of human nature. Getting this at least partially right is paramount to taking the next step. Recently, I've been particularly encouraged by this year's World Development Report 2015, which is entitled *Mind,*

Society, and Behavior. This World Bank document emphasizes how critical it is to recognize that people are automatic cultural learners, that we follow social norms, and that the cultural worlds we grow up in influence what we attend to, perceive, process, and value. This is a long way from my old paper's simple query, "Does Culture Matter in Economic Behavior?" The economists at the World Bank are apparently now convinced that it does.

In putting this book together, my intellectual and personal debts run deep. First and foremost, this book owes much to the running intellectual conversation between me and my wife, Natalie. Natalie read every chapter at least once and provided critical feedback throughout. It's only after she reads my work that anyone else can look at it.

Many at UCLA contributed greatly to this effort. Of course, my longtime collaborator, mentor, and friend, Rob Boyd, is central to this book and deserves many thanks for his decades of help and counsel. Rob Boyd read an early draft and provided great feedback. Similarly, Allen Johnson provided helpful comments on some of the very earliest drafts of several chapters. Allen brought me to UCLA, advised me, trained me in ethnography, and provided me with great freedom during my graduate school career. My thanks also go to Joan Silk, whose sage advice on many topics and deep knowledge of primates I still routinely rely on.

Over my career, I've been incredibly fortunate to have been a faculty member in the Department of Anthropology at Emory University (four years) and in both the Department of Psychology and Department of Economics at the University of British Columbia (nine years). I also spent two years in the Society of Scholars in the Business School at the University of Michigan, and one year at the Institute for Advanced Study in Berlin. This in-depth fieldwork gave me the rare opportunity to see the social sciences from the point of view of psychologists, sociologists, anthropologists, and economists. In particular, at the University of British Columbia (UBC), I developed a number of key collaborations with many of my psychology colleagues, including Steve Heine, Ara Norenzayan, Jessica Tracy, Sue Birch, and Kiley Hamlin, as well as learning much from Greg Miller and Edith Chen. Steve and Jess provided helpful feedback on early drafts of this book.

My students, former students, and members of UBC's Laboratory for Mind, Evolution, Cognition and Culture (MECC) deserve many thanks. In particular, thanks to Maciek Chudek, Michael Muthukrishna, Rita McNamara, James and Tanya Broesch, Cristina Moya, Ben Purzycki, Taylor Davis, Dan Hruschka, Rahul Bhui, Aiyana Willard, and Joey

Cheng. The products of our collaborations appear throughout this book. Michael and Rita provided useful comments on early drafts.

My co-directors at UBC's Centre for Human Evolution, Cognition and Culture played an important role in moving these ideas along. Conversations with the evolutionary anthropologist Mark Collard and sinologist-turned-cognitive-scientist Ted Slingerland are always stimulating, and Ted provided many useful comments on the near-final version of the manuscript.

During a crucial phase of my writing, New York University's Stern Business School generously provided me with a fellowship for 2013–2014. During this time, I learned a great deal from the psychologist Jon Haidt, the economist Paul Romer, and the philosopher Steve Stich. All three provided excellent feedback at various stages of my writing. I was also lucky to co-teach an MBA course with Jon, where I tried out several of the chapters below on future business leaders.

While working on this book I was fortunate to be a fellow at the Canadian Institute for Advanced Research (CIFAR), as a member of the Institutions, Organizations and Growth Group. This group was incredibly stimulating and supportive, and I learned much from many of its members. In particular, Suresh Naidu provided great comments on an early draft of this book.

As an ethnographer, I've been fortunate to have lived and worked among three quite different groups: Matsigenka in Peru, Mapuche in southern Chile, and Fijians on Yasawa Island in the South Pacific. In each place, many families shared their homes and lives with me, answered my endless questions, and deepened my understanding of human diversity. Special thanks goes to these folks.

In developing various ideas for this book, I often queried the authors or experts I drew on. Along the way, I got helpful replies from Daron Acemoglu, Siwan Anderson, Coren Apicella, Quentin Atkinson, Clark Barrett, Peter Blake, Monique Borgerhoff Mulder, Sam Bowles, Josep Call, Colin Camerer, Nicholas Christakis, Mort Christiansen, Alyssa Crittenden, Yarrow Dunham, Nick Evans, Dan Fessler, Jim Fearon, Ernst Fehr, Patrick Francois, Simon Gächter, Josh Greene, Avner Greif, Paul Harris, Ester Herrmann, Barry Hewlett, Kim Hill, Dan Hruschka, Erik Kimbrough, Michelle Kline, Kevin Laland, Jon Lanman, Cristine Legare, Hannah Lewis, Dan Lieberman, Johan Lind, Frank Marlowe, Sarah Mathew, Richard McElreath, Joel Mokyr, Tom Morgan, Nathan Nunn, David Pietraszewski, David Rand, Peter Richerson, James Robinson, Carel van Schaik, Joan Silk, Mark Thomas, Mike Tomasello, Peter Turchin, Carel van Schaik, Felix Warneken, Janet Werker, Annie Wertz, Polly Wiessner, David Sloan Wilson, Harvey Whitehouse, Andy Whiten, and Richard Wrangham (as well as many others, including those already mentioned above).

Over the years in planning and writing this book, I had conversations with an immense number of friends, coauthors, and colleagues who have shaped my thinking. This is my collective brain (see chapter 12).

Joe Henrich
22 January 2015
Vancouver, Canada

THE SECRET OF OUR SUCCESS

CHAPTER 1

A PUZZLING PRIMATE

You and I are members of a rather peculiar species, a puzzling primate.

Long before the origins of agriculture, the first cities, or industrial technologies, our ancestors spread across the globe, from the arid deserts of Australia to the cold steppe of Siberia, and came to inhabit most of the world's major land-based ecosystems—more environments than any other terrestrial mammal. Yet, puzzlingly, our kind are physically weak, slow, and not particularly good at climbing trees. Any adult chimp can readily overpower us, and any big cat can easily run us down, though we are oddly good at long-distance running and fast, accurate throwing. Our guts are particularly poor at detoxifying poisonous plants, yet most of us cannot readily distinguish the poisonous ones from the edible ones. We are dependent on eating cooked food, though we don't innately know how to make fire or cook. Compared to other mammals of our size and diet, our colons are too short, stomachs too small, and teeth too petite. Our infants are born fat and dangerously premature, with skulls that have not yet fused. Unlike other apes, females of our kind remain continuously sexually receptive throughout their monthly cycle and cease reproduction (menopause) long before they die. Perhaps most surprising of all is that despite our oversized brains, our kind are not that bright, at least not innately smart enough to explain the immense success of our species.

Perhaps you are skeptical about this last point?

Suppose we took you and forty-nine of your coworkers and pitted you in a game of Survivor against a troop of fifty capuchin monkeys

1

from Costa Rica. We would parachute both primate teams into the remote tropical forests of central Africa. After two years, we would return and count the survivors on each team. The team with the most survivors wins. Of course, neither team would be permitted to bring any equipment: no matches, water containers, knives, shoes, eyeglasses, antibiotics, pots, guns, or rope. To be kind, we would allow the humans—but not the monkeys—to wear clothes. Both teams would thus face surviving for years in a novel forest environment with only their wits, and their teammates, to rely on.

Who would you bet on, the monkeys or you and your colleagues? Well, do you know how to make arrows, nets, and shelters? Do you know which plants or insects are toxic (many are) or how to detoxify them? Can you start a fire without matches or cook without a pot? Can you manufacture a fishhook? Do you know how to make natural adhesives? Which snakes are venomous? How will you protect yourself from predators at night? How will you get water? What is your knowledge of animal tracking?

Let's face it, chances are your human team would lose, and probably lose badly, to a bunch of monkeys, despite your team's swollen crania and ample hubris. If not for surviving as hunter-gatherers in Africa, the continent where our species evolved, what are our big brains for anyway? How did we manage to expand into all those diverse environments across the globe?

The secret of our species' success lies not in our raw, innate, intelligence or in any specialized mental abilities that fire up when we encounter the typical problems that repeatedly challenged our hunter-gatherer ancestors in the Pleistocene. Our ability to survive and thrive as hunter-gatherers, or anything else, across an immense range of global environments is not due to our individual brainpower applied to solving complex problems. As you will see in chapter 2, stripped of our culturally acquired mental skills and know-how, we are not so impressive when we go head-to-head in problem-solving tests against other apes, and we certainly are not impressive enough to account for the vast success of our species or for our much larger brains.[1]

In fact, we have seen various versions of the human half of our Survivor experiment many times, as hapless European explorers have struggled to survive, stranded in seemingly hostile environments, from the Canadian Arctic to the Gulf Coast of Texas. As chapter 3 shows, these cases usually end in the same way: either the explorers all die, or some of them are rescued by a local indigenous population, which has comfortably been living in this "hostile environment" for centuries or millennia. Thus, the reason why your team would lose to the monkeys is that your species—unlike all others—has evolved an addiction to culture. By "culture" I mean the large body of practices, techniques, heuristics, tools, motivations, values, and beliefs that we all acquire while growing up, mostly by learning from other people. Your team's only hope is that you might bump into, and befriend, one of the groups of hunter-gatherers who live in the central African forests, like the Efe pygmies. These pygmy groups, despite their short stature, have been flourishing in these forests for a very long time because past generations have bequeathed to them an immense body of expertise, skills, and abilities that permit them to survive and thrive in the forest.

The key to understanding how humans evolved and why we are so different from other animals is to recognize that we are a *cultural species*. Probably over a million years ago, members of our evolutionary lineage began learning from each other in such a way that culture became cumulative. That is, hunting practices, tool-making skills, tracking know-how, and edible-plant knowledge began to improve and aggregate—by learning from others—so that one generation could build on and hone the skills and know-how gleaned from the previous generation. After several generations, this process produced a sufficiently large and complex toolkit of practices and techniques that individuals, relying only on their own ingenuity and personal experience, could not get anywhere close to figuring out over their lifetime. We will see myriad examples of such complex cultural packages, from Inuit snow houses, Fuegian arrows, and Fijian fish taboos to numerals, writing, and the abacus.

Once these useful skills and practices began to accumulate and improve over generations, natural selection had to favor individuals who were better cultural learners, who could more effectively tap in to and use the ever-expanding body of adaptive information available. The newly produced products of this cultural evolution, such as fire, cooking, cutting tools, clothing, simple gestural languages, throwing spears, and water containers, became the sources of the main selective pressures that genetically shaped our minds and bodies. This interaction between culture and genes, or what I'll call *culture-gene coevolution*, drove our species down a novel evolutionary pathway not observed elsewhere in nature, making us very different from other species—a new kind of animal.

However, recognizing that we are a cultural species only makes an evolutionary approach even more important. As you'll soon see in chapter 4, our capacities for learning from others are themselves finely honed products of natural selection. We are adaptive learners who, even as infants, carefully select when, what, and from whom to learn. Young learners all the way up to adults (even MBA students) automatically and unconsciously attend to and preferentially learn from others based on cues of prestige, success, skill, sex, and ethnicity. From other people we readily acquire tastes, motivations, beliefs, strategies, and our standards for reward and punishment. Culture evolves, often invisibly, as these selective attention and learning biases shape what each person attends to, remembers, and passes on. Nevertheless, these cultural learning abilities gave rise to an interaction between an accumulating body of cultural information and genetic evolution that has shaped, and continues to shape, our anatomy, physiology, and psychology.

Anatomically and physiologically, the escalating need to acquire this adaptive cultural information drove the rapid expansion of our brains, giving us the space to store and organize all this information, while creating the extended childhoods and long postmenopausal lives that give us the time to acquire all this know-how and the chance to pass it on. Along the way, we'll see that culture has left its marks all over our bodies, shaping the genetic evolution of our feet, legs, calves, hips, stomachs, ribs, fingers, ligaments, jaws, throats, teeth, eyes, tongues, and much more. It has also made us powerful throwers and long-distance runners who are otherwise physically weak and fat.

Psychologically, we have come to rely so heavily on the elaborate and complicated products of cultural evolution for our survival that we now often put greater faith in what we learn from our communities than in our own personal experiences or innate intuitions. Once we understand our reliance on cultural learning, and how cultural evolution's subtle selective processes can produce "solutions" that are smarter than we are, otherwise puzzling phenomena can be explained. Chapter 6 illustrates this point by tackling questions such as, Why do people in hot climates tend to use more spices and find them tastier? Why did aboriginal Americans commonly put burnt seashells or wood ash into their cornmeal? How could ancient divination rituals effectively implement game theoretic strategies to improve hunting returns?

The growing body of adaptive information available in the minds of other people also drove genetic evolution to create a second form of human status, called prestige, which now operates alongside the dominance status we inherited from our ape ancestors. Once we understand prestige, it will become clear why people unconsciously mimic more successful individuals in conversations; why star basketball players like LeBron James can sell car insurance; how someone can be famous for being famous (the Paris Hilton Effect); and, why the most prestigious participants should donate first at charity events but speak last in decision-making bodies, like the Supreme Court. The evolution of prestige came with new emotions, motivations, and bodily displays that are distinct from those associated with dominance.

Beyond status, culture transformed the environments faced by our genes by generating social norms. Norms influence a vast range of human action, including ancient and fundamentally important domains such as kin relations, mating, food sharing, parenting, and reciprocity. Over our evolutionary history, norm violations such as ignoring a food taboo, botching a ritual, or failing to give one's in-laws their due from one's hunting successes meant reputational damage, gossip, and a consequent loss of marriage opportunities and allies. Repeated norm violations sometimes provoked ostracism or even execution at the hands of one's community. Thus, cultural evolution initiated a process of *self-domestication*, driving genetic evolution to make us prosocial, docile, rule followers who expect a world governed by social norms monitored and enforced by communities.

Understanding the process of self-domestication will allow us to address many key questions. In chapters 9 to 11, we'll explore questions such as, How did rituals become so psychologically potent, capable of solidifying social bonds and fostering harmony in communities? How do marriage norms make better fathers and expand our family networks? Why is our automatic and intuitive response to stick to a social norm, even if that means paying a personal cost? Similarly, when and why does careful reflection cause greater selfishness? Why do people who wait for the "walk signal" at traffic lights also tend to be good cooperators? What was the psychological effect of World War II on America's Greatest Generation? Why do we prefer to interact with, and learn from, those who speak the same dialect as we do? How did our species become the most social of primates, capable of living in populations of millions, and at the same time, become the most nepotistic and warlike?

The secret of our species' success resides not in the power of our individual minds, but in the *collective brains* of our communities. Our collective brains arise from the synthesis of our cultural and social natures— from the fact that we readily learn from others (are *cultural*) and can, with the right norms, live in large and widely interconnected groups (are *social*). The striking technologies that characterize our species, from the kayaks and compound bows used by hunter-gatherers to the antibiotics and airplanes of the modern world, emerge not from singular geniuses but from the flow and recombination of ideas, practices, lucky errors, and chance insights among interconnected minds and across generations. Chapter 12 shows how it's the centrality of our collective brains that explains why larger and more interconnected societies produce fancier technologies, larger toolkits, and more know-how, and why when small communities suddenly become isolated, their technological sophistication and cultural know-how begins to gradually ebb away. As you'll see, innovation in our species depends more on our sociality than on our intellect, and the challenge has always been how to prevent communities from fragmenting and social networks from dissolving.

Like our fancy technologies and complex sets of social norms, much of the power and elegance of our languages come from cultural evolution, and the emergence of these communication systems drove much of our genetic evolution. Cultural evolution assembles and adapts our communicative repertoires in ways similar to how it constructs and adapts other aspects of culture, such as the making of a complicated tool or the performance of an intricate ritual. Once we understand that languages are products of cultural evolution, we'll be able to ask a variety of new questions such as, Why are languages from people in warmer climates more sonorous? Why do languages with larger communities of speakers have more words, more sounds (phonemes), and more grammatical tools? Why is there such a difference between the languages of small-scale societies and those that now dominate the modern world? In the longer run, the presence of such culturally evolved communicative repertoires created the genetic selective pressures that drove our larynx (the voice box) down, whitened our eyes, and endowed us with a birdlike propensity for vocal mimicry.

Of course, all these products of cultural evolution, from words to tools, do indeed make us individually smarter, or at least mentally better equipped to thrive in our current environments (so, "smarter"). You, for example, probably received a massive cultural download while growing up that included a convenient base-10 counting system, handy Arabic numerals for easy representation, a vocabulary of at least 60,000 words (if you are a native English speaker), and working examples of the concepts surrounding pulleys, springs, screws, bows, wheels, levers, and adhesives. Culture also provides heuristics, sophisticated cognitive skills like reading, and mental prostheses like the abacus that have evolved culturally to both fit, and to some degree, modify our brains and biology. However, as you'll see, we don't have these tools, concepts, skills, and heuristics because our species is smart; we are smart because we have culturally evolved a vast repertoire of tools, concepts, skills, and heuristics. Culture makes us smart.

Besides driving much of our species genetic evolution and making us (somewhat) "self-programmable," culture has woven itself into our biology and psychology in other ways. By gradually selecting institutions, values, reputational systems, and technologies over the eons, cultural evolution has influenced the development of our brains, hormonal responses, and immune reactions, as well as calibrating our attention, perceptions, motivations, and reasoning processes to better fit the diverse culturally constructed worlds in which we grow up. As we'll see in chapter 14, culturally acquired beliefs alone can change pain into pleasure, make wine more (or less) enjoyable, and, in the case of Chinese astrology, alter the length of believers' lives. Social norms, including those contained in languages, effectively supply training regimes that shape our brains in various ways, ranging from expanding our hippocampus to thickening our corpus callosum (the information highway that connects the two halves of our brains). Even without influencing genes, cultural evolution creates both psychological and biological differences between populations. You, for example, have been altered biologically by the aforementioned cultural download of skills and heuristics.

In the chapter 17, I'll explore how this view of our species changes how we think about several key questions:

1. What makes humans unique?
2. Why are humans so cooperative compared to other mammals?
3. Why do societies vary so much in their cooperativeness?
4. Why do we seem so smart compared to other animals?
5. What makes societies innovative, and how will the Internet influence this?
6. Is culture still driving genetic evolution?

The answers to these questions alter how we think about the interface of culture, genes, biology, institutions, and history and how we approach human behavior and psychology. This approach also has important practical implications for how we build institutions, design policies, address social problems, and understand human diversity.

CHAPTER 2

IT'S NOT OUR INTELLIGENCE

Humans have altered more than one-third of the earths' land surface. We cycle more nitrogen than all other terrestrial life forms combined and have now altered the flow of two-thirds of the earth's rivers. Our species uses 100 times more biomass than any large species that has ever lived. If you include our vast herds of domesticated animals, we account for more than 98% of terrestrial vertebrate biomass.[1]

Such facts leave little doubt that we are the ecologically dominant species on the planet.[2] Yet, they open the question of *why us?* What explains our species' ecological dominance? What is the secret of our success?

To pursue these questions let's begin by putting aside the hydroelectric dams, mechanized agriculture, and aircraft carriers of the modern world, along with the steel plows, massive tombs, irrigation works, and grand canals of the ancient world. We need to go back long before industrial technology, cities, and farming to understand how a particular tropical primate managed to spread across the globe.

Not only did ancient hunter-gatherers expand into most of the earth's terrestrial ecosystems, we probably also contributed to the extinction of much of its megafauna—that is, to the extinction of large vertebrates like mammoths, mastodons, giant deer, woolly rhinos, immense ground sloths, and giant armadillos, as well as some species of elephants, hippos, and lions. While climatic shifts were also likely contributors to these extinctions, the disappearance of many megafaunal species eerily coincides with the arrival of humans on different continents and large is-

8

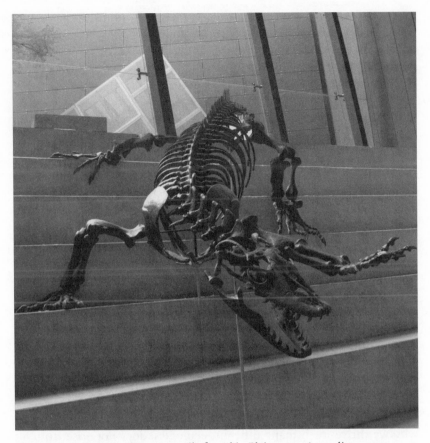

Figure 2.1. A giant, carnivorous reptile found in Pleistocene Australia.

lands. For example, before we showed up in Australia around 60,000 years ago, the continent was home to a menagerie of large animals, including two-ton wombats, immense meat-eating lizards (see figure 2.1), and leopard-sized marsupial lions. These, along with 55 other megafaunal species, went extinct in the wake of our arrival, resulting in the loss of 88% of Australia's big vertebrates. Tens of thousands of years later, when humans finally arrived in the Americas, 83 genera of megafauna went extinct, including horses, camels, mammoths, giant sloths, lions, and dire wolves, representing a loss of over 75% of the existing megafauna. Similar patterns emerged when humans arrived at different times in Madagascar, New Zealand, and the Caribbean.

Sharpening the point, the megafauna of Africa, and to a lesser extent of Eurasia, fared much better, probably because these species had long coevolved over hundreds of thousands of years with humans, including both our direct ancestors and our evolutionary cousins like the Neanderthals. African and Eurasian megafauna evolved to recognize that while we appear rather unintimidating, and perhaps easy prey given our lack of claws, canines, venom, and speed, we come with a dangerous bag of tricks, including projectiles, spears, poisons, snares, fire, and cooperative social norms that make us a top predator.[3] It's not just the fault of industrialized societies; our species' ecological impacts have a deep history.[4]

Other species have also spread widely and achieved immense ecological success; however, this success has generally occurred by speciation, as natural selection has adapted and specialized organisms to survive in different environments. Ants, for example, capture an equivalent biomass to modern humans, making them the most dominant of terrestrial invertebrates. To accomplish this, ant lineages have split, genetically adapted, and specialized into more than 14,000 different species with vast and complicated sets of genetic adaptations.[5] Meanwhile, humans remain a single species and show relatively little genetic variation, especially when the diverse range of environments we inhabit is considered. We have, for example, much less genetic variation than chimpanzees and show no signs of splitting into subspecies. By contrast, chimpanzees remain confined to a narrow band of tropical African forest and have already diverged into three distinct subspecies.[6] As will become clear in chapter 3, the manner in which we adapt to diverse environments, and why we thrive in so many different ecologies, does not arise from an array of environment-specific genetic adaptations, as with most species.

If not a dizzying array of genetic adaptations, what is the secret of our species' success? Most would agree that it traces, at least in part, to our ability to manufacture locally appropriate tools, weapons, and shelters, as well as to control fire and harness diverse food sources, like honey, game, fruit, roots, and nuts. Many researchers also point to our cooperative abilities and diverse forms of social organization.[7] Human hunter-gatherers all cooperate intensively within families and to some degree on larger scales ranging from a few families, called bands, to tribes of thousands. These forms of social organization vary on a bewildering array of dimensions, with differing rules for group membership/identity (e.g., tribal groups), marriage (e.g., cousin marriage, see chapter 9), exchange, sharing, ownership, and residence. Just considering hunter-gatherers, our species has more forms of social organization than the rest of the primate order combined.

While broadly on target, these observations succeed only in pushing the question back to how and why humans are capable of creating the necessary tools, techniques, and organizational forms to adapt to, and thrive in, such a diversity of environments. Why can't other animals achieve this?

The most common answer is that we are simply more intelligent. We have big brains with ample cognitive processing power and other souped-up mental abilities (e.g., greater working memory) that allow us to figure out how to solve problems creatively. The world's leading evolutionary psychologists, for example, have argued that humans evolved "improvisational intelligence," which allows us to formulate causal models of how the world works. These models then permit us to devise useful tools, tactics, and stratagems "on the fly." In this view, an individual faced with an environmental challenge, say hunting birds, throws his big primate brain into overdrive, figures out that wood can store elastic energy (a causal model), and then goes on to craft bows, arrows, and spring traps to get those birds.[8]

An alternative, though perhaps complementary, view is that our big brains are full of genetically endowed cognitive abilities that have emerged via natural selection to solve the most important and recurrent problems faced by our hunter-gatherer ancestors. These problems are often thought of as relating to specific domains, such as finding food, water, mates, and friends, as well as avoiding incest, snakes, and disease. When cued by environmental circumstances, these cognitive mechanisms take in problem-specific information and deliver solutions. The psychologist Steve Pinker, for example, has long argued that we are smarter and more flexible "not because we have fewer instincts than other animals; it is because we have more."[9] This view suggests that since our species has long relied on tracking and hunting, we might have evolved psychological specializations that fire up and endow us with tracking or hunting abilities (as cats have) when we are placed in the right circumstances.

A third common approach to explaining our species' ecological dominance focuses on our prosociality, our abilities to cooperate intensively across many different domains and extensively in large groups. Here the idea is that natural selection made us highly social and cooperative, and then by working together we conquered the globe.[10]

Thus, the three common explanations for our species' ecological success are (1) generalized intelligence or mental processing power, (2) specialized mental abilities evolved for survival in the hunter-gatherer environments of our evolutionary past, and/or (3) cooperative instincts or social intelligence that permit high levels of cooperation. All of these explanatory efforts are elements in building a more complete understanding of human nature. However, as I'll show, none of these approaches can explain our ecological dominance or our species' uniqueness without first recognizing the intense reliance we have on a large body of locally adaptive, culturally transmitted information that no single individual, or even group, is smart enough to figure out in a lifetime. To understand both human nature and our ecological dominance, we first need to explore how cultural evolution gives rise to complex repertoires of adaptive practices, beliefs, and motivations.

In chapter 3, lost European explorers will teach us about the nature of our vaulted intelligence, cooperative motivations, and specialized mental abilities. However, before setting sail with the explorers, I want to warm up by shaking your confidence on just how smart our species really is relative to other primates. Yes, we are intelligent, as earth creatures go; nevertheless, we are not nearly smart enough to account for our immense ecological success. Moreover, while we humans are good at certain cognitive feats, we are not so good at others. Many of both our mental skills and deficiencies can be predicted by recognizing that our brains evolved and expanded in a world in which the crucial selection pressure was our ability to acquire, store, organize, and retransmit an ever-growing body of cultural information. Like natural selection, our cultural learning abilities give rise to "dumb" processes that can, operating over generations, produce practices that are smarter than any individual or even group. Much of our seeming intelligence actually comes not from raw brainpower or a plethora of instincts, but rather from the accumulated repertoire of mental tools (e.g., integers), skills (differentiating right from left), concepts (fly wheels), and categories (basic color terms) that we inherit culturally from earlier generations.[11]

One quick terminological note before we face off against the apes. Throughout this book, *social learning* refers to any time an individual's learning is influenced by others, and it includes many different kinds of psychological processes. *Individual learning* refers to situations in which individuals learn by observing or interacting directly with their environment and can range from calculating the best time to hunt by observing when certain prey emerge, to engaging in trial-and-error learning with different digging tools. So, individual learning too captures many different psychological processes. Thus, the least sophisticated forms of social learning occur simply as a by-product of being around others and engaging in individual learning. For example, if I hang around you and you use rocks to crack open nuts, then I'm more likely to figure out *on my own* that rocks can be used to crack open nuts because I'll tend to be around rocks and nuts more frequently and can thus more easily make the relevant connections myself. *Cultural learning* refers to a more sophisticated subclass of social learning abilities in which individuals seek to acquire information from others, often by making inferences about their preferences, goals, beliefs, or strategies and/or by copying their actions or motor patterns. When discussing humans, I'll generally refer to *cultural learning*, but with nonhumans and our ancient ancestors, I'll call it *social learning*, since we often aren't sure if their social learning includes any actual cultural learning.

Showdown: Apes versus Humans

Let's begin by comparing the mental abilities of humans with two other closely related large-brained apes: chimpanzees and orangutans. As just mentioned, we "get smart" in part by acquiring a vast array of cognitive abilities via cultural learning. Cultural evolution has constructed a developmental world full of tools, experiences, and structured learning opportunities that harness, hone, and extend our mental abilities. This often occurs without anyone's conscious awareness. Consequently, to get a proper comparison with nonhumans, it might be misleading to compare apes to fully culturally equipped adults, who, for example, know fractions. Since its probably impossible, and certainly unethical, to raise children without access to these culturally evolved mental tools, researchers often compare toddlers to nonhuman apes (hereafter just "apes"). Admittedly, toddlers are already highly cultural beings, but they have had much less time to acquire additional cognitive endowments (e.g., knowing right vs. left, subtraction, etc.) and have had no formal education.

In a landmark study, Esther Herrmann, Mike Tomasello, and their colleagues at the Institute for Evolutionary Anthropology in Leipzig, Germany, put 106 chimpanzees, 105 German children, and 32 orangutans through a battery of 38 cognitive tests.[12] Their test battery can be broken down into subtests that capture abilities related to space, quantities, causality, and social learning. The space subtest includes tasks related to spatial memory and rotation in which participants have to recall the location of an object or track an object through a rotational movement. The quantities subtest measures participants' ability to assess relative amounts, or to account for additions and subtractions. The causality subtest assesses participants' abilities to use cues related to shape and sound to locate desirable things, as well as their ability to select a tool with the right properties to solve a problem (i.e., build a causal model). In the social learning subtest, participants are given opportunities to observe a demonstrator use a hard-to-discover technique to obtain a desirable object, such as extracting some food out of a narrow tube. Participants are then given the same task they just observed and can use what they just saw demonstrated to help them obtain the desired object.

Figure 2.2 is striking. On all the subtests of mental abilities, except social learning, there's essentially no difference between chimpanzees and two-and-a-half-year-old humans, despite the fact that the two-and-a-half-year-olds have much larger brains. Orangutans, who have slightly smaller brains than chimpanzees, do a bit worse, but not much worse. Even on the subtest that focused specifically on assessing the causal efficacy of tool properties (causal modeling), the toddlers got 71% correct, the chimps 61%, and the orangutans 63%. Meanwhile the chimps trounced the toddlers on tool use 74% to 23%.

By contrast, for the social learning subtest, the averages shown in figure 2.2 actually conceal the fact that most of the two-and-a-half-year-olds scored 100% on the test, whereas most of the apes scored 0%. Overall, these findings suggest that the only exceptional cognitive abilities possessed by young children in comparison to two other great apes relate to social learning, and not to space, quantities, or causality.

Crucially, if we gave this same battery to adult humans, they would blow the roof off the tests, performing at or near the ceiling (100% correct). This might lead you to think that the whole setup is unfair to the humans, because Esther, Mike, and their colleagues are comparing toddlers to older apes, who varied in ages from 3 to 21 years. Interestingly, however, older apes do not generally do better on these tests than younger apes—quite unlike humans. By age three, the cognitive performances of chimpanzees and orangutans—at least in these tasks—are about as good as they get.[13] Meanwhile, the young children will experience continuous, and eventually massive, improvements in their cognitive scores over at least the coming two decades of their lives. Just how good they will get will depend heavily on where, and with whom, they grow up.[14]

It's important to realize that chimpanzees and orangutans do have some social learning abilities, especially when compared to other animals, but when you have to design a test that is applicable to both apes

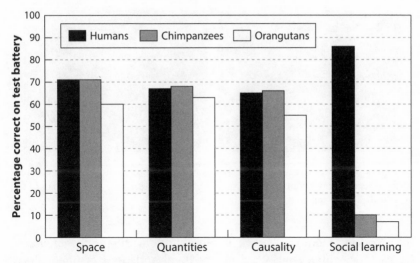

Figure 2.2. Average performance on four sets of cognitive tests with chimpanzees, orangutans, and toddlers.

and humans, the apes inevitably end up near the floor and the humans near the ceiling. In fact, we'll see later that when compared to other apes, humans are prolific, spontaneous, and automatic imitators, even willing to copy seemingly unnecessary or purely stylistic steps. When demonstrations include "extra" or "wasteful" steps, chimpanzee social learning emerges as superior to that of humans because we end up acquiring wasteful or inefficient elements whereas chimps filter these out.

Memory in Chimpanzees and Undergraduates

Despite the fact that our cognitive abilities improve as we grow up, especially in enriched cultural environments, we still do not end up with uniformly superior mental abilities compared to other apes. Let's consider first the available data comparing humans and chimpanzees in (1) working memory and information processing speed, and then in (2) games of strategic conflict. Both sets of findings bring into question the notion that our success as a species results from sheer brainpower or better mental processors. The second set of findings questions the notion that our minds are specialized for social maneuvering or strategizing in a Machiavellian world.

If you take an intelligence test, you may hear a list of numbers and then be asked to recall those numbers in reverse order. This measures

Figure 2.3. Working Memory Task. Participants saw a flash of numbers from 1 to 9 on the screen that were immediately covered by white squares. They then had to tap the number locations in order from memory.

your working memory. *Working memory*, along with *information processing speed*, are often considered two of the foundations of intelligence. Evidence indicates that greater informational processing speeds and working memory scores are associated with better problem solving and inductive reasoning (called fluid intelligence). Children and adolescents with greater working memory and information processing speed-scores at one age tend to have better problem solving and reasoning abilities at later ages.[15] Since working memory uses the neocortex, and humans have much larger neocortices than chimpanzees, we might expect adult humans to significantly outperform any chimpanzees in head-to-head competition.

Two Japanese researchers, Sana Inoue and Tetsuro Matsuzawa, set up just such a chimp-versus-human showdown. They trained three mother-offspring pairs of chimpanzees to identify numerals on a touch screen and to tap them in order (from 1 to 9). To test processing speed and working memory, they developed a task in which single digits flashed on the screen and were then covered by white squares (figure 2.3). The participants then had to tap on the squares (covering the numerals) in order from one to the highest digit. The time the participant had to view the numbers in their screen positions, before the white square covered them, ranged from 0.65 seconds down to one-fifth of a second.

These chimps faced off against university students.[16] For working memory, our species did well. In the easiest task, when six digits remained on the screen for a full 0.65 seconds, 7 of the 12 humans beat all the chimpanzees, even their star, the five-year-old Ayumu. On average, the humans tied Ayumu and handily beat the other chimps. And the tie is a bit deceptive, since the human score was dragged down by a weak (missing?) link, who got only a little over 30% of the sequences correct, making him worse than all the young chimps. However, when the numeral flashes got quicker and the task got tougher, Ayumu beat all the humans. Interestingly, as the flash of numbers sped up, Ayumu's performance remained consistent whereas the humans' performance, as well as that of the other chimps, rapidly degraded.

For informational processing speeds, which is the time from the end of the flash until the participant hits his first white square, the chimpanzees ruled. Every chimpanzee was faster than every human, and their speed did not vary with their performance. By contrast, faster human responses tended to be less accurate.

Typically, at this point, humans start making excuses for their uneven performance, arguing that the playing field was not level. For example, performances by chimpanzees were measured by their last 100 rounds after 400 rounds of practice. Human scores, however, were based on 50 rounds *with no practice*. Subsequent work indicates that university students, in fact, can be *trained* to tie or beat Ayumu in accuracy.[17]

However, such excuses can go both ways. The human team was stacked with young, educated adults who were likely at their human peak of working memory and processing speed. If the chimp team had a complex communicative repertoire, like humans, they would no doubt call for a rematch against five-year-old children to match the ages of the young chimpanzees. The young chimps, who actually outperformed their mothers, would probably defeat any group of young kids. The chimps would also charge that the students had a lifetime of experience with the bizarre Arabic numerals that they were forced to learn while in captivity.[18]

This dispute can go on and on and can't be resolved with the existing data. However, the point stands that the humans did not obviously dominate their fellow apes on either working memory or information processing speed, despite our much larger brains. Based on this evidence, it would seem hard to argue that our species' ecological dominance can be readily traced to our dazzling working memory or raw information processing speeds.

The True Machiavellians

Now, let's consider strategic conflict. We are a highly social species, so perhaps our global dominance has its origins in our social intelligence. One leading view on which selection pressures drove the expansion of

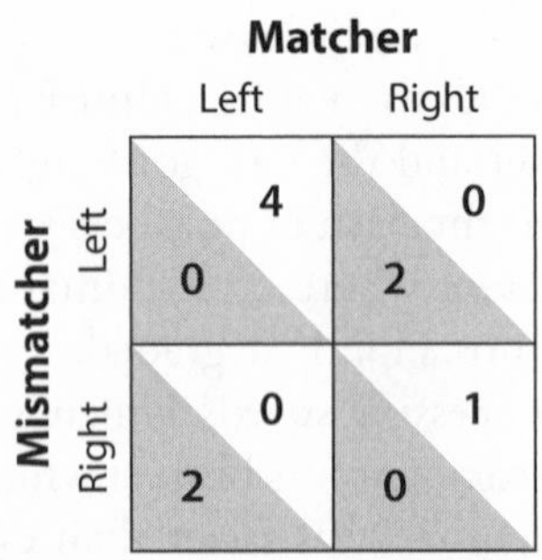

Matcher

Left   Right

Figure 2.4. The payoffs for the Matcher and Mismatcher in an Asymmetric Matching Pennies Game. Each player has to pick Left or Right. The Mismatcher's payoffs appear in the gray-shaded region of each cell while Matcher's payoffs appear in the white regions. The Matcher gets a higher payoff when he matches on Left than when he matches on Right (4 vs. 1). Meanwhile, the Mismatcher gets the same payoff regardless of how he mismatches.

human brains and created our fancy mental abilities is called the *Machiavellian intelligence hypothesis*. This view emphasizes that our brains and intelligence are specialized for dealing with other people and argues that our brain size and intelligence arose from an "arms race" in which individuals competed in an ever-escalating battle of wits to strategically manipulate, trick, exploit, and deceive each other. If this is so, we should be particularly good in games of strategic conflict compared to chimpanzees.[19]

Matching Pennies is a classic game of strategic conflict that has been played with both chimpanzees and humans. In the game, individuals are paired with another of their species for several rounds of interaction. Each player is placed into the role of either the Matcher or the Mismatcher. In each round, participants must select either Left or Right. The Matcher gets a reward only when his choice (Left or Right) *matches* the choice of his opponent. By contrast, the Mismatcher gets a reward only when his choice mismatches his opponent. The rewards, however, need not be symmetric, as illustrated in figure 2.4. In this asymmetric version, the Matcher gets 4 apple cubes (or cash for humans) when she successfully matches on Left, but only gets 1 cube when she matches on Right. Meanwhile, the Mismatcher gets only two cubes for any successful mismatches, no matter how they arise.

This kind of interaction can be analyzed using game theory. To win, the first thing to realize is that both players should be as unpredictable as possible. Nothing about your prior choices should allow your opponent to anticipate your next play—you have to randomize. To see this, put yourself into the shoes of the Matcher. Your opponent gets two cubes whether he plays Left (L) or Right (R), so you should essentially flip a coin with heads for R and tails for L. This means you'll play R and L each 50% of the time, and your opponent won't be able to predict your choices. If you deviate from 50%, your opponent will be able to exploit you more frequently. Now consider matters from the position of the Mismatcher: if you now similarly flip a coin, the Matcher will shift to play mostly L, since that gives him four instead of one. To compensate, as a Mismatcher you need to play R 80% of the time. Thus, the predicted winning strategy in a contest of intelligent rational actors is that Matchers should randomize their responses, playing L 50% of the time, while Mismatchers should randomize by playing L only 20% of the time. This outcome is called the Nash equilibrium. The fraction of the time that one should play L can be moved around by simply changing the payoffs for matching or mismatching on L or R.

A research team from Caltech and Kyoto University tested six chimpanzees and two groups of human adults: Japanese undergraduates and Africans from Bossou, in the Republic of Guinea. When chimpanzees played this asymmetric variant of Matching Pennies (figure 2.4), they zoomed right in on the predicted result, the Nash equilibrium. Humans, however, systematically and consistently missed the rational predictions, with Mismatchers performing particularly poorly. This deviation from "rationality," though it was in line with many prior tests of human rationality, was nearly seven times greater than the chimpanzees' deviation. Moreover, detailed analyses of the patterns of responses over many rounds of play show that the chimps responded more quickly to both their opponents' recent moves and to changes in their payoffs (i.e., when they switched from playing the Matcher to the Mismatcher). Chimpanzees seem to be better at individual learning and strategic anticipation, at least in this game.[20]

The performance of the apes in this setup was no fluke. The Caltech-Kyoto team also ran two other versions of the game, each with different payoffs. In both versions, the chimps zeroed in on the Nash equilibrium as it moved around from game to game. This means that chimps can develop what game theorists call a *mixed strategy*, which requires them to randomize their behavior around a certain probability. Humans, however, often struggle with this.

A final insight into the humans' poor performance comes from an analysis of participants' response times, which measures the time from the start of a round until the player selects his move. For both species, Mismatchers took longer than Matchers. However, the humans took *much longer* than the chimps. It's as if the humans were struggling to inhibit or suppress an automatic reaction.

This pattern may reflect a broader bug in human cognition: our automatic and unconscious tendency to imitate (to match). In Matching Pennies and other games like Rock-Paper-Scissors, one player sometimes accidentally reveals his or her choice a split-second before the other player. This flash look at an opponents' move could result in more victories for those who delay. And in Matching Pennies, experiments show that it does for Matchers, for whom copying leads to victories. For Mismatchers, however, it leads to more losses, because they sometimes fail to inhibit imitation. In Rock-Paper-Scissors, it results in more ties (e.g., rock-rock), because the slower player sometimes unconsciously imitates the choice of his or her opponent.[21] The reason is that we humans are rather inclined to copy—spontaneously, automatically, and often unconsciously. Chimpanzees don't appear to suffer from this cognitive "bug," at least not nearly to the same degree.

This is really just the beginning. So far, I have highlighted comparisons between human and ape cognition to suggest that although we are an intelligent species, we are not nearly smart enough to account for our species' ecological success. I could have also tapped the vast literature in psychology and economics, which tests the judgment and decision-making of undergraduates against benchmarks from statistics, probability, logic, and rationality. In many contexts, but not all, we humans make systemic logical errors, see illusory correlations, misattribute causal forces to random processes, and give equal weight to small and large samples. Not only do humans often fall systematically short of these standard benchmarks, we actually often don't do appreciably better than other species—like birds, bees, and rodents—on these tests. Sometimes, we do worse.[22] We, for example, suffer from the Gambler's fallacy, Concorde (or sunk cost) fallacy and Hot-hand fallacy, among others. Gamblers believe they are "due" at the craps tables (they're not), movie goers continue watching painfully bad movies even if they know they'd have more fun doing something else (e.g., sleeping), and basketball betters see certain players get the "hot-hand," even when they are actually seeing lucky streaks that are consistent with the player's typical scoring percentage. Meanwhile, rats, pigeons and other species don't suffer from such reasoning fallacies, and consequently often make the more profitable choices in analogous situations.

If our species is such a bunch of dimwits, how can we explain our species' success? And why do we seem so smart? I'll answer these questions in the next fifteen chapters. But before we get to all that, let's put my claims to the test. Stripped of our cultural know-how, can we humans fire-up our big brains and stoke up our fancy intellects enough to survive as hunter-gatherers?

CHAPTER 3

LOST EUROPEAN EXPLORERS

In June 1845 the HMS *Erebus* and the HMS *Terror*, both under the command of Sir John Franklin, sailed away from the British Isles in search of the fabled Northwest Passage, a sea channel that could energize trade by connecting western Europe to East Asia. This was the Apollo mission of the mid-nineteenth century, as the British raced the Russians for control of the Canadian Arctic and to complete a global map of terrestrial magnetism. The British admiralty outfitted Franklin, an experienced naval officer who had faced Arctic challenges before, with two field-tested, reinforced ice-breaking ships equipped with state-of-the-art steam engines, retractable screw propellers, and detachable rudders. With cork insulation, coal-fired internal heating, desalinators, five years of provisions, including tens of thousands of cans of food (canning was a new technology), and a twelve-hundred-volume library, these ships were carefully prepared to explore the icy north and endure long Arctic winters.[1]

As expected, the expedition's first season of exploration ended when the sea ice inevitably locked them in for the winter around Devon and Beechney Islands, 600 miles north of the Arctic Circle. After a successful ten-month stay, the seas opened and the expedition moved south to explore the seaways near King William Island, where in September they again found themselves locked in by ice. This time, however, as the next summer approached, it soon became clear that the ice was not retreating and that they'd remain imprisoned for another year. Franklin promptly died, leaving his crew to face the coming year in the pack ice with dwindling supplies of food and coal (heat). In April 1848, after nineteen months on the ice, the second-in-command, an experienced Arctic officer named Crozier, ordered the 105 men to abandon ship and set up camp on King William Island.

The details of what happened next are not completely known, but what is clear is that everyone gradually died. Both archaeological evidence and reports from Inuit locals gathered by the many explorers sent to rescue the expedition indicate that the crew fragmented, moved south, and cannibalism ensued. In one report, an Inuit band encountered one of the crew's parties. They gave the hungry men some seal meat but quickly departed when they noticed the crew transporting human limbs. Remains of the expedition have been located on several different parts of the island. There is also a rumor, never confirmed, that Crozier made it far enough south that he fell in with the Chippewa, where he lived out his days hiding from the shame of sustained and organized cannibalism.[2]

Why couldn't these men survive, given that some humans do just fine in this environment? King William Island lies at the heart of Netsilik territory, an Inuit population that spent its winters out on the pack ice and their summers on the island, just like Franklin's men. In the winter, they lived in snow houses and hunted seals using harpoons. In the summer, they lived in tents, hunted caribou, musk ox, and birds using complex compound bows and kayaks, and speared salmon using leisters (three-pronged fishing spears; see figure 3.1). The Netsilik name for the main harbor on King William Island is *Uqsuqtuuq*, which means "lots of fat" (seal fat).[3] For the Netsilik, this island is rich in resources for food, clothing, shelter, and tool-making (e.g., drift wood).

Franklin's men were 105 big-brained and highly motivated primates facing an environment that humans have lived in as foragers for over 30,000 years. They'd had three years in the Arctic, and nineteen months stuck in the ice, with their supplies slowly dwindling, to experience the environment and put those big brains to work. The men were all well known to each other after all this time, having worked together on the ship, so they should have been a highly cohesive group with a shared goal. At 105 persons, this group had roughly the same number of mouths to feed as a large Netsilik encampment, without the children or elderly to worry about. Yet the crew vanished, defeated by the hostile environment and only remembered in Inuit stories.

The reason Franklin's men could not survive is that humans don't adapt to novel environments the way other animals do or by using our

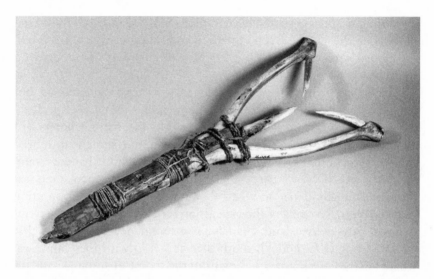

Figure 3.1. Head of a Netsilik leister used for fishing. The teeth are made from reindeer horn. Roald Amundsen collected this on King William Island during his visit in 1903–06.

individual intelligence. None of the 105 big brains figured out how to use driftwood, which was available on King William Island's west coast where they camped, to make the recurve composite bows, which the Inuit used when stalking caribou. They further lacked the vast body of cultural know-how about building snow houses, creating fresh water, hunting seals, making kayaks, spearing salmon and tailoring cold-weather clothing.

Let's briefly consider just a few of the Inuit cultural adaptations that you would need to figure out to survive on King William Island. To hunt seals, you first have to find their breathing holes in the ice. It's important that the area around the hole be snow covered—otherwise the seals will hear you and vanish. You then open the hole, smell it to verify that it's still in use (what do seals smell like?), and then assess the shape of the hole using a special curved piece of caribou antler. The hole is then covered with snow, save for a small gap at the top that is capped with a down indicator. If the seal enters the hole, the indicator moves, and you must blindly plunge your harpoon into the hole using all your weight. Your harpoon should be about 1.5 meters (5 ft) long, with a detachable tip that is tethered with a heavy braid of sinew line. You can get the antler from the previously noted caribou, which you brought down with your driftwood bow. The rear spike of the harpoon is made of extra-hard polar bear bone (yes, you also need to know how to kill polar bears; best to catch them napping in their dens). Once you've plunged your harpoon's head into the seal, you're then in a wrestling match as you reel him in, onto the ice, where you can finish him off with the aforementioned bear-bone spike.[4]

Now you have a seal, but you have to cook it. However, there are no trees at this latitude for wood, and driftwood is too sparse and valuable to use routinely for fires. To have a reliable fire, you'll need to carve a lamp from soapstone (you know what soapstone looks like, right?), render some oil for the lamp from blubber, and make a wick out of a particular species of moss. You will also need water. The pack ice is frozen salt water, so using it for drinking will just make you dehydrate faster. However, old sea ice has lost most of its salt, so it can be melted to make potable water. Of course, you need to be able to locate and identify old sea ice by color and texture. To melt it, make sure you have enough oil for your soapstone lamp.

These few examples are just the tip of an iceberg of cultural know-how that's required to live in the Arctic. I have not even alluded to the know-how for making baskets, fishing weirs, sledges, snow goggles, medicines, or leisters (figure 3.1), not to mention all the knowledge of weather, snow, and ice conditions required for safe travel using a sledge.

Nevertheless, while the Inuit are impressive, perhaps I am asking too much, and no one could have survived getting stuck in the ice for two years in the Artic. After all, we are a tropical primate, and the average temperatures during the winters on King William Island range between −25°C (−13°F) and −35°C (−31°F), and were even lower in the mid-nineteenth century. It happens, however, that two other expeditions have found themselves also stranded on King William Island, both before and after Franklin's expedition. Despite being much smaller and less well-equipped than Franklin's men, both crews not only survived but went on to future explorations. What was the secret of their success?[5]

Fifteen years before the Franklin Expedition, John Ross and a crew of twenty-two had to abandon the *Victory* off the coast of King William Island. During three years on the island, Ross not only survived but also managed to explore the region, including locating the magnetic pole. The secret of Ross's success is not surprising; it was the Inuit. Although not known as a "people person," he managed to befriend the locals, establish trading relations, and even fashion a wooden leg for lame Inuit man. Ross marveled at Inuit snow houses, multiuse tools, and amazing cold-weather attire; he enthusiastically learned about Inuit hunting, sealing, dogs, and traveling by dog sledge. In return, the Inuit learned from Ross's crew the proper use of a knife and fork while formally dining. Ross is credited with gathering a great deal of ethnological information, though in part this was driven by his practical need to obtain survival-crucial information and to maintain good relations. During their stay, Ross worried in his journals when the Inuit disappeared for long stretches and looked forward to the bounty they would return with—including packages such as 180 pounds of fish, fifty sealskins, bears, musk ox, venison, and fresh water. He also marveled at the health and vigor of the Inuit. Ross's sledge expeditions during this time always included parties of Inuit, who acted as guides, hunters, and shelter builders. After four years, during which time he was presumed dead by the British Admiralty, Ross managed to return to England with nineteen of his twenty-two men. Years later, in 1848, Ross would again deploy lightweight sledges, based on Inuit designs, in an overland search for Franklin's lost expedition. These sledge designs were adopted by many future British expeditions.

A little over a half century later, Roald Amundsen spent two winters on King William Island and three in the Arctic. In his refurbished fishing sloop, he went on to be the first European to successfully traverse the Northwest Passage. With knowledge of both Ross and Franklin, Amundsen immediately sought out the Inuit and learned from them how to make skin clothing, hunt seals, and manage dog sleds. Later, he would put these Inuit skills and technologies—clothing, sledges, and houses—to good use in beating Robert Scott to the South Pole. In praising the effectiveness of Inuit clothing at –63°F (–53°C), the Norwegian Amundsen wrote, "Eskimo dress in winter in these regions is far superior to our European clothes. But one must either wear it all or not at all; any mixture is bad. . . . You feel warm and comfortable the moment you put it on [in contrast with wool]." Amundsen made similar comments about Inuit snow houses (more on those in chapter 7). After finally deciding to replace the metal runners on his sledge with wooden ones, he noted, "One can't do better in these matters than copy the Eskimo, and let the runners get a fine covering of ice; then they slide like butter."[6]

The Franklin Expedition is our first example from the Lost European Explorer Files.[7] The typical case goes like this: Some hapless group of European or American explorers find themselves lost, cut off, or otherwise stuck in some remote and seemingly inhospitable place. They eventually run out of provisions and increasingly struggle to find food and sometimes water. Their clothing gradually falls apart, and their shelters are typically insufficient. Disease often follows, as their ability to travel deteriorates. Cannibalism frequently occurs, as things get desperate. The most instructive cases are those in which fate permits the explorers to gain exposure and experience in the "hostile" environment they will have to (try to) survive in before their supplies totally run out. Sadly, these explorers generally die. When some do survive, it's because they fall in with a local indigenous population, who provides them with food, shelter, clothing, medicine, and information. These indigenous populations have typically been surviving, and often thriving, in such "hostile" environments for centuries or millennia.

What these cases teach us is that humans survive neither by our instinctual abilities to find food and shelter, nor by our individual capacities to improvise solutions "on the fly" to local environmental challenges. We can survive because, across generations, the selective processes of cultural evolution have assembled packages of cultural adaptations—including tools, practices, and techniques—that cannot be devised in a few years, even by a group of highly motivated and cooperative individuals. Moreover, the bearers of these cultural adaptations themselves often don't understand much of how or why they work, beyond the understanding necessary for effectively using them. Chapter 4 will lay out the foundations of the processes that build cultural adaptations over generations.

Before moving on, however, let's again dip into the Lost European Explorer Files just to make sure the Arctic isn't a special case of an excessively challenging environment.

The Burke and Wills Expedition

In 1860, while returning from the first European trip across the interior of Australia, from Melbourne north to the Gulf of Carpentaria, four explorers found that they had nearly used up three months' worth of provisions and were increasingly forced to live off the land. The expedition leader, Robert Burke (a former police inspector) and his second-in-command, William Wills (a surveyor), along with Charles Gray (a 52-year-old sailor) and John King (a 21-year-old soldier), soon had to begin eating their pack animals, which included six camels that had been imported especially for this desert trip. The horse and camel meat extended their provisions but also meant they had to abandon their equipment as they traveled. Gray got increasingly weak, stole food, and soon died of dysentery. The remaining trio eventually made it back to their rendezvous point, an expedition depot at Coopers Creek, where they expected the rest of their large expedition party to be waiting with fresh supplies and provisions. However, this waiting party, who were also sick, injured, and running short on food, had departed earlier the same day. Burke, Wills, and King had just missed them, but the trio did manage to access some buried provisions. Still weak and exhausted, Burke decided not to try to catch the rest of their party, by heading south, but instead to follow Coopers Creek west toward Mount Hopeless (yes, really, Mount Hopeless), about 150 miles away, where there was a ranch and police outpost. While traveling along Coopers Creek, not long after departing the rendezvous depot, both of their two remaining camels died. This left them stuck along Coopers Creek because without either the camels to carry water or some knowledge of how to find water in the outback, the trio could not traverse the last open stretch of desert between the creek and the outpost at Mount Hopeless.[8]

Stranded, and now with their recent infusion of provisions running low, the explorers managed to make peaceful contact with a local aboriginal group, the Yandruwandrha. These aboriginal hunter-gatherers gave them gifts of fish, beans, and some cakes, which the men learned were made from a "seed" called nardoo (technically, it's a sporocarp, not a seed). Our trio clearly paid some attention when they were with the Yandruwandrha, but this didn't improve their success in fishing or trapping. However, impressed by the cakes, they did start searching for the source of the nardoo seeds, which they believed to be from a tree. After much searching, and running on empty, the trio finally wandered across a flat covered with nardoo—which turned out to be a cloverlike, semiaquatic fern, not a tree. Initially, the men just boiled the sporocarp, but later they found (not made) some grinding stones and copied the Yandruwandrha women whom they had observed preparing the cakes. They pounded the seeds, made flour, and baked nardoo bread.

This was an apparent boon in the men's plight, because it finally seemed they had a reliable source of calories. For more than a month, the men collected and consumed nardoo, as they all became increasingly fatigued and suffered from massive and painful bowel movements. Despite consuming what should have been sufficient calories (4–5 lbs. per day, according to Wills's journal), Burke, Wills, and King merely got weaker (see figure 3.2). Wills writes about what was happening to them by first describing the bowel movements caused by the nardoo:

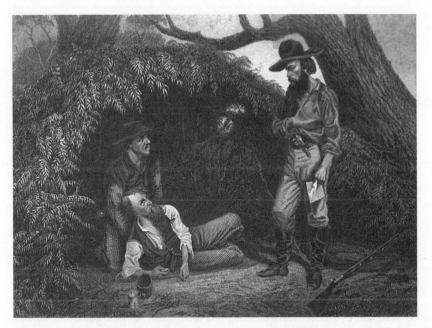

Figure 3.2. Painting of Burke, Wills, and King as they struggled to survive along Coopers Creek. Painted by Scott Melbourne and published in Wills's diary (Wills, Wills, and Farmer 1863).

I cannot understand this nardoo at all; it certainly will not agree with me in any form. We are now reduced to it alone, and we manage to get from four to five pounds per day between us. The stools it causes are enormous, and seem greatly to exceed the quantity of bread consumed, and is very slightly altered in appearance from what it was when eaten. . . . Starvation on nardoo is by no means unpleasant, but for the weakness one feels and the utter inability to move oneself, for as far as appetite is concerned, it gives me the greatest satisfaction.[9]

Burke and Wills died within a week of this journal entry. Alone, King managed to continue by appealing to the Yandruwandrha, who took him in, fed him, and taught him to construct a proper shelter. Three months later King was found by a relief expedition and returned to Melbourne.

Why did Burke and Wills die?

Like many plants used by hunter-gatherers, nardoo is indigestible and at least mildly toxic unless properly processed. Unprocessed nardoo passes through only partially digested and contains high levels of thi-aminase, which depletes the body's store of thiamine (vitamin B1). Low levels of thiamine cause the disease beriberi, resulting in extreme fa-tigue, muscle wasting, and hypothermia. To address this problem, the customary nardoo processing practices of aborigines appear to have multiple elements built in that make nardoo edible and nontoxic. First, they grind and leach the flour with copious amounts of water, which increases digestibility and decreases concentrations of the vitamin B1-destroying thiaminase. Second, in making the cakes, the flour is directly exposed to ash during heating, which lowers its pH and may break down the thiaminase. Third, nardoo gruel is consumed using only mus-sel shells, which may restrict the thiaminase's access to an organic sub-strate that is needed to fully initiate the B1-destroying reaction. Failure to deploy these local practices means that our unfortunate trio managed to starve and poison themselves while keeping their stomachs full.[10] Such subtle and nuanced detoxification practices are common in small-scale societies, and in later chapters we will see additional examples.

The effect of the nardoo, coupled with their lack of clothing, which was falling apart, and their inability to make a proper shelter, meant that the trio suffered greatly during the cold June winters. The effects of ex-posure probably accelerated their weakness and eventual demise. Their chances of learning from the locals, like Ross and Amundsen did, was diminished by Burke's flights of anger and impatience with the Yandru-wandrha. At one point, in response to their requests for gifts, he fired a shot over their heads, and they disappeared. Bad move.

If the Australian desert still seems too extreme, maybe our intelli-gence and/or evolved instincts might serve us better in subtropical cli-mates. Let's dip into the Lost European Explorer Files again.

The Narváez Expedition

In 1528, just north of Tampa Bay (Florida), Pánfilo Narváez made a cru-cial mistake. He split up his expedition, taking 300 conquistadors inland in search of the fabled cities of gold while sending his ships further up the coast for a later rendezvous at a new location. After wandering around the swamps and scrublands of northern Florida for two months (with no luck in finding the golden cities), and dealing treacherously with the locals, the mighty conquistadors attempted to head south to meet their ships. However, despite making great efforts through swampy terrain, they couldn't travel overland to their ships. Missing the rendezvous date, the remaining 242 men (50-some already dead) constructed five boats and planned to paddle along the Gulf coast to a Spanish port in Mexico.

Unfortunately, the conquistadors had dramatically underestimated the distance to Mexico, and the crude boats they constructed gradually stranded their crews on the barrier islands along the Gulf coast. The now-scattered Spanish parties starved, sometimes engaging in cannibalism, until they were aided by peaceful Karankawa hunter-gatherers, who had long lived along the Texas coast. Accounts suggest that with the Karankawa's help, the surviving parties were able to resume their journey to Mexico until starvation again marooned them in a wretched state. However, at least one of these groups got better at finding food, having learned from the locals to harvest seaweed and oysters. Interestingly, the floundering Spaniards, as well as later European travelers, always described the Karankawa as particularly tall, robust, and healthy-looking. So, this was a rich and bountiful environment for hunter-gatherers, if you knew what you were doing.

Though most died of starvation, a handful of Spaniards and one Moorish slave did reach the more densely populated heart of Karankawa territory. Having barely survived to arrive at this point, the remaining adventurers were promptly enslaved by these fiercer Karankawa and may have been forced into female gender roles. Among North American aboriginal populations, this male-to-female gender role switching was not uncommon. For our conquistadors, it meant a lot of toilsome work, such as carrying water, gathering firewood, and other unpleasant duties.

After several years of living among these scattered hunter-gatherers, four members of Narváez's crew were brought together by the annual prickly pear harvest season, during which time many local groups would congregate, feast, and celebrate. In the excitement, the foursome managed to slip away. After a long and rather circuitous route, staying among many different peoples in Mexico and Texas while operating as healers and shamans, they eventually found their way back to New Spain (colonial Mexico), eight years after they had begun in Florida.[11] Thus, the foursome had managed to survive by adopting a valued social role in these aboriginal American societies.

The Lone Woman

We can contrast these Lost European Explorer accounts, in which intrepid bands of hardy and experienced explorers find themselves struggling in novel environments, with another account, that of a lone young woman who found herself stranded for eighteen years in the place she grew up. Seventy miles off the coast of Los Angeles, and thirty miles from the nearest land, foggy, barren, and windswept San Nicolas Island was once inhabited by a thriving aboriginal society linked by trade to the other Channel Islands and the coast. However, by 1830 the island's population was dwindling, in part due to a massacre by Kodiak hunter-gatherers from the then Russian-controlled Aleutian Islands who had set up camp on San Nicolas to hunt otters. In 1835 Spanish missionaries from Santa Barbara sent a ship to transport the remaining island inhabitants to the mainland missions. During a rushed evacuation, one young native woman in her mid-twenties dashed off to search for her missing child. To evade a looming storm, the ship ended up leaving her behind on the island, and due to some unlucky quirks of fate, she was largely (but not entirely) forgotten.

Surviving for eighteen years, this lone castaway ate seals, shellfish, sea birds, fish, and various roots. She deposited dried meats on different parts of the island for times of sickness or other emergencies. She fashioned bone knives, needles, bone awls, shell fishhooks, and sinew fishing lines. She lived in whalebone houses and weathered storms in a cave. For transporting water, she wove a version of the amazing watertight baskets that were common among the California Indians. For clothing, she fashioned waterproof tunics by sewing together seagull skins with the feathers still on and wore sandals woven from grasses. When finally found, she was described as being in "fine physical condition" and attractive, with an "unwrinkled face." After overcoming an initial scare at being suddenly found, the lone woman promptly offered the search party dinner, which she was cooking at the time they arrived.[12]

The contrast with our lost European explorers could hardly be starker. One lone woman equipped with only the cumulative know-how of her ancestors survived for eighteen years whereas fully provisioned and well-financed teams of experienced explorers struggled in Australia, Texas, and the Arctic. These diverse cases testify to the nature of our species' adaptation. During eons of relying on large bodies of cumulative cultural knowledge, our species became addicted to this cultural input; without culturally transmitted knowledge about how to locate and process plants, fashion tools from available materials, and avoid dangers, we don't last long as hunter-gatherers. Despite the intelligence we acquire from having such big brains, we can't survive in the kinds of environments so commonly inhabited by our hunter-gatherer ancestors over our evolutionary history. While our attention, cooperative tendencies, and cognitive abilities have likely been shaped by natural selection to life in our ancestral environments, these genetically evolved psychological adaptations are entirely insufficient for our species. Neither our intelligence nor domain-specific psychological abilities fire up to distinguish edible from toxic plants or to construct watercraft, bone awls, snow houses, canoes, fishhooks, or wooden sledges. Despite the critical importance of hunting, clothing, and fire in our species' evolutionary history, no innate mental machinery delivered to our explorers information on locating snow-covered seal holes, making projectiles, or starting fires.

Our species' uniqueness, and thus our ecological dominance, arises from the manner in which cultural evolution, often operating over centuries or millennia, can assemble cultural adaptations. In the cases above, I've emphasized those cultural adaptations that involve tools and know-how about finding and processing food, locating water, cooking, and traveling. But as we go along, it will become clear that cultural adaptations also involve how we think and what we like, as well as what we can make.

In chapter 4, I will show how evolutionary theory can be successfully applied to build an understanding of culture. Once we understand how natural selection has shaped our genes and our minds to build and hone our abilities to learn from others, we will see how complex cultural adaptations—including tools, weapons, and food-processing techniques, as well as norms, institutions and languages—can emerge gradually without anyone fully apprehending how or why they work. In chapter 5, we will examine how the emergence of cultural adaptations began driving our genetic evolution, leading to an enduring culture-gene co-evolutionary duet that took us down a novel pathway, eventually making us a truly cultural species.

CHAPTER 4

HOW TO MAKE A
CULTURAL SPECIES

To understand why European explorers couldn't survive as hunter-gatherers while locals—even when stranded alone—could, we need to understand how populations generate cultural adaptations—the suites or packages of skills, beliefs, practices, motivations, and organizational forms that permit people to survive, and often thrive, in diverse and challenging environments. The process is—in some crucial sense—smarter than we are. Over generations, often outside of conscious awareness, individuals' choices, learned preferences, lucky mistakes, and occasional insights aggregate to produce cultural adaptations. These often-complex packages contain subtle and implicit insights that impress modern engineers and scientists (see chapter 7). We have glimpsed some of these cultural adaptations, from Inuit clothing to nardoo detoxification, and will study other such adaptations, ranging from food taboos that protect pregnant women from marine toxins to religious rituals that galvanize greater prosociality. Before getting to these, however, we need to build an understanding of cultural evolution, from the ground up, that can explain how it is that human populations end up with complexes of tools, tastes, and techniques that are honed to local environmental challenges.

This brings us to a central insight. Rather than opposing "cultural" with "evolutionary" or "biological" explanations, researchers have now developed a rich body of work showing how natural selection, acting on genes, has shaped our psychology in a manner that generates *nongenetic evolutionary processes* capable of producing complex cultural adaptations. Culture, and cultural evolution, are then a consequence of genetically evolved psychological adaptations for learning from other people. That is, natural selection favored genes for building brains with abilities to learn from others. These learning abilities, when operating in populations and over time, can give rise to subtly adaptive behavioral repertoires, including those related to fancy tools and large bodies of knowledge about plants and animals. These emergent products arose initially as unintended consequences of the interaction of learning minds in populations, over time. With this intellectual move, "cultural explanations" become but one type of "evolutionary explanation," among a potential host of other noncultural explanations.

In their now classic treatise, *Culture and the Evolutionary Process*, Rob Boyd and Pete Richerson laid the foundations of this approach by developing a body of mathematical models that explore our capacities for cultural learning as genetically evolved psychological adaptations. Once cultural learning is approached as a psychological adaptation or as a suite of adaptations, we can then ask how natural selection has shaped our psychology and motivations so as to allow us to most effectively acquire useful practices, beliefs, ideas, and preferences from others.[1] These are questions about *who* we should learn from, and *what* we should attend to and infer, as well as *when* input from cultural learning should overrule our own direct experience or instincts.

Evidence from diverse scientific fields is now revealing how finely tuned our psychological adaptations for cultural learning are. Natural selection has equipped our species with a wide range of mental abilities that allow us to effectively and efficiently acquire information from the minds and behaviors of other people. These learning instincts emerge early, in infants and young children, and generally operate unconsciously and automatically. In many circumstances, as we saw in the Matching Pennies Game and Rock-Paper-Scissors, we find it difficult to inhibit our automatic imitative instincts. As we'll see below, even when getting "right answers" is important, our cultural learning mechanisms fire up, to influence our practices, strategies, beliefs, and motivations. In fact, sometimes the more important getting the right answer is, the more we rely on cultural learning.

As a point of departure, it is worth considering how pervasive the effects of cultural learning are on our behavior and psychology. Box 4.1 lists just some of the domains where the influence of cultural learning has been studied.[2] The list includes domains of distinct evolutionary importance, such as food preferences, mate choice, technological adoptions, and suicide, as well as social motivations related to altruism and fairness. As we'll see in later chapters, cultural learning reaches directly into our brains and changes the *neurological values* we place on things and people, and in doing so, it also sets the standards by which we judge ourselves. One classic set of experiments shows that children acquire the performance standards by which they are willing or unwilling to reward themselves.[3] Children saw a demonstrator rewarding himself or herself with M&Ms only after exceeding either a relatively higher score in a bowling game or a relatively lower score. The children copied the rewarding standards of the demonstrator such that the kids exposed to the "high standards" model tended not to eat the M&Ms unless their score exceeded the higher threshold. As will become clear, culturally acquired standards or values guide our efforts and persistence at individual learning, training, and trial-and-error learning.

Box 4.1: Domains of Cultural Learning

- Food preferences and quantity eaten
- Mate choices (individuals and their traits)
- Economic strategies (investments)
- Artifact (tool) functions and use
- Suicide (decision and method)
- Technological adoptions
- Word meanings and dialect
- Categories ("dangerous animals")
- Beliefs (e.g., about gods, germs, etc.)
- Social norms (taboos, rituals, tipping)
- Standards of reward and punishment
- Social motivations (altruism and fairness)
- Self-regulation
- Judgment heuristics

Let's begin by exploring how thinking of our cultural learning abilities as genetically evolved psychological adaptations deepens our understanding of both how we adapt to our worlds as individuals and how populations adapt to their environments over generations. Our first question is: how should individuals figure out *whom* to learn from? This question is crucial because it will illuminate how cultural adaptations can emerge.

Suppose you are a boy living in a hunter-gatherer band. To survive, the men in your community hunt a wide range of game. How should you go about hunting? You could just start experimenting. Maybe try throwing rocks at gazelles or chasing some zebras. Or you could wait for some evolved instincts for hunting to fire up and tell you what to do. If you go this route, you'll probably wait a long time, as this is the situation that Franklin, Burke, and Narváez found themselves in. Franklin's men lived for nineteen months on the pack ice with starvation looming, but no one figured out how to harpoon a seal. In fact, since you are a member of a cultural species, your instincts will fire up, but instead of supplying specialized hunting abilities, they will cause you to start looking for people to copy—but not just any people. Aspiring young hunters first glean as much as they can from those to whom they have ready access, like their brothers, fathers, and uncles. Later, perhaps during adolescence, learners update and improve their earlier efforts by focusing on and learning from the older, most successful, and most prestigious hunters in their community. That is, learners should use three cultural learning cues to target their learning: age, success, and prestige. You might also use "being male" as a cue, since, let's assume, hunting is predominantly a male activity. These cues will not only help you zoom in on those in your community most likely to possess adaptive practices, routines, beliefs, and skills related to hunting, but will also allow you to gradually scaffold up your abilities and know-how while making the most of what those who care about you know. Moreover, since particular individuals may be successful or prestigious for idiosyncratic reasons (like possessing good genes), you'll want to sample several of the top hunters and use only those practices preferred by a plurality of them.[4]

More broadly, evolutionary reasoning suggests that learners should use a wide range of cues to figure out whom to selectively pay attention to and learn from. Such cues allow them to target those people most likely to possess information that will increase the learner's survival and reproduction. In weighting the importance of those they can potentially learn from (hereafter their *models*), individuals should combine cues related to the models' health, happiness, skill, reliability, competence, success, age, and prestige, as well as correlated cues like displays of confidence or pride. These cues should be integrated with others related to self-similarity, such as sex, temperament, or ethnicity (cued by, e.g., language, dialect, or dress). Self-similarity cues help learners focus on those likely to possess cultural traits (e.g., practices or preferences) that will be useful to the learner in their future roles. Same-sex cues, for example, reduce the time spent by teenage boys attending to the details of many female-specific activities, such as how to latch an infant, or what to do when a viscous, yellowish secretion (colostrum) first emerges from the nipple after birth. Together, all of these cues provide input to "model-based" cultural learning mechanisms because they help individuals figure out from whom to learn. Let's take a closer look at our selective cultural learning.

Skill and Success

Since many model-based cues, such as success and prestige, are only loosely tied to particular domains of behavior, such as hunting or golf, we expect them to influence learning about a broad spectrum of cultural traits, ranging from preferences for food, mates, wine, and linguistic labels to beliefs about invisible agents or forces, like gods, germs, angels, karma, and gravity. This is not to say that we should expect the same cues to have the same size impact on very different domains. What makes someone a successful hunter or basketball player is surely better predicted by their arrow-making techniques or their jump-shot form, respectively, but may also be impacted by the practice of eating carrots or by saying a brief prayer before heading out to hunt or play basketball. Eating carrots might improve a shooter's vision while a prayer ritual might calm and focus the mind (or, possibly bring supernatural aid).

Let's begin by looking more deeply at the impact of cues related to skill and success, which capture notions of competence and reliability. Skill cues are those that relate most directly to competence in a domain. For example, one might assess a writer's skill by reading his books. In a foraging society, an aspiring hunter might watch an older hunter adeptly stalk a giraffe, crouch in a tree's shadow, and fluidly release an accurately targeted arrow. By contrast, success cues are more indirect, but potentially more useful, because they aggregate information. Using success cues, you might evaluate a writer on the number of copies that his book sells, and a hunter by the frequency with which he brings home big game. Since many hunting societies have practices that facilitate easy accounting of prior kills, you might be able to look at how many monkey teeth a hunter has on his necklace or at the number of pig mandibles hung outside of his house.[5]

To see the power and pervasiveness of the use of success cues in cultural learning, consider the following experiment. MBA students participated in two different versions of an investment game. In the game, they had to allocate their money across three different investment options, labeled A, B, and C. They were told each investment's average monetary returns and its variation (sometimes you get more than average, other times, less). They were also told the relationships or correlations among the investments; for example, if investment A's value goes up, then B's value tends to go down. Participants could borrow money to invest. During each round of the game, each player would make his or her allocations and receive the returns. After each round, players could alter their investment allocations for the next round, and this went on for sixteen rounds. At the end of the game, each player's portfolio performance ranking relative to the other players heavily influenced their grade in the course, moving it up or down. If you know any MBAs, you'll know this is a serious incentive, and these players were thus strongly motivated to make the most money in the game.

The experimenters randomly assigned players to one of two different versions, or treatments. In one version, the MBAs made their decisions in isolation, receiving only the individual experience derived from their own choices over the sixteen rounds. The other version was identical except that the allocations chosen and performance rankings of all participants were posted between each round, using anonymous labels.

The difference in the results from each version surprised the economists who designed the experiment (though, admittedly, many economists are pretty easily surprised by human behavior[6]). Three patterns are striking. First, the MBAs didn't use the additional information available in the second treatment (with posted performances) in the complex and sophisticated way economic theory assumes. Instead, careful analysis shows that many participants were merely copying ("mimicking") the investment allocations made by the top performers in the previous round. Second, the environment of this experiment is simple enough that one can actually calculate the profit-maximizing investment allocations. This optimal allocation can be compared with where participants actually ended up in round 16 for each of the two versions. Left only to their own individual experience, the MBAs ended up very far away from the optimal allocation—thus, poor overall performance. However, in the second treatment, when they mimicked each other's investments, the group zeroed in on the optimal allocation by the end of the game. Here, the whole group made more money, which is interesting since there were no incentives for group performance, as grade assignments were all based on relative rankings. Finally, while opportunities to imitate each other had a dramatic effect on improving the overall group performance, it also led to some individual catastrophes. Sometimes top performers had taken large risks, which paid off in the short run—they got lucky. But their risky allocations, which often included massive borrowing, were copied by others. Since you can't copy the luck along with the allocation choices, an inflated number of bankruptcies resulted as a side effect.[7]

The central finding of this experiment, that people are inclined to copy more successful others, has been repeatedly observed in an immense variety of domains, both in controlled laboratory conditions and in real-world patterns.[8] In experiments, undergraduates rely on *success-biased* learning when real money is on the line—when they are paid for correct answers or superior performance. In fact, the more challenging the problem or the greater the uncertainty, the more inclined people are to rely on cultural learning, as predicted by evolutionary models. This tells us something about *when* individuals will rely on cultural learning over their own direct experience or intuitions.[9]

Interestingly, if you are a real stock market investor, this is now a formal strategy: you can purchase exchange-traded funds (ETFs) that match the picks of the market gurus (GURU), billionaire investors (iBillionaire), or the top money managers (ALFA).[10] But remember, you can't copy their luck.

Experimentally, economists have also shown that people rely on this skill- or success-biased cultural learning to (1) infer and copy others' beliefs about the state of the world, even when others have identical information, and (2) adapt to competitive situations, where copying others is far from the optimal strategy. In the real world, farmers from around the globe adopt new technologies, practices, and crops from their more successful neighbors.[11]

Running in parallel with the work in economics, decades of work by psychologists has also shown the importance of success and skill biases. This work underlines the point that these learning mechanisms operate outside conscious awareness and with or without incentives for correct answers.[12] One set of recent experiments by Alex Mesoudi and his collaborators are particularly relevant for our focus here on complex technologies.[13] In his arrowhead design task, participants engaged in repeated rounds of trial-and-error learning using different arrowhead designs to engage in virtual hunting on a computer. Whenever the opportunities were made available, students readily used success-biased cultural learning to help design their arrowheads. When cultural information was available, it rapidly led the group to the optimal arrowhead design and was most effective in more complex, more realistic environments.

In the last fifteen years, an important complementary line of evidence has become available as developmental psychologists have returned to focusing on cultural learning in children and infants. With new evolutionary thinking in the air, they have zoomed in on testing specific ideas about the *who*, *when*, and *what* of cultural learning. It's now clear that infants and young children use cues of competence and reliability, along with familiarity, to figure out from whom to learn. In fact, by age one, infants use their own early cultural knowledge to figure out who tends to know things, and then use this performance information to focus their learning, attention, and memory.

Infants are well known to engage in what developmental psychologists call "social referencing." When an infant, or young child, encounters something novel, say when crawling up to a chainsaw, they will often look at their mom, or some other adult in the room, to check for an emotional reaction. If the attending adult shows positive affect, they often proceed to investigate the novel object. If the adult shows fear or concern, they back off. This occurs even if the attending adult is a stranger. In one experiment, mothers brought one-year-olds to the laboratory at Seoul National University. The infants were allowed to play and get comfortable in the new environment, while mom received training for her role in the experiment. The researchers had selected three categories of toys, those to which infants typically react (1) positively, (2) negatively, and (3) with uncertain curiosity (an ambiguous toy). These different kinds of toys were each placed in front of the infants, one at a time, and the infant's reactions were recorded. Mom and a female stranger sat on either side of the baby and were instructed to react either with smiling and excitement or with fear.

The results of this study are strikingly parallel to studies of cultural learning among both young children and university students. First, the babies engaged in social referencing, looking at one of the adults, four times more often, and more quickly, when an ambiguous toy was placed in front of them. That is, under uncertainty, they used cultural learning. This is precisely what an evolutionary approach predicts for *when* individuals should use cultural learning (see note 9). Second, when faced with an ambiguous toy, babies altered their behavior based on the adults' emotional reactions: when they saw fear, they backed off, but when they saw happiness, they approached the toy and changed to regard it more positively. Third, infants tended to reference the stranger more than their moms, probably because mom herself was new to this environment and was thus judged less competent by her baby.[14]

By 14 months, infants are already well beyond social referencing and already showing signs of using skill or competence cues to select models. After observing an adult model acting confused by shoes, placing them on his hands, German infants tended not to copy his unusual way of turning on a novel lighting device: using his head. However, if the model acted competently, confidently putting shoes on his feet, babies tended to copy the model and used their heads to activate the novel lighting device.[15]

Later, by age three, a substantial amount of work shows that children not only track and use competence in their immediate cultural learning but retain this information to selectively target their future learning in multiple domains. For example, young children will note who knows the correct linguistic labels for common objects (like "ducks"), use this information for targeting their learning about both novel tools or words, and then remember this competence information for a week, using it to preferentially learn new things from the previously more competent model.[16]

Prestige

By observing whom others watch, listen to, defer to, hang-around, and imitate, learners can more effectively figure out from whom to learn. Using these "prestige cues" allows learners to take advantage of the fact that other people also are seeking, and have obtained, insights about who in the local community is likely to possess useful, adaptive information. Once people have identified a person as worthy of learning from, perhaps because they've learned about their success, they necessarily need to be around them, watching, listening, and eliciting information through interaction. Since they are trying to obtain information, learners defer to their chosen models in conversation, often giving them "the floor." And, of course, learners automatically and unconsciously imitate their chosen models, including by matching their speech patterns (see chapter 8). Thus, we humans are sensitive to a set of ethological patterns (bodily postures or displays), including visual attention, "holding the floor," deference in conversation, and vocal mimicry, as well as others. We use these prestige cues to help us rapidly zero in on whom to learn from. In essence, prestige cues represent a kind of second-order cultural learning in which we figure out who to learn from by inferring from the behavior of others who they think are worthy of learning from—that is, we culturally learn from whom to learn.

Despite the seeming ubiquity of this phenomenon in the real world, there is actually relatively little direct experimental evidence that people use prestige cues. There is an immense amount of indirect evidence that shows how the prestige of a person or source, such as a newspaper or celebrity, increases the persuasiveness of what they say or the tendency of people to remember what they say. This effect occurs even when the prestige of a person comes from a domain, like golf, that is far removed from the issue they are commenting on (like automobile quality). This provides some evidence, though it does not get at the specific cues that learners might actually use to guide them, aside from being told that someone is an "expert" or "the best."[17]

To address this in our laboratory, Maciej Chudek, Sue Birch, and I tested this prestige idea more directly. Sue is a developmental psychologist and Maciej was my graduate student (he did all the real work). We had preschoolers watch a video in which they saw two different potential models use the same object in one of two different ways. In the video, two bystanders entered, looked at both models, and then preferentially watched one of them. The visual attention of the bystanders provided a "prestige cue" that seemingly marked one of the two potential models. Then, participants saw each model select one of two different types of unfamiliar foods and one of two differently colored beverages. They also saw each model use a toy in one of two distinct ways. After the video, the kids were permitted to select one of the two novel foods and one of the two colorful beverages. They could also use the toy any way they wanted. Children were 13 times more likely to use the toy in the same manner as the prestige-cued model compared to the other model. They were also about 4 times more likely to select the food or beverage preferred by the prestige-cued model. Based on questions asked at the end of the experiment, the children had no conscious or expressible awareness of the prestige cues or their effects. These experiments show that young children rapidly and unconsciously tune into the visual attention of others and use it to direct their cultural learning. We are prestige biased, as well as being skill and success biased.[18]

Chapter 8 expands on these ideas to explore how selective cultural learning drove the evolution of a second form of social status in humans called Prestige, which in our species resides alongside the Dominance status we inherited from our primate ancestors. We'll see why, for example, it is possible to become famous for being famous in the modern world.

Self-Similarity: Sex and Ethnicity

Automatically and unconsciously, people also use cues of self-similarity, like sex and ethnicity, to further hone and personalize their cultural learning. Self-similarity cues help learners acquire the skills, practices, beliefs, and motivations that are, or were in our evolutionary past, most likely to be suitable to them, their talents, or their probable roles later in life. For example, many anthropologists argue that the division of labor between males and females is hundreds of thousands of years old in our species' lineage. If true, we should expect males to preferentially hang around, attend to, and learn from other males—and vice versa for females. This will result in novices learning the skills and expectations required for their likely roles later in life, as mothers, hunters, cooks, and weavers. Similarly, since individual differences, like height or personality, might influence one's success in various endeavors, learners may preferentially attend to those whom they resemble along these dimensions. In chapter 11, I will further detail the evolutionary logic underpinning the prediction that learners should preferentially attend to, and learn from, those who share their ethnic-group markers, such as language, dialect, beliefs, and food preferences. In short, these cues allow learners to focus on those most likely to possess the social norms, symbols, and practices that the learners will need over their life course for successful and coordinated social interactions.

There is ample evidence from psychological experiments—going back forty years—that both children and adults preferentially interact with, and learn from, same-sex models over opposite-sex models. In young children, these biases emerge even before they develop a gender identity and influence their learning from parents, teachers, peers, strangers, and celebrities. In fact, children learn their sex roles because they copy same-sex models, not vice versa. Evidence indicates that this learning bias influences diverse cultural domains, including musical tastes, aggression, postures, and object preferences. Later, we'll see that in the real world, it influences both students' learning (and performance) as well as patterns of copycat suicide.[19]

Recent work by the brain scientist Elizabeth Reynolds Losin, a former student of mine, and her colleagues at UCLA has begun to illuminate the neurological underpinnings of sex-biased cultural learning. Using fMRI technology, Liz focused on the difference between people's brains when they imitated a same-sex model versus an opposite-sex model. She asked both men and women in Los Angeles to first watch and then imitate the arbitrary hand gestures either of a same-sex or opposite-sex model. By comparing the same individuals' brain activity while watching versus watching and imitating both same- and opposite-sex models, Liz showed that women find mimicking other women more rewarding—neurologically speaking—than mimicking men. Men showed the opposite pattern. When people copied same-sex models, firing activity was higher in the nucleus accumbens, dorsal and ventral striatum, orbital frontal cortex, and left amygdala. Her analysis of the existing database of brain studies reveals that this pattern of brain activity emerges when people receive a reward, such as money, for getting the correct answer. This finding suggests that we experience copying same-sex others as internally more rewarding than copying opposite-sex others. We like it more, so naturally we are inclined to do it more.

While research on ethnicity biases in cultural learning is more limited, it's increasingly clear that infants, young children, and adults preferentially learn from co-ethnics—meaning that people attend to and selectively learn from those who share their ethnic markers. Young children preferentially acquire both food preferences and the functions of novel objects from those who share their language or dialect. This is true even when the potential models speak jabberwocky—English-sounding speech composed of nonsense words. That is, kids prefer to learn from those speaking nonsense *in their own dialect* rather than from someone speaking nonsense in a different dialect (suddenly, I'm reminded of much American political discourse). Infants preferentially copy the more unusual and difficult action—turning on a lamp with their head—from someone who speaks their language (German) as opposed to an unfamiliar language (Russian). And children and adults prefer to learn from those who already share some of their beliefs.[20]

These laboratory findings suggest that cues related to sex and ethnicity fire up our cultural learning psychology in ways that spark our interest in what the model is doing, or talking about, and focus our attention and memory. If true, students may learn more effectively from teachers or professors who match them on these dimensions, which may impact a person's grades, choice of major, or career preferences. Formal education is, after all, primarily an institution for intensive cultural transmission. Of course, identifying this learning bias as a causal influence is tricky in the real world because teachers have biases too, which may lead them to preferentially assist or reward those who share their sex or ethnic markers. Isolating causality in the real world is what economists are best at, so let's bring in some economists.

By exploiting large data sets of students, courses, and instructors, my UBC colleague Florian Hoffman and his collaborators unearthed real-world evidence consistent with the experimental findings discussed above: being taught by instructors whom you match on ethnicity/race reduces your dropout rate and raises your grades. In fact, for African-American students at a community college, being taught by an African-American instructor reduced class dropout rates by 6 percentage points and increased the fraction attaining a B or better by 13 percentage points. Similarly, using data from freshman (first-years) at the University of Toronto, Florian's team has also shown that getting assigned to a same-sex instructor increased students' grades a bit.

Unlike many researchers before them, Florian and his colleagues addressed concerns that these patterns are created by the instructors' biases by focusing on large undergraduate lectures where students (1) could not influence which instructors they got, (2) were anonymous to the professor, and (3) were graded by teaching assistants, not by the professors.[21] All this points to biases that influence whom learners readily tune in to and learn from.

Our cultural learning biases are why role models matter so much.

Older Individuals Often Do Know More

Both as an indirect measure of competence or experience, and as a measure of self-similarity, age cues may be important for cultural learning for two separate evolutionary reasons. For children, focusing on and learning from older children allows them to learn from more experienced individuals while at the same time providing a means of self-scaffolding, allowing them to bridge gradually from less to more complex skills. The idea here is that although a learner may be able to locate, and sometimes learn from, the most successful or skilled person in his community (say, the best hunter in a foraging band), many young learners will be too inexperienced or ill-equipped to take advantage of the nuances and fine points that distinguish the top hunters. Instead, by focusing on older children, young learners can isolate models who are operating at an appropriate increment of skill and complexity above their own. This creates a smoother and more continuous process of gradual skill acquisition, as learners move back and forth from observing older models to practicing, and repeat the process as they grow up. This is why, for example, younger children are often so desperate to hang around their "big cousins" or older siblings, and why mixed-age playgroups are the standard in small-scale societies.

Consistent with evolutionary expectations, young children do assess the age of potential models, perhaps by assessing physical size. Young children often prefer older models unless those individuals have proven unreliable. They trade off *age* against *competence* and in some cases will prefer younger but more competent models to older, less competent ones. For example, in one experiment second graders preferentially imitated the fruit choices of their fellow second graders over kindergarten models. However, when shown that some kindergarteners and second graders were superior puzzle solvers, many second graders shifted to the fruit choices of these good puzzle solvers, even if they were sometimes kindergarteners. In general, children and infants shift their food preferences in response to observing older, same-sex, models enjoying certain foods. Even infants, as young as 14 months, are sensitive to age cues.[22]

At the other end of the age spectrum, merely getting to be old was a major accomplishment in the societies of our evolutionary past. By the time ancient hunter-gatherers reached 65, and some did, natural selection had already filtered many out of their cohort. This means that not only were the senior members of a community the most experienced, but also that they had emerged from decades of natural selection acting selectively to shrink their age cohort. To see how this works, imagine you have a community with 100 people between the ages of 20 and 30. Of these 100, only 40 people routinely prepare their meat dishes using chili peppers. Suppose that using chili peppers, by virtue of their antimicrobial properties, suppresses food-borne pathogens and thereby reduces a person's chances of getting sick. If eating chili peppers year after year increases a person's chances of living past 65 from 10% to 20%, then a majority of this cohort, 57%, will be chili-pepper eaters by the time they reach 65. If learners preferentially copy the older cohort, instead of the younger cohort, they will have a greater chance of acquiring this survival-enhancing cultural trait. This is true even if they have no idea that chili peppers have any health impacts (see chapter 7). Age-biased cultural learning here can thus amplify the action of natural selection, as it creates differential mortality.[23]

Why Care What Others Think?
Conformist Transmission

Suppose you are in a foreign city, hungry, and trying to pick one of ten possible restaurants on a busy street. You can't read the menus because you don't know the local language, but you can tell that the prices and atmospheres of each establishment are identical. One place has 40 diners, six have 10 diners, and three are empty, except for the waitstaff. If you would pick the restaurant with 40 people (out of the 100 you've observed) *more than* 40% of the time then you are using conformist transmission—you are strongly inclined to copy the most common trait—the majority or the plurality.

Evolutionary models, which are built to mathematically capture the logic of natural selection, predict that learners ought to use what's called *conformist transmission* to tackle a variety of learning problems. As long as individual learning, intuitions, direct experience, and other cultural learning mechanisms tend to produce adaptive practices, beliefs, and motivations, then conformist transmission can help learners aggregate the information that is distributed across a group. For example, suppose long experience fishing will tend to cause anglers to prefer the blood knot to other potential knots (for connecting monofilament line) because the blood knot is objectively the best. However, individual experiences will vary, so suppose that long experience alone leads to only a 50% chance of an angler converging on the blood knot, a 30% chance of using the fisherman's knot, and a 20% chance of using one of five other knots. A conformist learner can exploit this situation and jump directly to the blood knot without experience. Thus, the wisdom of crowds is built into our psychology.

There is some laboratory evidence for conformist transmission, both in humans and sticklebacks (a fish), though there is not nearly as much as for the model-based cues discussed above. Nevertheless, when problems are difficult, uncertainty is high, or payoffs are on the line, people tend to use conformist transmission.[24]

Of course, we should expect learners to combine the learning heuristics I've described. For example, with regard to chili peppers, learners who apply conformist transmission to only the older cohort (sorting with age cues) will increase their chances of adopting this adaptive practice. If they are strong conformist learners, they will get the adaptive answer 100% of the time.

Culturally Transmitted Suicide

You probably know that committing suicide is prestige biased: when celebrities commit suicide there is a spike in suicide rates (celebrities: keep this in mind!). This pattern has been observed in the United States, Germany, Australia, South Korea, and Japan, among other countries. Alongside prestige, the cultural transmission of suicide is also influenced by self-similarity cues. The individuals who kill themselves soon after celebrities tend to match their models on sex, age, and ethnicity. Moreover, it's not just that a celebrity suicide vaguely triggers the suicide of others. We know that people are imitating because they copy not only the act of suicide itself but also the specific methods used, such as throwing oneself in front of a train. Moreover, most celebrity-induced copycat suicides are *not tragedies that would have occurred anyway*. If that were the case, there would be an eventual dip in suicide rates below the long-run average at some point after the spike, but there is not.[25] These are extra suicides that otherwise would not have occurred.

These effects can also be seen in suicide epidemics. Beginning in 1960, a striking pattern of suicide rippled through the pacific islands of Micronesia for about twenty-five years. As the epidemic spread, the suicides assumed a distinct pattern. The typical victim was a young male between 15 and 24 (modal age of 18) who still lived with his parents. After a disagreement with his parents or girlfriend, the victim experienced a vision in which past victims beckoned for him to come to them (we know this from attempted suicides). In heeding their call, the victim performed a "lean hanging," sometimes in an abandoned house. In a lean hanging, victims lean into the noose from a standing or kneeling position. This gradually depletes the victim's supply of oxygen, resulting in a loss of consciousness and then death. These suicides occurred in localized and sporadic outbreaks among socially interconnected adolescents and young men, a pattern common elsewhere. Sometimes these epidemics could be traced to a particular spark, such as the suicide of a 29-year-old prominent son of a wealthy family. In 75% of the cases, there was no prior hint of suicide or depression. Interestingly, these epidemics were restricted to only two ethnic groups within Micronesia, the Trukese and Marshallese.[26] Here we see that prestige and self-similarity, including both sex and ethnicity, shaped the diffusion of suicide.

While most people don't copy suicide, this domain illustrates just how potent our cultural learning abilities can be and how they influence broad social patterns. If people will acquire suicidal behavior via cultural learning, it's not clear what the boundaries are on the power of culture in our species. Copying suicide highlights the potency of our imitative tendencies and means that under the right conditions we can acquire practices via cultural learning that natural selection has directly acted to eliminate under most conditions. If humans will imitate something that is so starkly not in our self-interest, or that of our genes, imagine all the other less costly things we are willing to acquire by cultural transmission.

In addition to using model-based mechanisms for cultural learning, we should also expect natural selection to have equipped us with psychological abilities and biases for learning about certain predictable content domains, such as food, fire, edible plants, animals, tools, social norms, ethnic groups, and reputations (gossip), which have probably been important over long stretches of our species' evolutionary history. Here, natural selection may have favored attention and interest in these domains, as well as inferential biases, leading to ready encoding in memory and greater learnability. In the coming chapters, we'll explore how culture-gene coevolution drove the emergence of some of these specialized cognitive abilities, or *content biases*, and examine key lines of supporting evidence.

What's Mentalizing For?

If humans are a cultural species, then one of our most crucial adaptations is our ability to keenly observe and learn from other people. Central to our cultural learning is our ability to make inferences about the goals, preferences, motivations, intentions, beliefs, and strategies in the minds of others. These cognitive abilities relate to what is variously termed *mentalizing*, or *theory of mind*. Any learners who miss the boat on mentalizing and cultural learning, or get started too late, will be at a serious disadvantage because they won't have acquired all the norms, skills, and know-how necessary to compete with other, better, cultural learners. This logic suggests that the mental machinery we need for cultural learning ought to fire up relatively early in our development. It's this mental machinery that we will rely on to figure out what to eat, how to communicate, whom to avoid, how to behave, which skills to practice, and much more.

Evidence from young children and infants in Western populations combined with recent cross-cultural work in Fiji, Amazonia, and China suggests that these mentalizing abilities begin to develop early and reliably across diverse human societies. By the age of roughly eight months, infants in at least some societies have developed an ability to infer intentions and goals and to recognize who likely has knowledge and who does not. Infants, for example, will copy a model's goals or intentions, such as grasping a toy, even when the model fails to achieve his goal, but they won't copy unintentional actions that create the same physical results. By the time they are toddling, children are already making sophisticated judgments about others' mental states, for example, recognizing that a potential model mislabeled a familiar object and then subsequently devaluing what that model has to say. Similarly, toddlers can figure out what aspects of context are new for their model and use this to better target their learning, even when those same aspects are familiar to the learner.[27]

Though they agree mentalizing is important, many evolutionary researchers argue that these cognitive abilities evolved genetically in our lineage so we can better trick, manipulate, and deceive other members of our group—this is part of the Machiavellian intelligence hypothesis. The idea is that if Robin can infer Mike's goals, motivations, or beliefs, then Robin can exploit or manipulate Mike. He can outthink him and then outmaneuver him.[28]

However, another possibility is that mentalizing first evolved in our lineage, or perhaps was re-tasked away from trickery, deception, and manipulation, so that we could better learn from others by more effectively inferring our models' underlying goals, strategies and preferences for the purpose of copying them. Mentalizing also may help us teach more effectively, since good teaching requires us to assess what learners need to know. These expectations flow from the *cultural intelligence hypothesis*.[29]

In my psychology laboratory at UBC, our team has sought to allow these two hypotheses to face off against each other. In a novel situation, we gave young children the opportunity to deploy their mentalizing abilities either to copy the strategies of others or to exploit a hapless opponent. The results are stark: children strongly favor cultural learning over Machiavellian exploitation, even when their payoffs and personal experience point them away from copying others.

QUEEN MARGARET UNIVERSITY LRC

Of course, this doesn't mean that mentalizing isn't also deployed for social strategizing, as it clearly is in chimpanzees.[30] But what it does imply is that in humans you first need to acquire the social norms and rules governing the world you are operating in, and only then is strategic thinking useful. In our world, successful Machiavellians must first be skilled cultural learners. You can't bend, exploit, and manipulate the rules until you first figure out what the rules are.

Learning to Learn and to Teach

The evidence from infants and young children now suggest that humans rapidly develop a heavy reliance on carefully attending to, and learning from, other people, often using their mentalizing skills, and that they readily begin using cues such as success and prestige to figure out whom to learn from. However, it seems likely that both our degree of reliance on cultural learning over our own experience or innate intuitions, as well as how heavily we weight cues of prestige or gender vis-à-vis other cues may be itself tuned by both our own direct experience and our observations of others. That is, we need to be able to calibrate these systems for the contexts we encounter in the world.[31]

The importance of both direct experience and the observation of others is particularly clear for developing teaching abilities. Teaching is the flip side of cultural learning. It occurs when the model becomes an active transmitter of information. Later I'll discuss some evidence that suggests natural selection has operated to improve our transmission or communicative abilities, especially since the evolution of languages. Nevertheless, most people are still not particularly great teachers, especially of complex tasks, concepts, or skills, so cultural evolution has produced a wide range of strategies and techniques adapted for more effectively transmitting particular types of content, such as judo, algebra, or cooking. This is one way cultural transmission increases its own fidelity—learners acquire both the skills themselves as well as techniques for transmitting them.

In the early days, when our species was just beginning to rely on cultural learning, as cultural evolution was cranking up, it may be that attending to and copying others was acquired by experience, perhaps through trial-and-error learning, because it tended to get the best answer compared to other learning strategies.[32] Consistent with this idea, apes reared by humans, sometimes in human families, seem to be better at imitation compared to other apes. Importantly, however, though they improve relative to chimpanzees not reared by humans, they still pale in comparison to human children reared in the identical environments for the same time periods. Such evidence suggests that cultural learning may have initially developed as a response to the enriched environments created by the very earliest accumulations of cultural evolution (see chapter 16).[33] This learned increase in cultural learning would have permitted a greater accumulation of cultural know-how and further driven genetic evolution to make us better cultural learners. Now, the vast differences we observe between apes and human infants raised in the same environments suggests that the emergence of cultural learning is relatively canalized and rapid in our species, though, of course, it can still be modified by experience.[34]

CHAPTER 5

WHAT ARE BIG BRAINS FOR? OR, HOW CULTURE STOLE OUR GUTS

By selectively attending to certain types of cultural content, like food, sex, and tools, and to particular models based on cues related to prestige, success, and health, individuals can most efficiently equip themselves with the best available cultural know-how. This acquired repertoire can then be honed and augmented by an individual's experience in the world. Crucially, however, these individually adaptive pursuits have an unintended consequence, which we saw when the MBAs were allowed to copy each other—the whole group gradually zeroed in on the optimal investment allocation. As individuals go about their business of learning from others in their group, the overall body of cultural information contained and distributed across the minds in the group can improve and accumulate over generations.

To see more precisely how cumulative cultural evolution works, imagine a small group of forest-dwelling primates. Figure 5.1 represents this group along the top row, labeled Generation 0, with individuals represented by circles. One member of this generation has, on her own, figured out how to use a stick to extract termites from a termite mound (this trait is labeled T). Figuring this out is plausible for our ancestors, since modern chimpanzees do it. In Generation 1 (row 2), two of the offspring from Generation 1 copy the elder termite fisher because they note her success and are generally interested in "things related to food." However, while copying this termite-harvesting technique, one member of Generation 1 mistakenly infers that the stick his model was using had been sharpened (though actually, it just broke funny when the model grabbed it). When the learner made his stick, he used his teeth to sharpen it to match his model's stick (in figure 5.1, T^* marks the sharpened stick). Elsewhere, another member of Generation 1 realizes he can use hollow reeds to drink from the water that gathers deep in the troughs of large trees (this "straw" is labeled T2 in figure 5.1). He uses this technique to obtain water when he crosses the savannah, between patches of forest. In Generation 2, both the possessors of T2 and T^* were preferentially copied, so their practices spread a bit. One member of Generation 2 even managed to pick up both T2 and T^*, so she became particularly successful and was copied by three members of Generation 3. Then one day an inattentive member of Generation 3 plunged his fishing stick into an old, abandoned termite mound, not realizing the termites had long departed. Fortuitously, he happened to spear a rodent who had moved in after the termites left. Suddenly, the "termite fishing stick" became a general-purpose "hole-spear" (now labeled T^{**}), which allowed this lucky fellow to tap new food sources, as he started plunging his spear into every hole he could find. His success as a hole-spearer meant that several members of Generation 4 attended to and learned from him. Meanwhile, another member of Generation 3, while goofing off, just happened to notice a rabbit enter his hole after a rain. While looking at the rabbit's tracks in the mud and thinking how tasty that rabbit would be, it struck her that she could look for tracks like these and follow them to locate active rabbit holes (this "rabbit tracking" is labeled T3). This was interesting but not immediately useful since she had no way to get the rabbit out of the hole. Nevertheless, years later, this mom showed the tracks to her daughter after they saw a rabbit. This happenstance was crucial, because the daughter had already learned T^{**} ("hole-spear"). Now she could locate active rabbit holes and deploy her spear—a very useful technique. In Generation 5, while no one invented or lucked into anything, three members had learned T^{**}, T2, and T3. This package—a cultural adaptation—permitted them to spend more time on the savannah tracking rabbits to their holes, since they could also access water with T2 (the "straw"). Soon these primates began living at the edge of the forests, so they could hunt on the savannah. I'll call the combination of traits T^{**}, T2, and T3 the "savannah hunting package."

Keep in mind that this is a toy example meant to illustrate how selective cultural learning can generate a cumulative evolutionary process that generates cultural packages that are smarter than their bearers. My

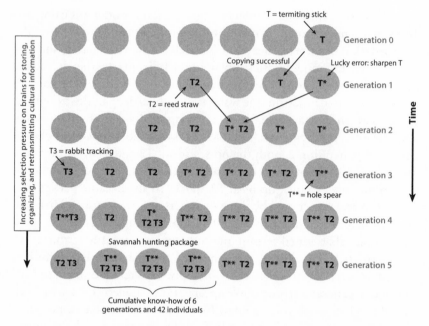

Figure 5.1. How learning from others generates cumulative cultural evolution.

imaginary primates are better cultural learners than any living primate except us. Nevertheless, even if I had made them worse learners, I could have achieved the same end point with a larger population or more generations. Like us (or at least me), these primates were just bumbling myopically around their daily lives. Sometimes their mistakes led to innovations, and sometimes a rare circumstance thrust an insight upon someone who was otherwise goofing off. The key is that these occasional insights and lucky errors were preferentially passed on, persisted, and were eventually recombined with other traits to create a savannah hunting package. Now consider this question: is Generation 5 smarter than Generation 0? The former has better tools and can obtain food more efficiently. Later, we will see a variety of evidence suggesting that Generation 5 may in fact be smarter than Generation 0, if "smartness" is defined as an individual's ability to solve novel problems. Of course, there will be caveats and cautions.

This imaginary ancestral primate crossed a crucial evolutionary threshold as it entered a regime of *cumulative cultural evolution*. This threshold is the point at which culturally transmitted information begins to accumulate over generations, such that tools and know-how get increasingly better fit to the local environments—this is the "ratchet effect."[1] It's this process that explains our cultural adaptations, and ultimately, the success of our species. As we'll see in chapter 7, individuals reliant on cultural adaptations often have little or no understanding of how or why they work, or even that they are "doing" anything adaptive.

The central argument in this book is that relatively early in our species' evolutionary history, perhaps around the origins of our genus (*Homo*) about 2 million years ago, we first crossed this evolutionary Rubicon, at which point cultural evolution became the *primary driver of our species' genetic evolution*. This interaction between cultural and genetic evolution generated a process that can be described as *autocatalytic*, meaning that it produces the fuel that propels it. Once cultural information began to accumulate and produce cultural adaptations, the main selection pressure on genes revolved around improving our psychological abilities to acquire, store, process, and organize the array of fitness-enhancing skills and practices that became increasingly available in the minds of the others in one's group. As genetic evolution improved our brains and abilities for learning from others, cultural evolution spontaneously generated more and better cultural adaptations, which kept the pressure on for brains that were better at acquiring and storing this cultural information. This process will continue until halted by an external constraint.

I call the threshold between typical genetic evolution and the regime of autocatalytic culture-driven genetic evolution the Rubicon. During the Roman Republic, the muddy red waters of the river Rubicon marked the boundary between the province of Cisalpine Gaul and Italy proper, which was administered directly by Rome itself. Provincial governors could command Roman troops outside of Italy, but under no circumstances could they enter Italy at the head of an army. Any commander who did this, and any soldiers who followed him, were immediately considered outlaws. This rule had served the old republic well until, in 49 BC, Julius Caesar crossed the Rubicon at the head of the loyal Legio XII Gemina. For Caesar and this legion, there was no turning back after crossing the Rubicon; civil war was inevitable and Roman history would be forever altered. Similarly, in crossing our evolutionary Rubicon, the human lineage embarked down a novel evolutionary path from which there was no turning back.

To see why there's no turning back, imagine being a member of Generation 6 in figure 5.1. Should you focus on inventing a new trait, or on making sure you accurately locate and copy those individuals with T**, T2, and T3? You might invent a good trait, as adaptive as T, but you'll never invent something as good as the savannah hunting package of T** + T2 + T3. Thus, if you don't focus on cultural learning, you will lose out to those who do.

As the process continues over generations, the selection pressures only increase: the more culture accumulates, the greater the selection pressures on genes for making one an adept cultural learner with a bigger brain capable of harnessing the ever-upward-spiraling body of cultural information. Figure 5.1 illustrates the point. Consider the memory—or brain storage space—required by our six different generations. In Generation 0, at most you could invent one trait in your lifetime, so you only need brain space for one trait. However, by Generation 5, you'd better have storage space in your brain for T**, T2, and T3—and you'd best know how they fit together. The memory space demanded in Generation 5, if one wants any chance of outsurviving and outreproducing others in the population, has increased threefold in only six generations. If genes spread that expand the brains of Generation 6, the selection pressure for bigger and better brains won't abate because cultural evolution will continue to expand the size of the cultural repertoire—of the body of know-how one could learn, if one were sufficiently well equipped. This culture-gene coevolutionary ratchet made us human.

We've already seen some of the evidence that culture drove human evolution. In chapter 2, we saw that when toddlers competed against other apes in a variety of cognitive tasks, the only domain in which they kicked butt was social learning. Otherwise, for quantities, causality, and space, it was pretty much a tie. That's precisely what you'd expect if culture drove the expansion of our brains, honed our cognitive abilities, and modified our social motivations. In chapter 3, by accompanying various hapless explorers, we saw that our species ability to live as hunter-gatherers depends on acquiring the local cultural knowledge and skills. And in chapter 4 we explored how natural selection has shaped our psychology to allow us to selectively target and extract adaptive information from our social milieu.

Moving forward, table 5.1 summarizes some of the products of culture-gene coevolution that I cover in this book. To begin, this chapter examines five ways in which cultural evolution has influenced, and interacted with, genetic evolution to shape our bodies, brains, and psychology. To understand table 5.1, look first at the column labeled "Culturally transmitted selection pressure." These are features of the

Table 5.1. Examples of How Cultural Evolution and Its Products Have Shaped Human Genetic Evolution

Chapters covered	Culturally transmitted selection pressure	Coevolved genetic consequences	Other implications
2–5, 7–8, 12, 13, 16	*Cumulative culture* Accumulating body of cultural knowledge creates dependence	Specialized cultural learning abilities for selectively acquiring adaptive information from others	Selection pressure for greater sociality. Difficult childbirth, due to oversized heads
		Long childhoods and larger brains prepared for cultural learning and practice with extensive brain "wiring" over decades	Demands for more child care
5–7, 12, 16	*Food processing* Cooking, leaching, pounding, chopping	Increasing dependence on processed food, including cooked foods. Results in small teeth, gapes, mouths, colons, and stomachs; possible interest in fire during childhood	Frees energy for brain building and favors the sexual division of labor
5, 15, 16	*Persistence hunting* Tracking, water containers, and animal behavioral know-how	Distance running facilitated by springy arches, slow-twitch muscle fibers, shock-enforced joints, a nuchal ligament, and innervated eccrine sweat glands	Human lineage becomes high-level predator
5, 7	*Folkbiology* Increasing knowledge about plants and animals	Folk biological cognition: hierarchical taxonomies with essentialized categories, category-based induction, and taxonomic inheritance	Universal treelike taxonomies for categorizing the natural world

Table 5.1. (*cont.*)

Chapters covered	Culturally transmitted selection pressure	Coevolved genetic consequences	Other implications
5, 12, 13, 15, 16	*Artifacts* Increasingly complex tools and weapons	Anatomical changes to hands, shoulders, and elbows. Direct cortical connections into spinal cord.	Greater manual dexterity and throwing abilities. Increased physical weakness
		Artifact cognition: functional stance	
4, 5, 8, 12, 15, 16	*Wisdom of age* Opportunities to use and transmit culture gleaned over a lifetime	Changes in human life history: extended childhood, adolescence, and a longer postreproductive lifespan (menopause)	Cooperation in child investment and rearing
4, 7, 12, 13	*Complex adaptations* Pressure for high fidelity cultural learning	Sophisticated abilities to infer others' mental states—theory of mind, or mentalizing, and overimitation	Dualism: a preparedness to understand minds without bodies
4, 8	*Information resources* Variation in skill or know-how among individuals	Prestige status: suite of motivations, emotions, and ethological patterns that produce a second type of status	Prestige-based leadership and greater cooperation
9–11	*Social norms* Enforced by reputations and sanctions	Norm psychology: concerns with reputation, internalization of norms, prosocial biases, shame and anger at norm violators, cognitive abilities for detecting violations	Strengthens effect of intergroup competition on cultural evolution

Table 5.1. (cont.)

Chapters covered	Culturally transmitted selection pressure	Coevolved genetic consequences	Other implications
11	*Ethnic groups* Culturally marked membership across social groups	Folksociology: in-group vs. out-group psychology that cues off phenotypic markers, which influence cultural learning and interaction	Tribal/ethnic groups, later nationalism and parochial religions
13	*Languages* Transmitted gestures and vocalizations	Changes in throat anatomy, audio processing, specialized brain regions, and tongue dexterity	Massive increase in the rate of cultural transmission
13	*Teaching* Opportunities to facilitate cultural transmission	Communicative or pedagogical adaptations: white sclera (whites of the eyes), eye contact, pedagogical inclinations, etc.	Higher-fidelity transmission and more rapid cultural evolution

world that were created by cultural evolution but that subsequently had consequences for genetic evolution (given in the column "Coevolved genetic consequence") and sparked a coevolutionary duet between genes and culture.

Now, let's examine how crossing the evolutionary Rubicon into a regime of cumulative cultural evolution helps explain several of our species' characteristics.

Big Brains, Fast Evolution, and Slow Development

Compared to those of other animals, our brains are big, dense, and groovy. While we don't have the biggest brains in the natural world—whales and elephants beat us—we do have the most cortical interconnections and the highest degree of cortical folding. Cortical folding produces that "crumpled wad of paper" (groovy) appearance that particularly characterizes human brains. But that's just the beginning of our oddities. Our brains evolved from the size of a chimpanzee's, at roughly 350 cm³, to 1350 cm³ in about 5 million years. Most of that expansion, from about 500 cm³ upward, took place only in about the last 2 million years. That's fast in genetic evolutionary terms.

This expansion was finally halted about 200,000 years ago, probably by the challenges of giving birth to babies with increasingly bulbous heads. In most species, the birth canal is larger than the newborn's head, but not in humans. Infant skulls have to remain unfused in order to squeeze through the birth canal in a manner that isn't seen in other species. It seems our brains only ceased expanding because we hit the stops set by our primate body plan; if babies' heads got any bigger, they wouldn't be able to squeeze out of mom at birth. Along the way, natural selection came up with numerous tricks to circumvent this *big-headed baby problem*, including intense cortical folding, high-density interconnections (which permit our brains to hold more information without getting bigger) and a rapid postbirth expansion. Specifically, newborn human brains continue expanding at the faster prebirth gestational rate for the first year, eventually tripling in size. By contrast, newborn primate brains grow more slowly after birth, eventually only doubling in size.[2]

After this initial growth spurt, our brains continue adding more connections for holding and processing information (new glial cells, axons, and synapses) over the next three decades of life and even beyond, especially in the neocortex. Consider our white matter and, specifically, the process of myelination. As vertebrate brains mature, their white matter increases as the (axonal) connections among neurons are gradually "burned in" and wrapped in a performance-enhancing coating of fat called myelin. This process of myelination makes brain regions more efficient, but less plastic and thus less susceptible to learning. To see how human brains are different, we can compare our myelination with that of our closest relatives, chimpanzees. For the cerebral cortex, figure 5.2[3] shows the fraction of myelination (as a percentage of the adult level) during three different developmental periods: (1) infancy, (2) childhood (called the "juvenile period" in primates), and (3) adolescence and young adulthood. Infant chimpanzees arrive in the world with 15% of their cortex already myelinized, whereas humans start with only 1.6% myelinized. For the neocortex, which has evolved more recently and is massive

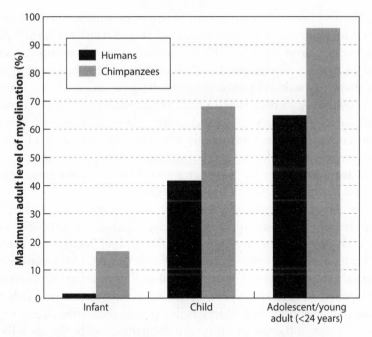

Figure 5.2. Myelination in chimpanzees and humans over development.

in humans, the percentages are 20% and 0%, respectively. During adolescence and young adulthood, humans still have only 65% of their eventual myelination complete, whereas chimpanzees are almost done, at 96%. These data suggest that, unlike chimpanzees, we continue substantial "wiring-up" into our third decade of life.

Human brain development is related to another unusual feature of our species, our extended childhoods and the emergence of that memorable period called adolescence. Compared to other primates, our gestational and infancy periods (birth to weaning) have shortened while our childhoods have extended and a uniquely human period of adolescence has emerged, prior to full maturity. Childhood is a period of intensive cultural learning, including playing and the practicing of adult roles and skills, during which time our brains reach nearly their adult size while our bodies remain small. Adolescence begins at sexual maturity, after which a growth spurt ensues. During this time, we engage in apprenticeships, as we hone the most complex of adult skills and areas of knowledge, as well as build relationships with peers and look for mates.[4]

The emergence of adolescence and young adulthood has likely been crucial over our evolutionary history, since in hunting and gathering populations, hunters do not produce enough calories to even feed themselves (let alone others) until around age 18 and won't reach their peak productivity until their late thirties. Interestingly, while hunters reach their peak strength and speed in their twenties, individual hunting success does not peak until around age 40, because success depends more on know-how and refined skills than on physical prowess. By contrast, chimpanzees—who also hunt and gather—can obtain enough calories to sustain themselves immediately after infancy ends, around age 5.[5] Consistent with our long period of wiring-up, this pattern and contrast with chimpanzees reveals the degree to which we humans are dependent on learning for our survival as foragers.

Our unusually big brains, with their slow neurological and behavioral development but rapid evolutionary expansion, is precisely what you'd expect if cumulative cultural evolution had become the driving selection pressure in the evolution of our species. Once cumulative cultural evolution began to produce cultural adaptations, like cooking and spears, individuals whose genes have endowed them with the brains and developmental processes that permit them to most effectively acquire, store, and organize cultural information will be the most likely to survive, find mates, and leave progeny. As each generation gets brains that are a little bigger and a little better at cultural learning, the body of adaptive know-how will rapidly expand to fill any available brain space. This process will shape the development of our brains, by keeping them maximally plastic and "programmed to receive," and our bodies, by keeping them small (and calorically inexpensive) until we've learned enough to survive. This culture-gene coevolutionary interaction creates an autocatalytic process such that no matter how big our brains get, there will always be much more cultural information in the world than any one of us can learn in a lifetime. The better our brains get at cultural learning, the faster adaptive cultural information accumulates, and the greater the pressure on brains to acquire and store this information.

This view also explains three puzzling facts about human infants. First, compared to other species, babies are altricial, meaning that they are weak, undermuscled, fat, and uncoordinated (sorry, babies, but it's true). By contrast, some mammals exit the womb ready to walk, and even primates rapidly figure out how to hang onto mom. Meanwhile, above the neck, human babies' brains are developmentally advanced at birth compared to those of other animals, having passed more of the mammalian neurological landmarks than other species. Fetuses are already acquiring aspects of language in the womb (see chapter 13), and babies arrive ready to engage in cultural learning. Before they can walk, feed themselves, or safely defecate, infants are selectively learning from others based on cues of competence and reliability (chapter 4) and can read others' intentions in order to copy their goals.[6] Third, despite being otherwise developmentally and cognitively advanced, babies' brains arrive highly plastic (unmyelinized) and continue to expand at their gestational rate. In short, while being otherwise nearly helpless, babies and toddlers are sophisticated cultural learning machines.

Natural selection has made us a cultural species by altering our development in ways that (1) slowed the growth of our bodies through a shortened infancy and extended childhood but added a growth spurt in adolescence, and (2) altered neurological development in complex ways that make our brain advanced at birth yet both highly expandable and enduringly plastic. As we go along, I will consider how our fast genetic evolution, big adult brains, slow bodily development, and gradual wiring-up are made possible only as part of a larger package of features that include the sexual division of labor, intensive parental investment in children, and the long postreproductive lives we associate with menopause. These features of our species will interact in crucial ways with cultural evolution.

Food Processing Externalizes Digestion

Compared to other primates, humans have an unusual digestive system. Starting at the top, our mouths, gapes, lips, and teeth are oddly small, and our lip muscles are weak. Our mouths are the size of a squirrel monkey's, a species that weighs less than three pounds. Chimpanzees can open their mouths twice as wide as we can and hold substantial amounts of food compressed between their lips and large teeth. We also have puny jaw muscles that reach up only to just below our ears. Other primates' jaw muscles stretch to the tops of their heads, where they sometimes even latch onto a bony central ridge. Our stomachs are small, having only a third of the surface area that we'd expect for a primate of our size, and our colons are too short, being only 60% of their expected mass. Our bodies are also poor at detoxifying wild foods. Overall, our guts—stomachs, small intestines, and colons—are much smaller than they ought to be for our overall body size. Compared to other primates, we lack a substantial amount of digestive power all the way down the line, from our mouth's (in)ability to breakdown food to our colon's capacity to process fiber. Interestingly, our small intestines are about the size they should be, an exception that we'll account for below.[7]

How can culture explain this strange physiological patterning in humans?

The answer is that our bodies, and in this case our digestive systems, have coevolved with culturally transmitted know-how related to food processing. People in every society process food using techniques that have accumulated over generations, including cooking, drying, pounding, grinding, leaching, chopping, marinating, smoking, and scraping. Of these, the oldest are probably chopping, scraping, and pounding with stone tools. Chopping, scraping, and pounding meat can go a long way, because they tenderize by slicing, dicing, and crushing the muscle fibers, partially replacing some of the functions of teeth, mouths, and jaws. Similarly, marinades initiate the chemical breakdown of foods. Acidic marinades, such as that used for the coastal South American dish ceviche, begin literally breaking down meat proteins before they reach your mouth, mimicking the approach taken by your stomach acid. And as we saw with nardoo, leaching is one of a host of techniques that hunter-gatherers have long used to process food and remove toxins.

Of all these techniques, cooking is probably the most important piece of cultural know-how that has shaped our digestive system. The primatologist Richard Wrangham has persuasively argued that cooking (and therefore fire) has played a crucial role in human evolution. Richard and his collaborators laid out how cooking, if done properly, does an immense amount of digestion for us. It softens and prepares both meat and plant foods for digestion. The *right* amount of heating tenderizes, detoxifies, and breaks down fibrous tubers and other plant foods. Heating also breaks down the proteins in meat, dramatically reducing the work for our stomach acid. Consequently, by contrast with meat-eating carnivores (e.g., lions), we do not often retain meat in our stomachs for hours, because it typically arrives partially digested by pounding, scraping, marinating, and cooking.

While all this food processing reduces the digestive workload of our mouths, stomachs, and colons, it does not alter the need to actually absorb the nutrients, which is why our small intestines are about the right size for a primate of our stature.

What is often underemphasized in this account is that food-processing techniques are primarily products of cultural evolution. Cooking, for example, is not something we instinctually know how to do, or even can easily figure out. If you don't believe me, go outside and make a fire without using any modern technology. Rub two sticks together, make a fire drill, find naturally occurring flint or quartz, etc. Put that big brain to work. Maybe some fire instincts designed by natural selection to solve this recurrent dilemma of our ancestral environments will kick in and guide you.

. . .

No luck? Unless you've had training—that is, received cultural transmission—it's very unlikely you were successful. Our bodies have been shaped by fire and cooking, but we have to learn from others how to make fire and cook. Making fire is so "unnatural" and technically difficult that some foraging populations have actually lost the ability to make fire. These include the Andaman Islanders (off the coast of Malaysia), Sirionó (Amazonia), Northern Aché, and perhaps Tasmanians. Now, to be clear, these populations couldn't have survived without fire; they retained fire but lost the ability to start new fires on demand. When one band's fire was inadvertently extinguished, say during a fierce storm, they had to head off to locate another band whose fire had not gone out (hopefully).[8] However, living in small and widely scattered groups in frosty Paleolithic Europe, the fires of our bigger-brained Neanderthal cousins probably sometimes went out and weren't to be reignited for thousands of years.[9] In chapter 12, we'll see how and why such important losses are not surprising.

It's likely that our species' reliance on fire began with the *control of fire*, perhaps obtained from naturally occurring sources. Nevertheless, just capturing, sustaining, and controlling fire requires some know-how. Keeping a fire going may sound easy, but you have to keep it going all the time, during rainstorms, high winds, and long journeys across rivers and through swamps. I learned something about this while living in the Peruvian Amazon among an indigenous group called the Matsigenka. After transporting what looked like a dead, charred log to her distant garden, I saw a Matsigenka woman breathe life back into a hidden ember using a combination of dried moss, which she brought with her, and thermal reflection from other logs. I was also embarrassed when another young Matsigenka woman, with the requisite infant slung at her side, stopped by my house in the village to rearrange my cooking fire. Her adjustments increased the heat, created a convenient spot for my pot, reduced the smoke (and my choking), and eliminated much of the need for my constant tending.[10]

Cooking is also difficult to learn through individual trial-and-error experience. For cooking to provide a digestive aid, it has to be done right. Bad cooking can actually make food harder to digest and increase its toxins. And what constitutes effective cooking depends on the type of food. With meat, doing the most obvious thing (to me) of placing pieces right in the flames can lead to a hard, charred outside and a raw interior—exactly what you don't want. Consequently, small-scale societies have complex repertoires of food-processing techniques that are specific to the food in question. For example, the best cooking technique for some foods involves wrapping them in leaves and burying them in the fire's ash for a long time (how long?). Meanwhile, many hunters eat the liver of their kill raw, on the spot. Livers, it turns out, are energy rich, soft, and delicious when eaten raw—except for those species in which eating the liver can be deadly (do you know which those are?).[11] Inuit hunters don't eat polar bear livers raw because they believe such livers are toxic (and they are correct, according to laboratory research on the question). The rest of the kill is typically butchered, sometimes pounded, possibly dried, and then cooked—though different parts of the kill are cooked in different ways.

The impact of this culturally transmitted know-how about fire and cooking has had such an impact on our species' genetic evolution that we are now, essentially, addicted to cooked food. Wrangham reviewed the literature on the ability of humans to survive by eating only raw foods. His review includes historical cases in which people had to survive without cooking, as well as studies of modern fads, such as the raw foods movement, The long and short of all this is that it's very difficult to survive for months without cooking. Raw-foodists are thin and often feel hungry. Their body fat drops so low that women often stop menstruating or menstruate only irregularly. This occurs despite the supermarket availability of a vast range of raw foods, the use of powerful processing technologies like blenders, and the consumption of some preprocessed foods. The upshot is that human foraging populations could never survive without cooking; meanwhile, apes do just fine without cooking, though they do love cooked foods.[12]

Our species' increasing dependence on fire and cooking over our evolutionary history may have also shaped our cultural learning psychology in ways that facilitated the acquisition of know-how about fire making. This is a kind of content bias in our cultural learning. The UCLA anthropologist, Dan Fessler, argues that during middle childhood (ages six to nine), humans go through a phase in which we are strongly attracted to learning about fire, by both observing others and manipulating it ourselves. In small-scale societies, where children are free to engage this curiosity, adolescents have both mastered fire and lost any further attraction to it. Interestingly, Fessler also argues that modern societies are unusual because so many children never get to satisfy their curiosity, so their fascination with fire stretches into the teen years and early adulthood.[13]

The influence of socially learned food-processing techniques on our genetic evolution probably occurred very gradually, perhaps beginning with the earliest stone tools. Such tools had likely begun to emerge by at least 3 million years ago (see chapter 15) and were probably used for processing meat—pounding, chopping, slicing, and dicing.[14] Drying meat or soaking plant foods may have emerged at any time, and probably repeatedly. By the emergence of the genus *Homo*, it's plausible that cooking began to be used sporadically but with increasing frequency, especially where large fibrous tubers or meat were relatively abundant.

Our repertoire of food-processing methods altered the genetic selection pressures on our digestive system by gradually supplanting some of its functions with cultural substitutes. Techniques such as cooking actually increase the energy available from foods and make them easier to digest and detoxify. This effect allowed natural selection to save substantial amounts of energy by reducing our gut tissue, the second most expensive tissue in our bodies (next to brain tissue), and our susceptibility to various diseases associated with gut tissue. The energy savings from the externalization of digestive functions by cultural evolution became one component in a suite of adjustments that permitted our species to build and run bigger and bigger brains.

How Tools Made Us Fat Wimps

Responding to posters that read "Wanted, athletic men to earn $5 per second by holding 85-pound ape's shoulders to the floor," beefy linebacker types would line up at *Noell's Ark Gorilla Show*, part of a circus that travelled up and down the eastern seaboard of the United States from the 1940s to the 1970s. Inspired to impress the crowds at this star attraction, no man in thirty years ever lasted more than five seconds pinning down a juvenile chimpanzee. Moreover, the chimpanzees had to be seriously handicapped, as they wore "silence of the lambs" masks to prevent them from using their preferred weapon, their large canine teeth. Later, the show's apes were forced to wear large gloves because a chimp named Snookie had rammed his thumbs up an opponent's nose, tearing the man's nostrils apart. The organizers of Noell's Ark Gorilla Show were wise to use young chimps, because a full-grown male chimpanzee (150 lbs.) is quite capable of breaking a man's back. The authorities did finally put an end to this spectacle, but it wasn't clear whether they were concerned about the young apes or the brawny wrestlers who voluntarily entered the ring with them.[15]

How did we become such wimps?

It was culture. As cumulative cultural evolution generated increasingly effective tools and weapons, like blades, spears, axes, snares, spear-throwers, poisons, and clothing, natural selection responded to the changed environment generated by these cultural products by shaping our genes to make us weak. Manufactured from wood, flint, obsidian, bone, antler, and ivory, effective tools and weapons can replace big molars for breaking down seeds or fibrous plants and big canines, strong muscles, and dense bones for fighting and hunting.

To understand this, realize that big brains are energy hogs. Our brains use between a fifth and a quarter of the energy we take in each day, while the brains of other primates use between 8% and 10%. Other mammals use only 3% to 5%. Even worse, unlike muscles, you can't shut down a brain to save energy; it takes almost as much energy to sustain a resting brain as it does an active one. Our cultural knowledge about the natural world combined with our tools, including our food-processing techniques, allowed our ancestors to obtain a high-energy diet with much less time and effort than other species. This was crucial for brain expansion in our lineage. However, since brains need a constant supply of energy, periods of food scarcity—such as those initiated by floods, droughts, injuries, and disease—pose a serious threat to humans. To deal with this threat, natural selection needed to trim our body's energy budget and create a storehouse for times of scarcity. The emergence of tools and weapons allowed natural selection to trade expensive tissues for fat, which is cheaper to maintain and provides an energy-storage system crucial for sustaining big brains through periods of scarcity.[16] This is why infants, who devote 85% of their energy to brain building, are so fat—they need the energy buffer to sustain neurological development and optimize cultural learning.

So, if you are challenged to wrestle a chimpanzee, I recommend that you decline and instead suggest a contest based on (1) threading a needle (a sewing contest?), (2) fast-ball pitching or (3) long-distance running.[17] While natural selection traded strength for fat, increasingly complex tools and techniques drove another key genetic change: the human neocortex sends corticospinal connections deeper into the motor neurons, spinal cord, and brain stem than in other mammals. It is the depth of these connections—in part—that facilitates our fine dexterity for learned motor patterns (recall the plasticity of the neocortex mentioned above). In particular, these motor neurons directly innervate our hands, allowing us to thread a needle or throw accurately, as well as our tongues, jaws, and vocal cords, facilitating speech (see chapter 13). Improved motor control was favored once cumulative cultural evolution began delivering more and finer tools. Such tools also created genetic evolutionary pressures that shaped the anatomy of our hands, giving us wider fingertips, more muscular thumbs, and a "precision grip." Cultural evolution may also have produced packages for throwing, including techniques, artifacts (wooden spears and throwing clubs), and strategies, suitable for using projectiles in hunting, scavenging, raiding, or community policing. The emergence of these, along with the ability to learn to practice throwing by observing others, may have fostered some of the anatomical specializations in our shoulders and wrist, while at the same time explaining why many children are so keenly interested in throwing (more on this in chapter 15).[18]

Alongside these anatomical changes, our species' long history with complex tools has also likely shaped our learning psychology. We are cognitively primed to categorize "artifacts" (e.g., tools and weapons) as separate from all other things in the world, like rocks and animals. Unlike plants, animals, and other nonliving things like water, we think about function when we think about artifacts. For example, when young children ask about artifacts they ask "What's it for" or "What does it do?" instead of "What kind is it?" which is their initial query when seeing a novel plant or animal. This specialized thinking about artifacts, as opposed to thinking about other nonliving things, requires, first, that there be some complex artifacts with nonobvious, or *causally opaque*, functions in the world that one needs to learn about.[19] Cumulative cultural evolution will readily generate such cognitively opaque artifacts, a point I'll make in spades in chapter 7.

How Water Containers and Tracking Made Us Endurance Runners

Traditional hunters throughout the world have shown that we humans can run down antelopes, giraffes, deer, steenboks, zebras, waterbucks, and wildebeests. These pursuits often take three or more hours, but eventually the prey animal drops over, either from fatigue or heat exhaustion. With the exception of domesticated horses,[20] which we have artificially selected for endurance, our species' main competition for the mammalian endurance champion comes from some of the social carnivores, like African wild dogs, wolves, and hyenas, that also engage in persistence hunting and habitually run 6 to 13 miles (10 to 20 km) per day.

To beat these species, we only need to turn up the heat, literally, because these carnivores are much more susceptible to warmer temperatures than we are. In the tropics, dogs and hyenas can only hunt at dawn and dusk, when it's cooler. So, if you want to race your dog, plan a 25-kilometer race on a hot summer day. He'll conk out. And the hotter it is, the more you'll beat him by. Chimpanzees aren't even in our league in this domain.[21]

Comparisons of human anatomy and physiology with those of other mammals, including both living primates and hominins (our ancestor species and extinct relatives), reveal that natural selection has likely been at work shaping our bodies for serious distance running for over a million years. We have a full suite of specialized distance running adaptations, from toe to head. Here's a sampling:

- Our feet, unlike other apes, possess springy arches that store energy and absorb the shock of repeated impacts. This is provided that we learn proper form, and avoid landing on our heels.
- Our comparatively longer legs possess extended springlike tendons, including the crucial Achilles, that connect to short muscle fibers. This setup generates efficient power and provides us with the ability to increase speed by taking longer, energy-saving, strides.[22]
- Unlike animals built for speed, which possess mostly fast-twitch muscle fibers, frequent distance running can shift the balance in our legs upward from 50% slow-twitch muscle fibers to as high as 80%, yielding much greater aerobic capacity.
- The joints in our lower body are all reinforced to withstand the stresses of endurance running.
- To stabilize our trunk while running, our species sports a distinctively enlarged gluteus maximus, along with substantial muscles—the erector spinae—that run up our backbone.
- Coupled with our notably broad shoulders and short forearms, arm swinging creates a compensatory torque that balances us while running. And unlike other primates, the musculature in

our upper back allows our head to twist independently from our torso.

- The nuchal ligament, connecting our heads and shoulders, secures and balances our skulls and brains against running-related shocks. Other running animals also have a nuchal ligament, but other primates do not.

Perhaps most impressive of all are our thermoregulatory adaptations—we are certainly the sweatiest species. Mammals must maintain their body temperature in a relatively narrow range, from roughly 36°C (96.8°F) to 38°C (100°F). The lethal core temperature of most mammals ranges from 42°C (107.6°F) to 44°C (111.2°F). Since running can generate a tenfold increase in heating, the inability of most mammals to run long distances arises from their inability to manage this heat buildup.

To overcome this adaptive challenge, natural selection favored the (1) nearly complete loss of hair, (2) proliferation of eccrine sweat glands, and (3) emergence of a "head-cooling" system. The idea here is that sweat coats and cools the skin through evaporation, which is fostered by the airflow generated by running. To appreciate what happened, note that sweat glands come in two varieties, apocrine and eccrine. At puberty, apocrine glands start producing a viscous pheromone-containing secretion, which is often processed by bacteria to create a strong aroma. These glands are confined to our armpits, nipples, and groin (guess what they are for?). By contrast, eccrine glands, which secrete clear salty water and some other electrolytes, can be found all over our bodies and are much more numerous on us than on other primates. The highest densities of these glands occur in the scalp and feet, the two locations most in need of cooling during running. Measured over body surfaces, no other animal can sweat faster than we do. Moreover, our eccrine glands are "smart glands," because they contain nerves that may permit centralized control from the brain (in other animals sweating is controlled locally). It was these innervated eccrine glands, and not the apocrine glands, that proliferated to cover our bodies during human evolution.

Because brains are particularly susceptible to overheating, natural selection also engineered a special *brain cooling system* in our ancestors. This system involves a network of veins that run near the surface of the skull, where they are first cooled by the ample sweat glands on the face and head. They then flow into the sinus cavities, where they absorb heat from the arteries responsible for transporting blood to the brain. This cooling system may be why humans, unlike so many mammals, can sustain core temperatures above the 44°C (111.2°F) limit.[23]

At this point, you might be thinking that all these features of our bodies are clearly adaptive, so why would I think that it was cultural evolution that created the conditions that led to the evolution of our species' running adaptations. To get at this point, let's look at three aspects of this adaptive design more closely. First, really putting our endurance abilities into action, where they give us the biggest survival advantage, requires running for hours in the heat of the day in the tropics. When our evaporative cooling system kicks into overdrive, a prime athlete will begin sweating out 1 to 2 liters per hour, with 3 liters being well within our bodies' capacity. This system can run, and keep us running, for many hours provided it doesn't run short of a critical ingredient—water. So, where's the genetically evolved water storage system or tank?

Horses, which as I mentioned can compete with us for distance, do have the ability to store large amounts of water. By contrast, not only are humans unable to consume and store large amounts of water, but we are actually relatively poor hydrators compared to other animals. While a donkey can drink 20 liters in 3 minutes, we top out at 2 liters in 10 minutes (camels do 100 liters in the same time). How can this crucial element be missing from our thermoregulatory system? Is our otherwise elegant running design fatally flawed?[24]

The answer is that cultural evolution supplied water containers and water-finding know-how. Among ethnographically known foraging populations, hunters carry water in gourds, skins, and ostrich eggs. Such containers are used in conjunction with detailed, local, culturally transmitted knowledge about where and how to locate water. In the Kalahari Desert of southern Africa, foragers use ostrich eggs as canteens (water containers), which keep the water refreshingly cool, or occasionally use the stomach sacs of small antelopes. They also use long reed straws to suck water from hollow tree trunks, where it collects, and they can readily locate water-bearing roots by spotting certain dry wispy vines. In Australia, hunter-gatherers created water containers using a technique that involved turning small mammals "inside-out" (see figure 5.3). Like the Kalahari foragers, they also used surface signs to locate hidden underground water sources. These techniques are nonobvious: recall that Burke and Wills became trapped along Coopers Creek for want of such know-how.

This reasoning suggests that the evolution of our fancy sweat-based thermoregulatory systems could take off only *after* cultural evolution

Figure 5.3. Water containers used by hunter-gatherers in Australia.

generated the know-how for making water containers and locating water sources in diverse environments. The suite of adaptations that make us stunning endurance runners is actually part of a coevolutionary package into which culture delivered a critical ingredient, water.

Supplied with water, any good marathoner probably has the endurance to chase down a zebra, antelope, or steenbok. However, there's more to persistence hunting than endurance, a lot more. Endurance hunters need to be able to recognize specific target prey and then track that specific individual over long distances. Almost any animal we might want to pursue is much faster than we are in a sprint and will immediately disappear into the distance. To exploit our endurance edge, we need to be able to track a specific individual for several hours, by identifying and reading their spoor and anticipating their actions. The ability to distinguish the target, say a zebra, from other zebras is crucial since many herd animals have a defensive strategy in which they circle back to their herd and try to disappear by blending back into the group. If you can't selectively target the one you've been chasing—the tired one— then you could end up chasing a fresh zebra (a disaster). Thus, persistence hunters must be able to track and identify individuals.

Although many species engage in some form of tracking, no other animal tracks the way we do. Studies of tracking among modern hunter-gatherers reveal that it is an arena of intensive cultural knowledge that is acquired through a kind of apprenticeship, as adolescents and young men watch the best hunters in their group interpret and discuss spoor. From the spoor, skilled trackers can deduce an individual's age, sex, physical condition, speed, and fatigue level, as well as the time of day it passed by. Such feats are accomplished, in part, by knowledge of particular species' habits, feeding preferences, social organization, and daily patterns.[25]

A number of culturally transmitted tricks further aid persistence hunters. The most interesting of these bits of strategy highlights the subtle, adaptive edge that culture-gene coevolution has isolated and exploited. This is a little complicated, so stay with me.

Many four-legged animals are saddled with a design disadvantage. Game animals thermoregulate by panting, like a dog. If they need to release more heat, they pant faster. This works fine unless they are running. When they run, the impact of their forelimbs compresses their chest cavities in a manner that makes breathing during compressions inefficient. This means that, ignoring oxygen and thermoregulation requirements, running quadrupeds should breathe only once per locomotor cycle. But since the need for oxygen goes up linearly with speed, they will be breathing too frequently at some speeds and not frequently enough at other speeds. Consequently, a running quadruped must pick a speed that (1) demands only one breath per cycle but (2) supplies enough oxygen for his muscle-speed demands (lest fatigue set in), and (3) delivers enough panting to prevent a meltdown (heat stroke), which depends on factors unrelated to speed, such as the temperature and breeze. The outcome of these constraints is that quadrupeds have a discrete set of optimal—or preferred—speed settings (like the gears on a stick-shift car) for different styles of locomotion (e.g., walking, trotting, and galloping). If they deviate from these preferred settings, they operate less efficiently.

Humans lack these restrictions because (1) our lungs do not compress when we stride (we're bipedal) so (2) our breathing rates can vary independent of our speed, and (3) our thermoregulation is managed by our fancy sweating system, so the need to pant does not constrain our breathing. Because of this, within our range of aerobic running speeds (not sprinting), energy use doesn't vary too much. That means we can change speeds within this range without paying much of a penalty. As a result, a skilled endurance hunter can strategically vary his speed in order to force his prey to run inefficiently. If his prey picks an initial speed to escape that is just faster than the hunter's, the hunter can speed up. This forces the prey to shift to a much faster speed, which will cause rapid overheating. The animal's only alternative is to run inefficiently, at a slower speed, which will exhaust his muscles more quickly. The consequence is that hunters force their prey into a series of sprints and rests that eventually result in heat stroke. The overheated prey collapses and is easily dispatched. Tarahumara, Paiute, and Navajo hunters report that they then simply strangle the collapsed deer or pronghorn antelope.[26]

Persistence hunters can also take advantage of a wide range of other tricks that increase their edge. In the Kalahari, where this aspect has been most studied, hunters tend to pursue game during midday, when temperatures are hottest, between 39°C (102°F) and 42°C (107.6°F). They adjust their prey choice depending on the seasonally varying health status of their target species, pursuing duiker, steenbok, and gemsbok in the rainy season and zebra and wildebeest in the dry season. They hunt in the morning after a bright full moon (no clouds) because many species will be tired after remaining active on well-lit nights. When chasing a herd, hunters watch for "dropouts" since these will be the weakest members. Nonhuman predators tend to follow the herd, not the loners, since they rely on scent, not sight and spoor. Perhaps not surprisingly, foragers can spot heat stroke in other people and know how to treat it, as they did with one anthropologist who tried to keep up with the locals (an occupational hazard).[27]

Finally, to achieve a running form that maximizes both performance and freedom from injury, humans need to rely on some cultural learning, on top of much individual practice. The evolutionary biologist and anatomist Dan Lieberman has studied long-distance barefoot and minimally shod running in communities around the globe. When he asks runners of all ages how they learned to run, they never say they "just knew how." Instead, they often name or point to an older, highly skilled, and more prestigious member of their group or community and say they just watch him and do what he does. We are such a cultural species that we've come to rely on learning from others even to figure out how to run in ways that best harness our anatomical adaptations.[28]

Thinking and Learning about
Plants and Animals

Over generations, cultural evolution generates a large, and potentially ever expanding, body of knowledge about plants and animals. This knowledge, as we saw with our lost European explorers, is crucial for survival. Given the criticality of this knowledge, we should expect humans to be equipped from a young age with psychological abilities and motivations to acquire, store, organize, extend (via inference), and retransmit this information. In fact, we humans have an impressive *folkbiological cognitive system* for dealing with information about plants and animals. Much research by anthropologists and psychologists, such as the dynamic duo of Scott Atran and Doug Medin, working in diverse human populations have shown that these cognitive systems have several interesting properties. Children rapidly organize information about plants and animals into (1) *essentialized categories* (e.g., "cobras" and "penguins") embedded in (2) *hierarchical (treelike) taxonomies* that permit inferences using (3) *category-based induction* and (4) *taxonomic inheritance*.

These are fancy cognitive science terms for rather intuitive ideas. In using *essentialized categories*, learners implicitly assume that membership in a category (say, "cats") results from some hidden essence deep inside that all members share. This essence cannot be removed by superficial changes to an individual. For example, suppose you operate on a cat and then paint it so that this individual now looks exactly like a skunk. Is it a cat or a skunk? Or, something new, like a "skat" or "cunk"? Children and adults will typically say that it is still a cat which currently looks like a skunk. By contrast, if a table is dismantled and reconstructed as a chair, no one thinks it's still a table. It "is" what it "does." Using *category-based induction* learners can readily extend information learned about one particular cat to all cats—if you see Felix go crazy over catnip, you readily infer that all cats will likely similarly respond to catnip. These essentialized categories are assembled, over development and cul-

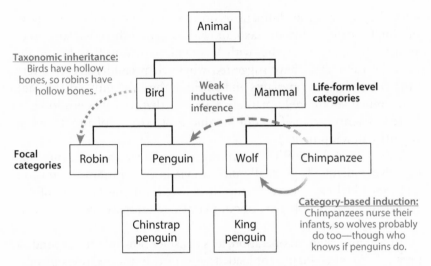

Figure 5.4. Diagram illustrating aspects of folkbiological thinking.

tural evolution, into increasingly complex hierarchical taxonomies, as shown in figure 5.4. With such taxonomies in mind, category-based induction allows people to use their knowledge of one category, say "chimpanzees," to make inferences about other categories. The confidence one puts in these inferences depends on the relationship in one's mental taxonomy. For example, knowing a fact about chimpanzees (e.g., that they nurse their infants), one can readily infer that wolf moms also probably nurse baby wolves, since both are types of mammals. The tree of relationships also allows us to use *taxonomic inheritance*: learners may find out that one of their higher-level categories, like "birds," possesses particular traits, such as laying eggs or having hollow bones. When they encounter a new type of bird, say a robin, they can readily infer that it likely lays eggs and has hollow bones without explicitly learning these facts about robins.[29]

Though these patterns of thinking are quite consistent across diverse small-scale societies, it's worth noting that Western, urban populations appear to have some anomalies in their folkbiological psychology. In small-scale societies, people typically use, and children first learn, focal categories (see figure 5.4) like "robin," "wolf," and "chimpanzee." But the urban-living children and university students that psychologists typically study use what are called "life-form-level" categories, such as "bird" and "fish." Moreover, these urbanites seem to preferentially reason from what they know about humans to other species, rather than just positioning humans within the taxonomy and treating them like any other animal. Comparative studies with Maya children, as well as Americans living in rural areas, have suggested why: urban children receive very little cultural input on plants and animals; thus, the only critters they know much about are humans. In essence, urbanized Western folkbiological systems are malfunctioning due to a poverty of input during cognitive development.[30]

This powerful cognitive system organizes the vast body of information that individuals gradually pick up via both cultural transmission and individual experience during their lives.[31] Of course, most of the knowledge that people have about plants and animals comes to them via cultural transmission.

To see this system in operation, let's consider how infants respond to unfamiliar plants. Plants are loaded with prickly thorns, noxious oils, stinging nettles, and dangerous toxins, all genetically evolved to prevent animals like us from messing with them. Given our species' wide geographic range and diverse use of plants as foods, medicines, and construction materials, we ought to be primed to both learn about plants and avoid their dangers. To explore this idea in the lab, the psychologists Annie Wertz and Karen Wynn first gave infants who ranged in age from 8 to 18 months, an opportunity to touch novel plants (basil and parsley) and artifacts, including both novel objects and common ones, like wooden spoons and small lamps.

The results were striking. Regardless of age, many infants flatly refused to touch the plants at all. When they did touch them, they waited substantially longer before doing so than they did with the artifacts. By contrast, even with the novel objects, infants showed none of this reluctance. This suggests that well before one year of age, infants can readily distinguish plants from other things and are primed for caution with plants. But how do they get past this conservative predisposition?

The answer is that infants keenly watch what other people do with plants and are only inclined to touch or eat plants that other people have already touched or eaten. In fact, once they get the "go-ahead via cultural learning, they are suddenly interested in eating plants. To explore this, Annie and Karen exposed the infants to models who picked fruit from plants and also picked fruitlike things from an artifact of similar size and shape to the plant. The models put both the fruit and the fruit-like things in their mouths. Next, the infants were given a choice to go for the fruit (picked from the plant) or the fruit-like things picked from the object. More than 75% of the time the infants went for the fruit, not the fruit-like things, since they'd gotten the go-ahead via cultural learning.

As a check, the infants were also exposed to models putting the fruit or fruit-like things behind their ears (not in their mouths). In this case, the infants went for the fruit or fruit-like things in equal measure. It seems that plants are most interesting if you can eat them, but only if you have some cultural learning cues that they aren't toxic.[32]

After Annie first told me about her work while I was visiting Yale in 2013, I went home to test it on my six-month-old son, Josh. Josh seemed very likely to overturn Annie's hard empirical work, since he immediately grasped anything you gave him and put it rapidly in his mouth. Comfortable in his mom's arms, I first offered Josh a novel plastic cube. He delighted in grabbing it and shoving it directly into his mouth without any hesitation. Then I offered him a sprig of arugula. He quickly grabbed it, but then paused, looked with curious uncertainty at it, and then slowly let it fall from his hand while turning to hug his mom.

It's worth pointing out how rich the psychology is here. Not only do infants have to recognize that plants are different from objects of similar size, shape, and color, but they also need to create categories for types of plants, like basil and parsley, and distinguish "eating" from just "touching." It does them little good to code their observation of someone eating basil as "plants are good to eat" since that might cause them to eat poisonous plants as well as basil. But it also does them little good to narrowly code the observation as "that particular sprig of basil is good to eat" since that particular sprig has just been eaten by the person they are watching.[33] This is another content bias in cultural learning.

The genetic evolution of our big brains, long childhoods, short colons, small stomachs, tiny teeth, flexible nuchal ligaments, long legs, arched feet, dexterous hands, lightweight bones, and fat-laden bodies was driven by cumulative cultural evolution—by the growing pool of information available in the minds of other people. Beyond our bodies, culture has also shaped the genetic evolution of our minds and psychology, as we've just seen in how people learn about artifacts, animals, and plants. In chapter 7, we'll examine how eons of adapting to a world full of complex and nuanced cultural adaptations, including tools, practices, and recipes, has led our species to be capable of placing immense faith on cultural information, often trumping our own direct experience or innate intuitions. Later chapters will then explore how cultural evolution shaped the genetic evolution of our status psychology, communicative abilities and sociality, eventually domesticating us into the only ultra-social mammal. However, before departing on this journey, I want to alleviate any doubts you have that culture can really drive genetic change.

CHAPTER 6

WHY SOME PEOPLE HAVE BLUE EYES

If you create a global map of eye color but put aside the migrations of peoples in the last few hundred years, you will see that light eyes—blue and green—are common only in a region centered on the Baltic Sea in Northern Europe. Meanwhile, almost everyone else in the world has brown eyes, and there's good reason to believe that brown eyes were universal, or nearly so, prior to the emergence of this pattern of eye color. Here's the puzzle: why are light eyes distributed in this peculiar way?[1]

To understand this, we need first to consider how culture has shaped genes for skin color over the last 10 millennia. Much evidence now indicates that the shades of skin color found among different populations—from dark to light—across the globe represent a genetic adaptation to the intensity and frequency of exposure to ultraviolet light, including both UVA and UVB. Near the equator, where the sun is intense year-round, natural selection favors darker skin, as seen in populations near the equator in Africa, New Guinea, and Australia. This is because both UVA and UVB light can dismantle the folate present in our skin if not impeded or blocked by melanin. Folate is crucial during pregnancy, and inadequate levels can result in severe birth defects like spina bifida. This is why pregnant women are told by their physicians to take folic acid. In men, folate is important in sperm production. Preventing the loss of this reproductively valuable folate means adding protective melanin to our epidermis, which has the side effect of darkening our skin.[2]

The threat from intense UV light to our folate diminishes for populations farther from the equator. However, a new problem pops up, because darker skinned people face a potential vitamin D deficiency. Our bodies use UVB light to synthesize vitamin D. At higher latitudes, the protective melanin in dark skin can block too much of the UVB light and thereby inhibit the synthesis of vitamin D. This vitamin is important for the proper functioning of the brain, heart, pancreas, and immune system. If a person's diet lacks other significant sources of this vitamin, then having dark skin and living at high latitudes increases one's chances of experiencing a whole range of health problems, including, most notably, rickets. A terrible condition, especially in children, rickets causes muscle weakness, bone and skeletal deformities, bone fractures, and muscle spasms. Thus, living at high latitudes will often favor genes for lighter skin. Not surprising for a cultural species, many high-latitude populations of hunter-gatherers (above 50°–55° latitude), such as the Inuit, culturally evolved adaptive diets based on fish and marine animals, so the selection pressures on genes to reduce the melanin in their skin were not as potent as they would have been in populations lacking such resources. If these resources were to disappear from the diet of such northern populations, selection for light skin would intensify dramatically.

Among regions of the globe above 50°–55° latitude (e.g., much of Canada), the area around the Baltic Sea was almost unique in its ability to support early agriculture. Starting around 6000 years ago, a cultural package of cereal crops and agricultural know-how gradually spread from the south and was adapted to the Baltic ecology. Eventually, people became primarily dependent on farmed foods and lacked access to the fish and other vitamin-D-rich food sources that local hunter-gatherer populations had long enjoyed. As a consequence of this combination of living at high latitude and a lack of vitamin D, natural selection kicked in to favor genes for really light skin, so as to maximize whatever vitamin D could be synthesized using UVB light.

Natural selection would have operated on many different genes to favor very light skin among cereal-eating Baltic peoples, because there are many genetic routes to reduce melanin in our skin. One of those genes is called *HERC2* which is located on chromosome 15. *HERC2* inhibits, or suppresses, the production of a protein by a nearby gene called *OCA2*. Suppression of the synthesis of this protein, which occurs through a long and complicated set of biochemical pathways, results in less melanin in people's skin. However, unlike other genes that influence skin color at other places in those pathways, *HERC2* usually causes light eyes because it also reduces the melanin in irises. Blue and green eyes, then, are a side effect of natural selection favoring genes for lighter skin among cereal-dependent populations living at high latitudes. If cultural evolution hadn't produced agriculture, and specifically techniques and technologies suitable for higher latitudes, then there would be no blue or green eyes.[3] In all likelihood then, this genetic variant only started spreading within the last six millennia, after agriculture arrived in the Baltic region.

The point of this example is this: cultural evolution can shape our environments, and consequently, it can drive genetic evolution. In cases of recent culture-gene coevolution, in which the relevant genes have not spread to replace all or most competing genetic variants, we can isolate the causes and effects and sometimes even finger the specific genes being favored. This is important because some researchers have argued that culture could never be strong enough for long enough to drive genetic evolution. Recently, however, new mathematical models and mounting evidence from the human genome provide a clear, if only preliminary, answer. Not only has culture driven specific genes to high frequency in some populations in the last ten millennia, but in fact, sometimes cultural evolution can create selection pressures *more* powerful than seen elsewhere in nature. Sometimes, culture catalyzes and drives more rapid genetic evolution.

To be clear, this book is about how culture drove genetic evolution during the emergence of our species. It's about human nature, *not about the genetic differences among current populations* in our species now. However, I'm going to use the fact that culture-gene coevolution continues today, with many culture-gene interactions still in progress in our species, to illustrate the power of culture to shape the genome. Aside from this chapter, I'll only occasionally be able to link specific genes to the culture-gene coevolutionary processes described. This is for several reasons. First, many of the coevolutionary processes I'm focused on are "completed," such that the traits under selection don't vary across our species. This means that we can't exploit the variation among populations, or what is known about the movements of populations around the world, to infer the underlying causes of the spread of particular genetic variants. Second, many human traits are influenced by many genes at different locations in our chromosomes. This makes it quite difficult to finger specific genetic variants, since any one variant contributes only tiny effects. Finally, this enterprise is really just beginning, so while the broad outlines can be discerned, much more work needs to be done.

Let's consider another example.

Rice Wine and *ADH1B*

In mammals, the alcohol from rotting fruit and other sources is broken down by enzymes produced by alcohol dehydrogenase (ADH) genes and eventually processed into energy and metabolites in the liver. However, if the rate of inflow of alcohol (ethanol) into the liver is too high, then it "overflows," going into the heart and then spreading throughout the body. Intoxication ensues. Most primates aren't particularly good at processing alcohol. However, about 10 million years ago, when our common ancestor with gorillas came down from the trees to spend more time on the ground, rotting fruit probably became a more important food source, so our ape lineage evolved a higher tolerance for consuming alcohol.[4] This ancient adaptation appears to have set the stage for much more recent culture-gene coevolution, as humans have experienced a great deal of evolutionary action on various alcohol-processing genes since the origins of agriculture.

Let's consider just one of those genetic changes. Between 7000 to 10,000 years ago, the DNA on one of these *ADH* genes (*ADH1B*) on chromosome 4 flipped a bit, causing it to code for the amino acid histidine instead of arginine. The evidence seems to suggest that this new version of the *ADH1B* gene metabolizes alcohol much more efficiently in the liver. Perhaps, more important, the rapid breakdown of alcohol produces high levels of acetaldehyde, which causes dizziness, increased heart rate, nausea, weakness, overheating, and a flushing of the skin. The unpleasantness of this flushing reaction reduces people's susceptibility to alcoholism and parallels the effects created by drugs used to treat alcoholism. Estimates vary, but possessing the booze-inhibiting variant *ADH1B* reduces the likelihood of alcohol dependence by a factor of between two to nine times, and both heavy and excessive drinking by a factor of about five. The *more efficient* alcohol breakdown performed by this variant also likely protects the body against drinking binges but may make hangovers worse.[5] Have you ever noticed anyone flushing after drinking a relatively small amount of alcohol? Who was it?

Data on *ADH1B* has been gathered from around the world. It turns out that the booze-inhibiting variant of this gene is distributed rather nonrandomly. Check out figure 6.1. The hottest spot is in southeastern

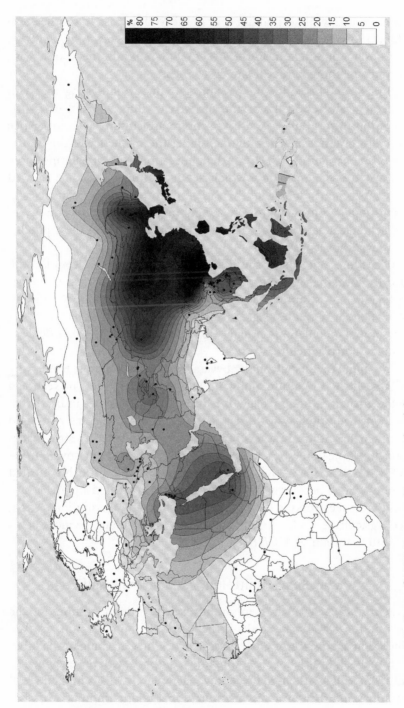

Figure 6.1. Distribute of *ADH1B* gene variant across the globe.

China, with a weak second hot spot in the Middle East. In southeastern China, the frequency of the booze-inhibiting gene goes as high as 99%, with several populations in the 70% to 90% range. In the Middle East, the rates are more in the 30% to 40% range.[6]

Bing Su and his colleagues have brought these findings together with archeological data on the origins of rice agriculture in East Asia—the transition from hunting and gathering to agriculture. The earlier rice agriculture started in a region, the higher the frequency of the booze-inhibiting variant of *ADH1B* in the populations now inhabiting those regions. Knowing the date at which rice agriculture began then allowed them to account for 50% of the variation in this gene's frequencies in Asian populations, which is amazingly high given the uncertainty in archaeological dates and all the other factors at work on these populations over thousands of years.[7]

Okay, fine, but what's the link from agriculture to alcohol? Well, broadly speaking, agriculture and the making of fermented beverages go together. Most hunter-gatherer populations do not have the means, know-how, or resources (e.g., cereals) to make beer, wine, or spirits. Yet agricultural populations usually do, even small-scale, semi-nomadic, slash-and-burn agriculturalists.

In China, the first alcoholic beverages date back almost to the very origins of rice agriculture, along the Yellow River. About 9000 years ago in the ancient farming village of Jaihu, chemical analyses indicate that someone had stored away thirteen pottery jars of a fermented rice-based beverage that probably also contained honey and fruit.[8] It seems that as soon as people domesticated rice, they quickly figured out how to make rice wine. Based on other historical episodes, this probably created alcohol-related problems for rice farmers, which favored any *ADH* variants that make drinking less fun. Without the cultural evolution of, first, rice agriculture and, second, rice wine, there might be no booze-inhibiting variant of *ADH1B*.

Why Some Adult Humans Can Drink Milk

As with most mammals, drinking milk does little or nothing for the nutrition of 68% of the adults in the world. If you are a milk drinker, you are in the minority. Of course, whether they are humans or other mammals, all healthy babies come fully equipped with the enzyme lactase, which allows them to breakdown lactose milk sugars in their small intestines and thereby access the nutritional bounty in milk. Milk is a

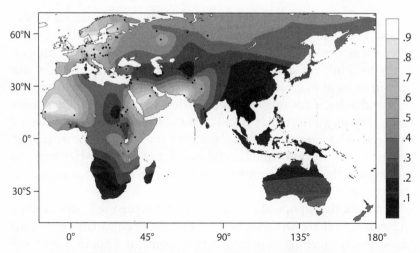

Figure 6.2. Distribution of lactase persistence. Shading indicates the percentages of people in those regions who can digest milk in adulthood.

bonanza of calcium, vitamins, fat, proteins, carbs, and even water. In most people, the production of lactase wanes after nursing ends. By age five, most people can no longer break down the lactose sugars in milk. Even worse, drinking milk often—though not always—causes diarrhea, cramping, gas, nausea, and even vomiting. This is lactose intolerance. In populations without access to medical care, such diarrhea can be deadly.[9]

However, in scattered populations around the world, including groups in Europe, Africa, and the Middle East, people can digest milk throughout adulthood. This *lactase persistence* allows older children, adolescents, and adults to access milk's nutrition. Figure 6.2 shows the global distribution of lactase persistence. Among indigenous populations from the British Isles and Scandinavia, over 90% of people are lactase persistent, while rates in eastern and southern Europe range from 62% to 86%. In India, the rate is 63% in the north but 23% in the south. In Africa, the patterns are strikingly patchy. Some groups have high frequencies of lactase persistence whereas neighboring groups do not. In the Sudan alone, rates range from around 20% up to 90%, depending on the ethnic group. In East Asia, lactase persistence is rare and often nonexistent.

Lactase persistence is under fairly direct genetic control and involves genes that inhibit the typical mammalian shutdown of the production of lactase after nursing ends. Though many factors have shaped the distribution of these regulatory genes, there are two key culturally evolved packages that have driven this bit of genetic evolution. First, it's only in the last 12 millennia that humans have domesticated animals like cows, sheep, camels, horses, and goats, which can potentially provide milk for adults to drink. So some populations adopted cultural practices that permitted them to keep animals and milk them. Such animals provide meat and hides, if nothing else. Initially, the extra milk would have been useful only for young children and infants. However, its presence would have created a genetic selection pressure for extending this lactose-processing ability into middle childhood and beyond. Herding and milking are the first elements in the cultural package that selects for these genes.

Crucially, these populations must also have persisted with herding and milking *but not have* adopted the practices, or culturally evolved the know-how, for turning milk into cheese, yogurt, and kumis. Kumis is a fermented beverage made from mare's milk. While fresh whole milk from cows is 4.6% lactose by weight, cheddar cheese is 0.1%, and tzatziki is 0.3% (tzatziki is a traditional Middle Eastern yogurt-based dish). Some fancy cheeses like Gouda and Brie have only trace amounts of lactose. Thus, cheese and yogurt making are, at least in part, cultural adaptations for reducing lactose, which permit everyone to access much of the otherwise unavailable nutrition from milk. If populations had developed these bodies of technical know-how too soon, the selection pressure on genes for doing the same job would have been weakened. Thus, understanding who is, and is not, lactase persistent requires understanding both how cultural evolution can drive, and how it can inhibit, genetic evolution.

Of course, many other factors influence when and where people can engage in herding and thereby influence the strength of selection for lactase persistence genes. As with blue eyes, the region of northern Europe suitable for agriculture, but with limited UVB, may have created an especially strong selection pressure because of the calcium, proteins, and small amounts of vitamin D in milk. Calcium may inhibit the breakdown of vitamin D in the liver. And where it's cold, fresh milk can be stored for longer periods without having to turn it into cheese.

Elsewhere, such as in the arid deserts of the Middle East and Africa, selection pressures for lactose persistence may have been increased by the water available in milk. Herders capable of drinking camel's milk, for example, might have had an advantage in travelling through arid regions or in surviving droughts. In some African regions, where herding might otherwise be rare or impossible due to extreme temperatures and diseases that plague herd animals, some societies developed cultural adaptations that involved systematically moving the animals to avoid temperature extremes and spacing their herds to suppress pathogen transmission. These populations, despite living in regions full of challenges to a herding lifestyle, are lactose persistent too, perhaps because of their locally specialized herding packages.[10]

Particularly interesting in this case is that natural selection found different ways to create lactase persistence in different populations. It appears that as herding animals became a centerpiece of economies around Eurasia and Africa, natural selection independently found five different genetic variants to suppress the shutdown of lactase production in different populations. In Europe, there was a flip on chromosome 2 just a short distance upstream from the gene that codes for the protein lactase (*LCT*). The DNA base cytosine got swapped for thymine. This flip may simply have thrown a monkey wrench into the otherwise mammalian standard "off switch" for the making of lactase proteins after weaning. Elsewhere, in Africa and the Middle East, the DNA swaps are different, though they are all between 13,000 and 15,000 bases upstream from *LCT*.[11]

Dating of the spread of these genes suggests that one of the African variants may be the oldest, with the European variants in the middle, dating to between 7450 and 10,250 years ago. The variant centered in the Arabian Peninsula is probably more recent, between 2000 and 5000 years old. This timing suggests that the domestication of the Arabian camel may have created the selection pressures that favored this particular variant. The speed of this culture-driven genetic evolution is noteworthy. These selection pressures drove genes for lactose persistence to 32% of the global population in less than 10 millennia. That's really fast compared to the rates observed elsewhere in nature, and even in the human genome.

Before moving on, it's worth highlighting the downside to not recognizing that such culture-gene coevolutionary processes have occurred, and are occurring. The ill effects of promoting milk drinking among peoples without lactase persistence didn't begin to dawn on American researchers until 1965. Prior to this, Americans assumed that if drinking cow's milk was good for "our" (European-descent) children it must be good for everyone's children. In 1946, the National School Lunch program required that fluid milk be part of any student lunch funded by the program. Despite the growing scientific literature, government efforts continued to promote milk drinking for everyone into the 1990s.

As late as 1998, the then Secretary of Health and Human Services even appeared in the well-known Got Milk? commercials. Over the years, many of the famous sports and music stars who appeared in these commercials donning milk moustaches were not, in all likelihood, actually able to digest the very beverage they were promoting.[12]

Culture-Gene Revolutions

These cases of culture-driven genetic evolution are three of the best documented examples we have, though there's every reason to suspect that they represent only the tip of an iceberg. The evolutionary biologist Kevin Laland and his collaborators have already fingered over 100 genes that have likely been under selection, based on analyses of the genome, and have at least plausible cultural origins. These genes influence an immense spectrum of traits ranging from dry earwax and malaria resistance to skeletal development and the digestion of plant toxins.[13] What these cases illustrate for our purposes is as follows:

1. Culture can exert a powerful force on genes, driving genetic evolution. Gene-culture packages emerge and spread rapidly, as with milk-drinking, blue eyes, and booze avoidance.
2. In fact, the selection pressures created by culture can be among the most powerful observed in nature, and broad genetic sweeps can occur in tens of thousands of years. Culture-gene coevolution can be remarkably fast.
3. We can point to specific genes on particular chromosomes, and sometimes even know which molecular base changed. Once-hypothetical genes have now been pinpointed.
4. Once cultural evolution creates the selection pressures, natural selection often manages to find and favor several different genetic variants to address the challenge.
5. However, sometimes cultural evolution can sap the strength of selection, as we saw for populations who rapidly developed cheese- and yogurt-making technology.

One concern with the above examples is that all stem from the emergence of food production—from agriculture and animal domestication. Perhaps this major revolution in human history is a unique event from which we cannot draw general conclusions? To the contrary, my view is that the agricultural revolution just happens to be the best-timed revolution in order for us to detect its causes and effects in our genome. The industrial revolution is too recent, and the revolutions that preceded food production are much older and thus harder to study. Nevertheless, there's every reason to suspect that there was a cooking-and-fire revolution, a projectile-weapons revolution, and a spoken-language revolution, among many others. And as you will see in later chapters, technologically driven revolutions are probably underpinned by revolutions in forms of social organization or institutions. The agricultural revolution is just the one in the temporal sweet spot for today's science.

To see something of this, consider that chimpanzees have two copies of the *AMY1* gene, but humans have, on average, six copies. The gene codes for a protein—amylase—found in saliva that helps break down starch. The extra copies mean that humans, on average, end up having six to eight times more amylase in their saliva than chimpanzees. All other things equal, this means we are better at starch processing than chimps. So, after you beat that chimp in a marathon, you should challenge him to a potato-digesting contest.

Human populations, however, vary in the number of *AMY1* copies they have. Populations who have long been dependent on eating high-starch diets have between 6.5 and 7 copies, on average. The Hadza, African hunter-gatherers who live in savannah woodlands and rely on starchy roots and tubers, have the most, at almost 7 copies, on average, and some Hadza have as many as 15 copies. European-Americans and Japanese are not far behind, at 6.8 and 6.6 copies. By contrast, populations long dependent on low-starch diets have copy counts around 5.5. These include other African hunter-gathers who live in the tropical forest of the Congo basin and herders in both Africa and central Asia, who depend primarily on some combination of meat, blood, fish, fruit, insects, seeds, and honey.[14]

These differences are likely part of a long and meandering evolutionary story and emerged as our ancestors shifted to a heavy reliance on underground roots and tubers over a million years ago. However, just how much populations have depended on starch since then has been influenced by a combination of ecology and cultural evolution, including the practices, preferences, technologies, and know-how of different populations. As we see from the examples above, groups can live relatively close by, in similar ecologies, but still maintain different numbers of *AMY1* genes because they operate with different economic packages.

There is also evidence that culturally prescribed forms of social organization can shape our genome. This is important since some have argued that the forms of social organization created by cultural evolution are too weak or unstable to affect our genes. One important aspect of human social organization is what anthropologists call *postmarital residence*. In many human societies, especially until recently, local norms specified that a newly married couple went to live either with the husband's family or with the wife's family. The first is called *patrilocal residence*, and the second, *matrilocal residence*. Working in three patrilocal and three matrilocal farming populations in Northern Thailand, Hiroki Oota and his colleagues examined the variation in people's mitochondrial DNA and Y chromosomes. Both sons and daughters get their mitochondrial DNA from mom, and only mom. Sons get their Y chromosomes from dad, whereas daughters don't get Y chromosomes at all. If social organization is stable enough to influence the genome, than patrilocal communities should have relatively low variation in their Y chromosomes compared to mitochondrial DNA because sons always stick with their fathers. Similarly, because daughters stick with their mothers, matrilocal communities should show the opposite pattern, low variation in mitochondrial DNA and higher variation in Y chromosomes. This is precisely what Oota's team found, showing that culturally evolved social norms can shape the genome.[15]

Overall, cultural evolution can, and has, powerfully shaped the human genome in a variety of important ways. As we saw in chapter 5, this culture-gene coevolutionary interaction goes well back into our species' history, where culturally transmitted know-how about fire, water containers, tracking, and projectiles were some of the key selection pressures favoring aspects of our anatomy and physiology. Moving forward, I'll begin to focus on how culture created selection pressures on genes that influence our psychology and sociality. In chapter 7, we'll take another step by going deeper into the subtle and nuanced ways cultural evolution can build adaptations without the culture-bearers themselves having any idea what's going on.

Genes and Races

Before moving on, it's worth highlighting a point about genes and race. Anthropologists have long argued that race is not a biological concept. What we mean by this is that the racial categories developed historically by Europeans—such as Caucasian, Negroid, and Mongoloid—do not convey or contain much, if any, useful genetic information, aside from capturing something of the migration patterns of ancient peoples.[16] Detailed studies of the genome, including the research highlighted above, has only served to further underline this point. As we saw, skin-color genes are heavily influenced by a combination of UV radiation and diet, because they effect vitamin D and folate. This means that people in New Guinea and Africa are both very dark skinned despite being from opposite ends of our species' family tree. And very light-skinned Europeans are evolutionarily recent, being mostly a product of agriculture at high latitudes. Other genes have quite different distributions for distinct reasons. For example, we saw that lactase persistent genes are common among indigenous populations in Britain and some African groups, exist at moderate frequencies among Eastern Europeans and Middle Easterners, and remain at low frequencies in other African groups and many Asian populations. Similarly, amylase genes are more common among Japanese, European-Americans, and Tanzanian foragers but less common among Congo forgers and herders in both Tanzania and Central Asia. What does race tell us about these genetic differences?

Nothing. Traditional racial categories just don't tell us anything about this important variation. In fact, the processes I've described above actually make classical racial categories even less informative, since they operate in diverse and nonconcordant ways *within* races to make local groups less similar (e.g., lactase persistent and nonpersistent Africans) while at the same time making different continental races more similar (e.g., amylase genes in Japanese and Americans). The current evidence indicates that natural selection operates in diverse ways on scales much smaller than races and simultaneously on different continents.

Moreover, figures 6.1 and 6.2 show that even using *categories*, racial or otherwise, often distorts the picture. The genetic distributions on these maps vary continuously, so it's best to forget about discrete boundaries. Overall, traditional racial categories capture only about 7% of the total genetic variation in our species, which reveals that races are nothing like the subspecies found in chimpanzees.[17] Given our global distribution and range of environments, our species genetic variation is actually rather limited. Of course, this is not surprising when you realize that in addition to sometimes driving genetic evolution, cultural evolution can also inhibit genetic responses by more rapidly generating cultural adaptations of the kind discussed in chapter 7.[18]

For good historical reasons, many people are sensitive to scientific and evolutionary research on genetic variation, especially variation among populations. In the last century, pseudoscientific efforts to formalize folk concepts of race were used to justify much violence, oppression, and even genocide. However, two rapidly developing areas of research ought to (somewhat) allay concerns about the return of pseudoscientific racism. First, our understanding of human genetic variation, derived from studying actual genes, completely dismantles any remaining shreds of the old racial notions, as the examples above show. The best antidote for pseudoscience is real science. Second, psychologically-oriented researchers have come to increasingly understand how and why humans are susceptible to lumping people together in labeled groups and slapping stereotypes on those groups. As you'll see in chapter 11, racial and ethnic categories arise when cultural evolution taps our human universal tribal psychology to carve up the social world in particular ways. Though these categories are not usually rooted in any important genetic variation, these categories are learned unconsciously and can affect our perceptions, automatic intuitions, and rapid judgments. Increasingly, we understand what triggers prejudice and what the implications of this is for health, education, economics, conflict, and social life.[19] What we need is more evolution-grounded science on genes, culture, ethnicity, and race, not less.

These insights will continue to fuel the spread of a new social construct: the view that all people, and perhaps some other species as well, are endowed with certain inalienable rights—we call these *human rights*. No new facts about genes, biology, or culture can alienate a person from these rights.

CHAPTER 7

ON THE ORIGIN OF FAITH

As one of the world's staple crops, manioc (or cassava) is a highly productive, starch-rich tuber that has permitted relatively dense populations to inhabit drought-prone tropical environments. I've lived on it, both in Amazonia and in the South Pacific. It's tasty and filling. However, depending on the variety of manioc and the local ecological conditions, the tubers can contain high levels of cyanogenic glucosides, which release toxic hydrogen cyanide when the plant is eaten. If eaten unprocessed, manioc can cause both acute and chronic cyanide poisoning. Chronic poisoning, because it emerges only gradually after years of consuming manioc that tastes fine, is particularly insidious and has been linked to neurological problems, developmental disorders, paralysis in the legs, thyroid problems (e.g., goiters), and immune suppression. These so-called "bitter" manioc varieties remain highly productive even in infertile soils and ecologically marginal environments, in part due to their cyanogenic defenses against insects and other pests.[1]

In the Americas, where manioc was first domesticated, societies who have relied on bitter varieties for thousands of years show no evidence of chronic cyanide poisoning. In the Colombian Amazon, for example, indigenous Tukanoans use a multistep, multiday processing technique that involves scraping, grating, and finally washing the roots in order to separate the fiber, starch, and liquid. Once separated, the liquid is boiled into a beverage, but the fiber and starch must then sit for two more days, when they can then be baked and eaten. Figure 7.1 shows the percentage

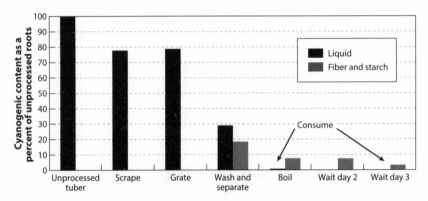

Figure 7.1. Effects of each major step in the Tukanoan manioc processing technique. Percentages are relative to the raw tuber.

of cyanogenic content in the liquid, fiber, and starch remaining through each major step in this processing.[2]

Such processing techniques are crucial for living in many parts of Amazonia, where other crops are difficult to cultivate and often unproductive. However, despite their utility, one person would have a difficult time figuring out the detoxification technique. Consider the situation from the point of view of the children and adolescents who are learning the techniques. They would have rarely, if ever, seen anyone get cyanide poisoning, because the techniques work. And even if the processing was ineffective, such that cases of goiter (swollen necks) or neurological problems were common, it would still be hard to recognize the link between these chronic health issues and eating manioc. Most people would have eaten manioc for years with no apparent effects. Low cyanogenic varieties are typically boiled, but boiling alone is insufficient to prevent the chronic conditions for bitter varieties. Boiling does, however, remove or reduce the bitter taste and prevent the acute symptoms (e.g., diarrhea, stomach troubles, and vomiting). So, if one did the common-sense thing and just boiled the high-cyanogenic manioc, everything would *seem* fine. Since the multistep task of processing manioc is long, arduous, and boring, sticking with it is certainly nonintuitive. Tukanoan women spend about a quarter of their day detoxifying manioc, so this is a costly technique in the short term.[3]

Now consider what might result if a self-reliant Tukanoan mother decided to drop any seemingly unnecessary steps from the processing of her bitter manioc. She might critically examine the procedure handed down to her from earlier generations and conclude that the goal of the procedure is to remove the bitter taste. She might then experiment with alternative procedures by dropping some of the more labor-intensive or time-consuming steps. She'd find that with a shorter and much less labor-intensive process, she could remove the bitter taste. Adopting this easier protocol, she would have more time for other activities, like caring for her children. Of course, years or decades later her family would begin to develop the symptoms of chronic cyanide poisoning.[4]

Thus, the unwillingness of this mother to take on faith the practices handed down to her from earlier generations would result in sickness and early death for members of her family. Individual learning does not pay here, and intuitions are misleading. The problem is that the steps in this procedure are *causally opaque*—an individual cannot readily infer their functions, interrelationships, or importance. The causal opacity of many cultural adaptations had a big impact on our psychology.

Wait. Maybe I'm wrong about manioc processing. Perhaps it's actually rather easy to individually figure out the detoxification steps for manioc? Fortunately, history has provided a test case.

At the beginning of the seventeenth century, the Portuguese transported manioc from South America to West Africa for the first time. They did not, however, transport the age-old indigenous processing protocols or the underlying commitment to using those techniques. Because it is easy to plant and provides high yields in infertile or drought-prone areas, manioc spread rapidly across Africa and became a staple food for many populations. The processing techniques, however, were not readily or consistently regenerated. Even after hundreds of years, chronic cyanide poisoning remains a serious health problem in Africa. Detailed studies of local preparation techniques show that high levels of cyanide often remain and that many individuals carry low levels of cyanide in their blood or urine, which haven't yet manifested in symptoms. In some places, there's no processing at all, or sometimes the processing actually increases the cyanogenic content. On the positive side, some African groups have in fact culturally evolved effective processing techniques, but these techniques are spreading only slowly.[5]

The point here is that cultural evolution is often much smarter than we are. Operating over generations as individuals unconsciously attend to and learn from more successful, prestigious, and healthier members of their communities, this evolutionary process generates *cultural adaptations*. Though these complex repertoires appear well designed to meet local challenges, they are not *primarily* the products of individuals ap-

99

plying causal models, rational thinking, or cost-benefit analyses. Often, most or all of the people skilled in deploying such adaptive practices do not understand how or why they work, or even that they "do" anything at all. Such complex adaptations can emerge precisely because natural selection has favored individuals who often place their *faith* in cultural inheritance—in the accumulated wisdom implicit in the practices and beliefs derived from their forbearers—over their own intuitions and personal experiences. In many crucial situations, intuitions and personal experiences can lead one astray, as we saw with our lost explorers (the nardoo was satisfying). To see this more clearly, let's look at some more cultural adaptations.

Taboos during Breast-Feeding and Pregnancy?

We were eating a large, tasty moray eel when I noticed that Mere was not eating any of the eel, only the manioc. I asked her why she was not eating the eel. I recall Mere saying something like "A tabu; qi sa bukete," which translates as, "It's taboo; I'm pregnant." "Interesting," I thought; this suggested to me that there may be some taboos against consuming certain foods during pregnancy. I had noticed Mere not eating because I'd been worried about eating the moray eel myself, since I'd read that this species is known to carry high levels of ciguatera toxin. Of course, following the ethnographers' axiom, I pressed on eating the eel since no one else seemed at all worried. Many folks were even enthusiastic about the eel, since it has a richer flavor than the typical white fish. This incident, early in my fieldwork in Fiji, sparked my interest and led me to investigate pregnancy practices and food taboos more deeply over the next several years.[6]

To tap her experience with public health research, pregnancy, and breastfeeding, I teamed with my wife, Natalie, on this project. Here's what we found: during both pregnancy and breast-feeding, women on Yasawa Island (Fiji) adhere to a series of food taboos that selectively excise the most toxic marine species from their diet. These large marine species, which include moray eels, barracuda, sharks, rock cod, and several large species of grouper, contribute substantially to the diet in these communities; but all are also known in the medical literature to be associated with ciguatera poisoning. Ciguatera toxin is produced by a marine microorganism that thrives on dead coral reefs. The toxin accumulates up the food chain to achieve dangerous levels in some large and long-living members of these species. The acute symptoms of poisoning, which last about a week, involve diarrhea, vomiting, headache, itchiness, and a distinctive hot-cold reversal on the skin. My village friends say they know they've been poisoned when they bathe. Bathing is always done with cool water, and when poisoned, the water provokes a burning sensation on their skin. These symptoms sometimes return periodically, weeks or even months later. Little is known about the effects of ciguatera toxin on fetuses, though we know that pregnant women have reduced resistance to toxins, and I found cases in the medical literature showing that fetuses can be highly disturbed by ciguatera poisoning. Like other toxins, it seems likely that ciguatera can accumulate in mother's milk and endanger nursing infants. For adults, ciguatera poisoning results in death in a small percentage of cases. While you have probably never heard of ciguatera toxin, it's the most common form of fish poisoning and creates a health problem for any population that routinely consumes tropical reef species.[7]

This set of taboos represents a cultural adaptation that selectively targets the most toxic species in women's usual diets, just when mothers and their offspring are most susceptible. To explore how this cultural adaptation emerged, we studied both how women acquire these taboos and what kind of causal understandings they possess. As adolescents and young women, these taboos are first learned from mothers, mothers-in-law, and grandmothers. However, this initial repertoire is then updated by a substantial portion of women who learn more taboos from village elders and prestigious local *yalewa vuku* (wise women), who are known for being knowledgeable about birthing and medicinal plants. Here we see Fijian women using cues of age, success or knowledge, and prestige to figure out from whom to learn their taboos. As explained in earlier chapters, such selectivity alone is capable of generating an adaptive repertoire over generations, without anyone understanding anything.

We also looked for a shared underlying mental model of why one would not eat these marine species during pregnancy or breast-feeding—a causal model or set of reasoned principles. Unlike the highly consistent answers on *what* not to eat and *when*, women's responses to our *why* questions were all over the map. Many women simply said they did not know and clearly thought it was an odd question. Others said it was "custom." Some did suggest that the consumption of at least some of the species might result in harmful effects to the fetus, but what precisely would happen to the fetus varied greatly, though a nontrivial segment of the women explained that babies would be born with rough skin if sharks were eaten and smelly joints if morays were eaten.

Unlike most of our interview questions on this topic, the answers here had the flavor of post-hoc rationalization: "Since I'm being asked for a reason, there must be a reason, so I'll think one up now." This is extremely common in ethnographic fieldwork, and I've personally experienced it in the Peruvian Amazon with the Matsigenka and with the Mapuche in southern Chile.[8] Of course, it's not particularly difficult to get similar responses from educated Westerners, but there remains a striking difference: educated Westerners are trained their entire lives to think that behaviors must be underpinned by explicable and declarable reasons, so we are more likely to have them at the ready and feel more obligated to supply "good" reasons upon request. Saying "it's our custom" is not considered a good reason. The pressure for an acceptable, clear, and explicit reason for doing things is merely a social norm common in Western populations, which creates the illusion (among Westerners) that humans generally do things based on explicit causal models and clear reasons.[9] They often do not.

Finally, our evidence from Yasawa suggests that these taboos, while causally opaque, do actually work. We compared women's chances of getting fish poisoning during pregnancy and breast-feeding with the rest of their adult lives. Our analyses show that rates of fish poisoning are cut by a third during pregnancy and breast-feeding. Thus, the taboos are cultural prescriptions that reduce fish poisoning.

Why Put Ash in the Corn Mix?

One morning in 1998, when I was living in rural southern Chile and working with the indigenous Mapuche, I arrived at my friend Fonso's farmhouse to find him preparing what he called *mote*, a traditional Mapuche corn dish. He showed me how you have to scoop fresh ash out of the wood stove and put it into the corn mix for soaking, before heating it. I thought that was curious, so I asked him why he mixed the wood ash in with the corn. His answer was, "It's our custom." And a wise custom it is.

In the Americas before 1500 CE, corn was the stable crop for many farming societies. However, relying heavily on corn presents some tricky nutritional issues. A diet based on corn can leave one short on niacin (vitamin B3). Failure to get enough niacin results in a disease called pellagra, a horrible condition characterized by diarrhea, lesions, hair loss, tongue inflammation, insomnia, dementia, and then death. There is actually niacin in corn, but it's chemically bound and cannot be freed by normal cooking. To release this niacin, populations throughout the New World culturally evolved practices that introduced an alkali (a base) into their corn preparations. In some places, the alkali came from burning seashells (generating calcium hydroxide) or the ash of certain kinds of wood. Elsewhere, there were natural sources of lye (providing potassium hydroxide). Mixing the alkali into the recipe in the right way chemically releases the otherwise unavailable niacin in the corn, which stops pellagra in its tracks, and allowed corn-based agricultural populations to grow and spread.[10]

Perhaps mixing nonfood substances, like wood ash or burned seashells, with foods during cooking is easy for a big-brained ape like us to figure out?

History, again, provides us with a natural experiment, because corn was brought from the New World to Europe after 1500. By 1735, some populations in Italy and Spain had already become reliant on cornmeal as a staple, and pellagra had emerged. The condition was theorized to be a form of leprosy or somehow caused by spoiled corn. Pellagra spread across Europe with this new staple crop into Romania and Russia but remained confined mostly to poor populations, who relied on it almost exclusively through the winter—making pellagra the "springtime disease." Experiments were done, and laws were passed to address the problem, by prohibiting the sale of spoiled or moldy corn. This did little to reduce pellagra, since spoilage is not the issue—the Europeans developed the wrong causal model.[11]

Later, pellagra also emerged in the southern United States during the late nineteenth and early twentieth century and spread in epidemic fashion until the 1940s. Millions died, because poor people and institutions, including prisons, sanitariums, and orphanages, had come to rely heavily on diets of cornmeal and molasses. Despite alarms raised by the Surgeon General, special commissions, medical conferences, and private donations to find a cure, the plague raged on for thirty years.

One man, Dr. Joseph Goldberger, investigated orphanages, performed controlled experiments on prisoners, and had begun to construct the right causal model by 1915. However, at the time, the medical community was convinced that pellagra must be an infectious disease, so Goldberger was ineffective and his ideas thought "absurd." Goldberger even injected his wife and friends with blood from people suffering from pellagra to demonstrate the noninfectious nature of the condition. These studies were dismissed by asserting that Goldberger's staff must have been "constitutionally resistant" to the disease.[12]

Thus, not only did people—Europeans and Americans in this case—not figure out the right causal model, but they actively resisted it when it was presented to them by Goldberger. Instead, they preferred to hold firmly to the wrong causal model, probably because the right model was rather less intuitive. Spoiled food and contamination were, and are, relatively "easy to think" about with regard to food compared to the concept of chemical reactions initiated by the introduction of nonfoods, like burnt seashells, into culinary recipes. Cultural evolution had produced a rather nonintuitive fix for the pellagra challenge.

Note, if you are educated and Western, you might be thinking that my numerous examples of toxic plants and animals are merely special cases, because you might be under the impression that few plants need detoxification and that nature's bounty is pure and safe. For many Westerners, "it's natural" seems to mean "it's good." This view is wrong and comes from shopping in supermarkets and living in landscaped environments. Plants evolved toxins to deter animals, fungi, and bacteria from eating them. The list of "natural" foods that need processing to detoxify them goes on and on. Early potatoes were toxic, and the Andean peoples ate clay to neutralize the toxin. Even beans can be toxic without processing. In California, many hunter-gatherer populations relied on acorns, which, similar to manioc, require a labor intensive, multiday leaching process. Many small-scale societies have similarly exploited hardy, tropical plants called cycads for food. But cycads contain a nerve toxin. If not properly processed, they can cause neurological symptoms, paralysis, and death. Numerous societies, including hunter-gatherers, have culturally evolved an immense range of detoxification techniques for cycads.[13] By contrast with our species, other animals have far superior abilities to detoxify plants. Humans, however, lost these genetic adaptations and evolved a dependence on cultural know-how, just to eat.

Divination and Game Theory

Remember from chapter 2 when the chimpanzees and humans each played Matching Pennies? Game theory tells us that the optimal rational strategy involves randomizing, by playing Left (L) or Right (R) with some fixed probability. For example, a player's optimal strategy might be to play R 80% of the time. The humans lost to the chimpanzees because we are bad randomizers, and probably because we tend to copy each other automatically. As I noted, much work in psychology shows that people (well, at least educated Westerners) are subject to the Gambler's Fallacy, in which we perceive streaks in the world where none exist or we believe that we are "due" after an extended losing streak. In fact, we struggle to recognize a sequence of hits and misses as random—instead, we find phony patterns in the randomness. One famous version of this is the hot-hand fallacy in basketball, in which people perceive a player as suddenly better than his long-term scoring average would suggest (it's an illusion). This is a problem for us, since the best strategies in life sometimes require randomizing. We are just not good at shutting down our mental pattern recognizers.[14]

When hunting caribou, Naskapi foragers in Labrador, Canada, had to decide where to go. Common sense might lead one to go where one had success before or to where friends or neighbors recently spotted caribou. However, this situation is like Matching Pennies in chapter 2. The caribou are mismatchers and the hunters are matchers. That is, hunters want to match the locations of caribou while caribou want to mismatch the hunters, to avoid being shot and eaten. If a hunter shows any bias to return to previous spots, where he or others have seen caribou, then the caribou can benefit (survive better) by avoiding those locations (where they have previously seen humans). Thus, the best hunting strategy requires randomizing. Can cultural evolution compensate for our cognitive inadequacies?

Traditionally, Naskapi hunters decided where to go to hunt using divination and believed that the shoulder bones of caribou could point the way to success.[15] To start the ritual, the shoulder blade was heated over hot coals in a way that caused patterns of cracks and burnt spots to form. This patterning was then read as a kind of map, which was held in a prespecified orientation. The cracking patterns were (probably) essentially random from the point of view of hunting locations, since the outcomes depended on myriad details about the bone, fire, ambient temperature, and heating process. Thus, these divination rituals may have provided a crude randomizing device that helped hunters avoid their own decision-making biases. The undergraduates in the Matching Pennies game could have used a randomizing device like divination, though the chimps seem fine without it.[16]

This is not some obscure, isolated practice, and other cases of divination provide more evidence. In Indonesia, the Kantus of Kalimantan use bird augury to select locations for their agricultural plots. The anthropologist Michael Dove argues that two factors will cause farmers to make plot placements that are too risky. First, Kantu ecological models contain the Gambler's Fallacy and lead them to expect that floods will be less likely to occur in a specific location after a big flood in that location (which is not true).[17] Second, as with the MBAs' investment allocations in chapter 4, Kantus pay attention to others' success and copy the choices of successful households, meaning that if one of their neighbors has a good yield in an area one year, many other people will want to plant there in the next year.

Reducing the risks posed by these cognitive and decision-making biases, the Kantu rely on a system of bird augury that effectively randomizes their choices for locating garden plots, which helps them avoid catastrophic crop failures. The results of divination depend not only on seeing a particular bird species in a particular location, but also on what type of call the bird makes (one type of call may be favorable, and another unfavorable).[18]

The patterning of bird augury supports the view that this is a cultural adaptation. The system seems to have evolved and spread throughout this region since the seventeenth century when rice cultivation was introduced. This makes sense, since it is rice cultivation that is most positively influenced by randomizing garden locations. It's possible that, with the introduction of rice, a few farmers began to use bird sightings as an indication of favorable garden sites. On average, over a lifetime, these farmers would do better—be more successful—than farmers who relied on the Gambler's Fallacy or on copying others' immediate behavior. Whatever the process, within 400 years, the bird augury system had spread throughout the agricultural populations of this Borneo region. Yet it remains conspicuously missing or underdeveloped among local foraging groups and recent adopters of rice agriculture, as well as among populations in northern Borneo who rely on irrigation. So, bird augury has been systematically spreading in those regions where it is most adaptive.

This example makes a key point: not only do people often not understand what their cultural practices are doing, but sometimes it may even be important that they don't understand what their practices are doing or how they work. If people came to understand that bird augury or bone divination didn't actually predict the future, the practice would probably be dropped or people would increasingly ignore ritual findings in favor of their own intuitions.

Manufacturing complex technologies is also causally opaque. Consider just one element of the archery package found among hunter-gatherers, the arrow. Let's also pick a society known to possess one of the least complex toolkits known, the hunter-gatherers of Tierra del Fuego, who entered the historical record when they encountered Ferdinand Magellan and, later, Charles Darwin. Among the Fuegians, making an arrow requires a fourteen-step procedure that involves using seven different tools to work six different materials. Here are some of the steps:

- The process begins by selecting the wood for the shaft, which preferably comes from chaura, a bushy, evergreen shrub. Though strong and light, this wood is a nonintuitive choice since the gnarled branches require extensive straightening. (Why not start with straighter branches?)
- The wood is heated, straightened with the craftsman's teeth, and eventually finished with a scraper. Then, using a preheated and grooved stone, the craftsman presses the shaft into the grooves and rubs it back and forth, pressing it down with a piece of fox skin. The fox skin becomes impregnated with the dust, which prepares it for the polishing stage. (Does it have to be fox skin?)
- Bits of pitch, gathered from the beach, are chewed and mixed with ash. (What if you don't include the ash?)
- The mixture is then applied to both ends of a heated shaft, which must then be coated with white clay. (What about red clay? Do you have to heat it?) This prepares the ends for the fletching and arrowhead.
- Two feathers are used for the fletching, preferably from upland geese. (Why not chicken feathers?)
- Right-handed bowman must use feathers from the left wing of the bird, and vice versa for lefties. (Does this really matter?)
- The feathers are lashed to the shaft with sinews from the back of the guanaco, after they are smoothed and thinned with water and saliva. (Why not sinews from the fox that I had to kill for the aforementioned skin?)

Next is the arrowhead, which must be crafted and then attached to the shaft, and of course, there is also the bow, quiver, and archery skills. But I'll leave it there, since I think you get the idea.[19] It's an extensively causally opaque process.

"Overimitation" in the Laboratory

Crucial to making cultural adaptations like manioc, corn, or nardoo processing work is not only faithfully copying all the steps, but also *sometimes* actually avoiding putting much emphasis on causal understandings that one might build on the fly, on one's own. As shown above, dropping seemingly unnecessary steps from one's cultural repertoire can result in neurological disorders, paralysis, pellagra, reduced hunting success, pregnancy problems, and death. In a species with cumulative cultural evolution, but only in such a species, faith in one's cultural inheritance often favors greater survival and reproduction.

Dovetailing with the above field observations, experimental work with children and adults on the fidelity of cultural learning allows us to put a microscope on the cultural transmission process. Recently, psychologists have studied the when and why of people's willingness to copy the seemingly irrelevant steps used by another to get to a reward. In a typical experiment, a participant sees a model engage in a multistep procedure that involves using simple tools to push, pull, lift, poke, and tap an "artificial fruit" (often a large box with doors and holes). The procedure usually results in obtaining some desirable outcome, such as a toy or snack. Some of the steps in the procedure are not apparently required to achieve the goal of getting the reward. Sometimes people even copy steps with no evident material-physical connection to the outcome. Notorious for inappropriately naming of behavioral patterns, psychologists have labelled this not-particularly-shocking phenomenon *overimitation*.

Let's examine a specific experiment that has been tested and replicated with children, adults, and chimpanzees. In the experiment, participants first observe a model engage in a series of steps using a slender rod to access a reward in an "artificial fruit." The fruit is a large opaque box with two entry points. The first entry point is sealed by bolts, which can be (a) *pushed* or (b) *dragged* out of the away—using the rod—to provide access to the tube. This tube, however, merely dead ends—it's a decoy and is irrelevant to obtaining the reward. The second entry point is concealed by a doorway, which can be (a) *slid* or (b) *lifted*. The rod, which has a Velcro tip, can then be maneuvered down the tube to obtain the reward, a sticker for the kids or food for the chimps.[20]

The robust results from these kinds of experiments are that children and adults are rather inclined to copy whatever the model does to obtain the reward. People even copy the irrelevant actions when they are alone, after they think the experiment is over, and when they've been told explicitly *not to copy* any irrelevant actions.[21] However, as we'd expect from chapter 4, people are more likely to copy irrelevant actions when the model is older and higher in prestige. This is also not merely some tendency of little children: assuming the problem is sufficiently opaque, the magnitude of "overimitation" *increases* with age.[22] This also isn't just educated Western peoples. Research in the Kalahari Desert in southern Africa, whose populations lived as foragers until recent decades, show them to be *at least* as inclined to high-fidelity cultural transmission as Western undergraduates.[23]

As you may anticipate, the chimpanzees outperformed their big-headed cousins once again. In this work, the comparative psychologists Vicki Horner and Andy Whiten used the same opaque "artificial fruit" used above and also a clear version of the fruit in which one could readily see that the top slot was not connected to the area with the reward. When the causality was more transparent, with the clear box, the chimpanzees immediately dropped all irrelevant actions, whereas the three- to four-year-old Scottish kids copied the irrelevant actions as much as with the opaque fruit. Chimpanzees did learn some stuff by watching the model work on the fruit: it helped them assess the affordances of the apparatus. They learned how different parts of the fruit could move. But once they had visual evidence that these actions would not do anything, they dropped them.[24] Though chimps clearly have some culture, they aren't a cultural species.[25]

Nevertheless, there's much more to overimitation than this. As we saw in chapter 2, humans possess a certain degree of low-level automatic mimicry—this is one reason why the chimpanzees can zoom in on the optimal Matching Pennies solution but we can't. Second, as we'll see in chapter 8, humans have also evolved to use mimicry to build social relationships and to cue status differences. So we also mimic others to say, "Hey, I wanna relate to you; you're swell." Finally, beginning in chapter 9, we'll see how cultural evolution generates social norms that, if violated, can result in a bad reputation or other punishment. So sometimes people may also "overimitate" to avoid getting a bad reputation as a deviant. Culture-gene coevolution generates many reasons why our species is inclined to copy all the steps or closely follow the local protocols.[26]

Our reliance on cultural transmission, however, goes much deeper. In addition to acquiring practices and beliefs, which may violate our intuitive understandings, we can also acquire tastes, preferences, and motivations. These too can be acquired in the face of our instinctual or innate inclinations. Such acquisitions do not mean we lack instincts or innate inclinations, but merely that natural selection has endowed our cultural learning systems with the ability to, under the right conditions, overwrite or work around them.

Overcoming Instinct:
Why Chili Peppers Taste Good

Why do we use spices in our foods? In thinking about this question keep in mind that (1) other animals don't spice their foods, (2) most spices contribute little or no nutrition to our diets, and (3) the active ingredients in many spices are actually aversive chemicals that evolved to keep insects, fungi, bacteria, mammals, and other unwanted critters away from the plants that produce them.

Several lines of evidence indicate that spicing may represent a class of cultural adaptations to the problem of food-borne pathogens. Many spices are antimicrobials that can kill pathogens in foods. Globally, the common spices are onions, pepper, garlic, cilantro, chili peppers (capsicum), and bay leaves. Here's the idea: the use of many spices represents a cultural adaptation to the problem of pathogens in food, especially in meat. This challenge would have been most important before refrigerators came on the scene. To examine this, two biologists, Jennifer Billing and Paul Sherman, collected 4578 recipes from traditional cookbooks from populations around the world. They found three distinct patterns.[27]

1. Spices are, in fact, antimicrobial. The most common spices in the world are also the most effective against bacteria. Some spices are also fungicides. Combinations of spices have synergistic effects, which may explain why ingredients like chili powder (a mix of red pepper, onion, paprika, garlic, cumin and oregano) are so important. And ingredients like lemon and lime, which are not on their own potent antimicrobials, appear to catalyze the bacteria-killing effects of other spices.

2. People in hotter climates use more spices, and more of the most effective bacteria killers. In India and Indonesia, for example, most recipes used many antimicrobial spices, including onions, garlic, capsicum, and coriander. Meanwhile, in Norway, recipes use some black pepper and occasionally a bit of parsley or lemon, but that's about it.

3. Recipes appear to use spices in ways that increase their effectiveness. Some spices, like onions and garlic, whose killing power is resistant to heating, are deployed in the cooking process. Other spices, like cilantro, whose antimicrobial properties might be damaged by heating are added fresh in recipes.[28]

Thus, many recipes and preferences appear to be cultural adaptations that are suited to local environments and that operate in subtle and nuanced ways not understood by those of us who love spicy foods. Billing and Sherman speculated that these evolved culturally, as healthier, more fertile, and more successful families were preferentially imitated by less successful ones. This is quite plausible given what we know about our species' evolved psychology for cultural learning, including specifically cultural learning about foods and plants.

Among spices, chili peppers are an ideal case. Chili peppers were the primary spice of New World cuisines prior to the arrival of Europeans and are now routinely consumed by about a quarter of all adults globally. Chili peppers have evolved chemical defenses, based on capsaicin, that make them aversive to mammals and rodents but desirable to birds. In mammals, capsicum directly activates a pain channel (TrpV1), which creates a burning sensation in response to various specific stimuli, including acid, high temperatures, and allyl isothiocyanate (which is found in mustard and wasabi). These chemical weapons aid chili pepper plants in their survival and reproduction, because birds provide a better dispersal system for the plants' seeds than other options (like mammals). Consequently, chilies are innately aversive to nonhuman primates, babies, and many human adults. Capsaicin is so innately repellent that nursing mothers are advised to avoid chili peppers lest their infants reject their breast milk, and in some societies, capsicum is even put on a mother's breasts to initiate weaning. Yet adults who live in hot climates regularly incorporate chilies into their recipes. And those who grow up among people who enjoy eating chili peppers not only eat chilies but love eating them. How do we come to like the experience of burning and sweating—the activation of pain channel TrpV1?[29]

Research by the psychologist Paul Rozin shows that people come to enjoy the experience of eating chili peppers mostly by reinterpreting the pain signals caused by capsicum as pleasure or excitement. Based on work in the highlands of Mexico, children acquire this preference gradually, without being pressured or compelled.[30] They want to learn to like chili peppers, to be like those they admire. This fits with what we've already seen: children readily acquire food preferences from older peers. In chapter 14, I will further examine how cultural learning can alter our bodies' physiological response to pain, and specifically to electric shocks. The bottom line is that culture can overpower our innate mammalian aversions when necessary and without us knowing it.

As a product of this long-running duet between cumulative cultural evolution and genes, our brains have genetically adapted to a world in which information crucial to our survival was embedded implicitly in a vast body of knowledge that we inherit culturally from previous generations. This information comes buried in daily cooking routines (manioc), taboos, divination rituals, local tastes (chili peppers), mental models, and tool-manufacturing scripts (arrow shafts). These practices and beliefs are often (implicitly) MUCH smarter than we are, as neither individuals nor groups could figure them out in one lifetime. As you'll see in later chapters, this is also true of some institutions, religious beliefs, rituals, and medical practices. For these evolutionary reasons, learners first decide if they will "turn on" their causal-model builders at all, and if so, they have to carefully assess how much mental effort to put into them. And if cultural transmission supplies a prebuilt mental model for how things work, learners readily acquire and adhere to those.

Of course, people can, and do, attempt to break down complex procedures and protocols in order to understand the causal links between them and to engineer better versions. They also alter practices through experimentation, errors in learning, and idiosyncratic actions. Nevertheless, as a cultural species, we have an instinct to faithfully copy complex procedures, practices, beliefs, and motivations, including steps that may appear causally irrelevant, because cultural evolution has proved itself capable of constructing intricate and subtle cultural packages that are far better than we could individually construct in one lifetime. Often, people don't even know what their practices are actually doing, or that they are "doing" anything. Spicy-food lovers in hot climates don't know that using recipes involving garlic and chili peppers protect their families from meat-borne pathogens. They just culturally inherited the tastes and the recipes, and implicitly had faith in the wisdom accumulated by earlier generations.

Finally, we humans do, of course, construct causal models of how the world works. However, what's often missed is that the construction of these models has long been sparked and fostered by the existence of complex culturally evolved products. When people have accurately speculated on why they do something, this realization often occurs after the fact: "Why do we always do it this way? There must be a reason.... Maybe it's because ..." However, just because some people have speculated accurately as why they themselves, or their groups, do something in a particular way does not mean that this is the reason why they do it. An enormous amount of scientific causal understanding, for example, has developed in trying to explain existing technologies, like the steam engine, hot air balloon, or airplane. A device or technology often pre-existed the development of any causal understanding, but *by existing*, such cultural products opened a window on the world that facilitated the development of an improved causal understanding. That is, for much of human history until recently, cumulative cultural evolution drove the emergence of deeper causal understandings much more than causal understanding drove cultural evolution.[31]

This historical observation is consistent with experimental studies in young children. Research by the developmental psychologists Andrew Meltzoff, Alison Gopnik, and Anna Waismeyer suggests that its exposure to people using artifacts and trying to "do things" that most effectively sparks the causal inference machinery in our minds. Toddlers, for example, will more accurately infer the causal connection between a particular means and a particular end when they observe a person using an artifact than when they observe exactly the same "naturally occurring" physical movements and environmental correlations. That is, children switch their causal inference machinery on in the presence of people actually operating cultural artifacts; the causal models they build help learners better operate the artifact, or engage in the practice, in culturally prescribed ways.[32] More on this later.

Move Over, Natural Selection

Famous evolutionary psychologists, from Steve Pinker to David Buss, are fond of claiming that natural selection is the only process capable of creating complex adaptations that are functionally well designed to meet environmental challenges or the demands of organisms' lives.[33] They are impressed by the fact that products of natural selection—like eyes, wings, hearts, spider webs, nests, and polar bear snow caves—seem well suited or fit to the problems they solve. Save for certain telltale imperfections, these adaptations look well designed, even engineered. Eyes seem crafted for seeing and wings for flying; yet, there's no engineer or designer, and no agents had intentions to create them or a mental model of how they work. I largely agree with this view and certainly share their sense of awe at the stunning power of natural selection. However, I part ways with them on the word "only." At least since the rise of cumulative cultural evolution, natural selection has lost its status as the *only* "dumb" process capable of creating complex adaptations well fit to local circumstances. As this chapter aimed to show, cultural evolution, through the selective attention and learning processes discussed in chapters 4 and 5, as well as others I've not mentioned, is fully cable of generating these complex adaptive products, which no one designed or had a causal mental model of *before* they emerged.

To see this, let's compare two types of houses—two artifacts—one built by natural selection and one built by cumulative cultural evolution. In Africa, male village weaverbirds construct strong, kidney-shaped nests with downward-facing tubular entrances that effectively protect two to three eggs from larger predators. Each species of weaver uses a stereotyped set of techniques to build the house in the same step-by-step pattern. Weavers first create an attachment and then construct a ring, roof, egg chamber, antechamber, and entryway (see figure 7.2). Weaving, in different parts of the house, involves one of three knots (overhand, half hitch, and slipknot) and three different weave patterns. To build the house, weavers must locate and harvest particularly stout strips from tall grasses or palm fronds. The shape of the interior combined with the downward-facing entry tunnel means that predators will have a hard time getting at the eggs. The thickness and layered construction of the woven floor means that the eggs can even survive a fall, should the nest be knocked off its branch. None of these techniques or layouts are learned from other birds. Weavers either just know them innately or are geared to reliably figure them out on the fly, on their own. Natural selection has constructed many such complex artifacts, and invertebrates such as termites, wasps, and spiders make many such beautiful structures without any mental model of their final form.[34]

Inuit snow houses are also a complex adaptation for living in many parts of the Arctic (see figure 7.2). Architecturally, these snow houses are unique in that they are constructed from snow blocks cut from drifts created during a single snowfall to form an aerodynamic dome-shape that can stand against strong Arctic winds. Properly constructed, with blocks cut to fit, this dome is strong enough for a person to stand on without danger of collapse. Heated by small soapstone lamps that are fueled with rendered fat from marine mammals, the insulating properties of snow mean that inside temperatures are 10°C (50°F). This internal warmth slightly melts the snow, thereby allowing the walls and ceil-

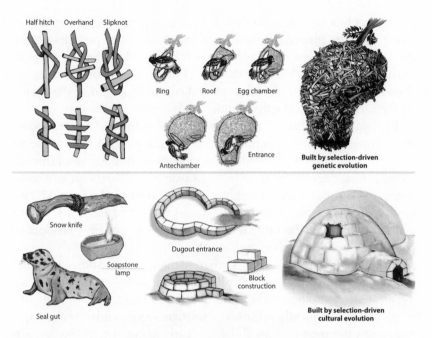

Figure 7.2. Products of different selective processes operating in genetic (*top*) and cultural evolution (*bottom*). Top: In Africa, male village weaverbirds construct strong, kidney-shaped, nests with downward facing tubular entrances that effectively protect two to three eggs from larger predators. Bottom: In the Arctic, Inuit hunters traditionally constructed snow houses using specially crafted bone knives for cutting blocks. These houses were heated with soap stone lamps, often fueled with rendered seal fat.

ing to freeze together even more solidly. Properly oriented, the long tunnel entrance not only blocks the wind, but also uses pressure differences to create a heat trap. Windows, created from translucent membranes cut from seal guts or sheets of ice, provide light inside, and small holes maintain air circulation.[35]

Like village weaver nests, Inuit snow houses look designed and are clearly functionally well fit to life in the Arctic. In fact, they appear to call for a team of engineers with knowledge of aerodynamics, thermodynamics, material science, and structural mechanics. Not surprisingly, facing the real threat of freezing to death in their tents, Franklin's men didn't figure out how to make snow houses. No single individual or even a group of a hundred highly motivated men in this case, could figure this out. It's a product of cumulative cultural evolution and contains features that many or most Inuit builders just learn as "that's the way you do it" without any big causal model. Of course, there's little doubt that bits and pieces of causal models were culturally transmitted along with the procedures, rules, and protocols, since partial or mini-models help builders make sure the parts are working and to adapt to changing or unusual circumstances. However, most of these causal mini-models are themselves transmitted culturally as part of the overall package, not built on the fly by individuals.

Recognizing the power of cultural evolution to produce such adaptive complexity has serious implications for studying humans. It means that when we observe something functionally well suited to address an adaptive challenge outside of conscious awareness, whether it be a snow house or complex cognitive ability (like subtracting 16 from 17), we can't assume that the complexity comes from either natural selection acting on genes or intentional construction. It might be a product of cumulative cultural evolution.

Overall, cultural evolution is smarter than we are, and our species evolved genetically in a world full of cultural stuff—ranging from sophisticated technologies like snow houses to nuanced protocols like using ash to chemically release key nutrients from corn—that people had to just put their faith in. Relatively early in our species lineage, surviving by one's wits alone without leaning on any cultural know-how from prior generations meant getting outcompeted by better cultural learners, who put their efforts into focusing selectively on what and from whom to learn. However, even if you can figure out what to learn and from whom, it doesn't mean that those who possess the most valuable cultural know-how will be motivated to permit you to hang around them and freely tap their accumulated wisdom. It's this evolutionary challenge that gave us prestige.

CHAPTER 8

PRESTIGE, DOMINANCE, AND MENOPAUSE

In *Into Thin Air*, the author Jon Krakauer described the influence of the famed mountaineer Rob Hall at Everest Base Camp. Base Camp provides an interesting situation in which a diverse set of people have been plucked from the modern world and plopped down at 17,598 feet, where they must figure out how to organize themselves at least well enough to complete a challenging task. At the time, Hall was recognized as perhaps the best mountain climber in the world, having summited Everest more times than any other non-Sherpa climber. Krakauer sketched the scene:

> Base Camp bustled like an anthill. In a certain sense, Rob Hall's Adventure Consultants compound served as the seat of government for the entire Base Camp, because nobody on the mountain commanded more respect than Hall. Whenever there was a problem—a labor dispute with the Sherpas, a medical emergency, a critical decision about climbing strategy—people trudged over to our mess tent to seek Hall's advice. And, he generously dispensed his accumulated wisdom to the very rivals who were competing with him for clients.[1]

Rob Hall's ability to influence life at Base Camp arose from his *prestige*. Even in dealing with his rivals in both climbing and business, he was the first among equals. He held this position not because of any formal or official position, but because of a shared sense of respect or admiration for him among those at Base Camp. People sought him out and deferred to his judgments in many domains, including domains that have little to do with climbing skill per se, such as Sherpa labor disputes. Hall responded generously, which only enhanced his influence. Not long after the scene described by Krakauer, Hall froze to death on Mt. Everest as he stayed behind, still high on the mountain, in an effort to save a weakened climber.

Such patterns are not some peculiarity of late twentieth-century Westerners. The same phenomenon emerges all over the world. Consider the isolated inhabitants of the Andaman Islands, a population of egalitarian hunter-gatherers, who were studied by the renowned British social anthropologist A. R. Radcliffe Brown from 1906 to 1908. Radcliffe Brown observed:

> Besides the respect for seniority, there is another important factor in the regulation of social life, namely the respect for certain personal qualities. These qualities are skill in hunting and warfare, generosity and kindness, and freedom from bad temper. A man possessing them inevitably acquires a position of influence in the community. His opinion on any subject carries more weight than that of another even older man. The younger men "attach" themselves to him, are anxious to please him by giving him any presents that they can, or by helping him in such work as cutting a canoe, and to join him in hunting parties or turtle expeditions. . . . In each local group there was usually to be found one man who thus by his influence could control and direct others.[2]

Radcliffe Brown was describing prestige, which creates similar patterns across the globe and back in time. Great climbers and highly skilled hunters, as well as those that excel in other locally valued domains, are sought out, deferred to, and naturally emerge as influential across a wide range of domains. Such respected individuals are rarely ill-tempered or erratic, and instead are often renowned for their generosity. This phenomenon occurs even in societies that are highly egalitarian, possessing no formal leadership roles or hierarchy. Across human societies, prestige is consistently associated with great skill, knowledge, and success in activities or tasks people care about. This prestige status readily forms a foundation for leadership in egalitarian societies.[3]

To understand the psychology that underlies these patterns, let's examine how *prestige psychology* evolved in our species' lineage.[4] The key is recognizing that once humans became good cultural learners, they needed to locate and learn from the best models. The best models are those who seem to possess the information most likely to be valuable to learners, now or later in their lives. To be effective, learners must hang around their chosen models for long periods and at crucial times. Learners also benefit if their models are willing to share nonobvious aspects of their practices, or at least not actively conceal the secrets of their success. As a consequence, humans reliably develop emotions and motivations to seek out particularly skilled, successful, and knowledgeable models and then are willing to pay deference to those models in order to gain their cooperation (pedagogy), or at least acquiescence, in cultural transmission. This deference can come in many forms, including giving assistance (e.g., helping with chores), gifts and favors (e.g., watching their children), as well as speaking well of them in public (thus broadcasting their prestige). Without some form of deference, prestigious individuals have little incentive to allow unrelated learners to be around them and would not be inclined to provide any preferential access to their skills, strategies, or know-how.

This patterning, as it emerged over human evolutionary history, created another opportunity for natural selection to sharpen our cultural learning abilities. When you are just learning a complex skill, it may often be difficult to distinguish a truly great performance from just a good one, or even a mediocre performance (e.g., playing the violin). To solve this problem, a young or naïve learner can watch other more experienced individuals to see who they pay attention to, defer to, and mimic. Our naïve learner can then use this to figure out whom they should begin learning from.[5] As we saw in chapter 4, this represents a kind of second-order cultural learning, in which we figure out from whom to learn by assessing who others think are worthy models. From this, prestige status is born. Individuals who receive this kind of attention, imitation, and deference are prestigious, even if they turn out not to be very knowledgeable or skilled. These cues of prestige, such as visual attention, cause learners to preferentially target their learning efforts. Individuals become increasingly prestigious as other members of their community come to believe they are worthy of respect, deference, and admiration. This occurs even if most of those who come to respect the person cannot—or do not—directly evaluate the person's success, knowledge, or skill themselves. This respect and admiration are the emotions that drive prestige deference. It's why the young Andaman Islanders would "attach" themselves to certain individuals and why they would help him cut a canoe or hunt a turtle.

To understand prestige as a social phenomenon, it's crucial to realize that it's often difficult to figure out what precisely makes someone successful. In modern societies, the success of a star NBA basketball player might arise from his (1) intensive practice in the off-season, (2) sneaker preference, (3) sleep schedule, (4) pregame prayer, (5) special vitamins, or (6) taste for carrots. Any or all of these might increase his success. A naïve learner can't tell all the causal links between an individual's practices and his success (see chapter 7). As a consequence, learners often copy their chosen models broadly across many domains. Of course, learners may place more weight on domains that for one reason or other seem more causally relevant to the model's success. This copying often includes the model's personal habits or styles as well as their goals and motivations, since these may be linked to their success. This "if in doubt, copy it" heuristic is one of the reasons why success in one domain converts to influence across a broad range of domains.[6]

The immense range of celebrity endorsements in modern societies shows the power of prestige. For example, the NBA star LeBron James, who went directly from high school to the pros, gets paid millions to endorse State Farm Insurance. Though a stunning basketball talent, it's unclear why Mr. James is qualified to recommend insurance companies. Similarly, Michael Jordan famously wore Hanes underwear and apparently Tiger Woods drove Buicks. Beyoncé drinks Pepsi (at least in commercials). What's the connection between musical talent and sugary cola beverages? Finally, while new medical findings and public educational campaigns only gradually influence women's approach to preventive medicine, Angelina Jolie's single op-ed in the *New York Times*, describing her decision to get a preventive double mastectomy after learning she had the "faulty" *BRCA1* gene, flooded clinics from the U.K. to New Zealand with women seeking genetic screenings for breast cancer.[7] Thus, an unwanted evolutionary side effect, the influence of prestige across many domains, turns out to be worth millions and represents a powerful and underutilized public health tool.

Having evolved alongside cultural learning in the human lineage, prestige was a latecomer to our status psychology. We humans also possess a *dominance psychology*, which was inherited from our primate ancestors and is thus much older than prestige. In both primates and humans, individuals attain dominance status when others fear them and believe they will use physical violence or other means of coercion if they do not receive deference in the form of appeasement displays and preferred access to mates and resources (e.g., foods). In these hierarchies, subordinates signal their acceptance of a lower rank with displays involving diminutive body positions, including narrowed shoulders and a downward gaze. Dominant individuals remind subordinates of who the boss is with expansive body positions, upright torsos, widely spread limbs, and broadened chests. In some primates, high rank is achieved purely through fighting ability, which is based mostly on size and strength, though coalition partners and kinship also play a role. In chimpanzees, alliances are also often crucial, as pairs or trios establish coalitions in order to secure the top spots in the dominance hierarchy. These rankings are not the unstable products of continuous fighting but often provide a relatively stable social order that is established after periods of fierce conflict. High dominance rank in both males and females generally leads to greater reproductive success, as measured by numbers of surviving offspring.[8]

Thus, because of culture-gene coevolution, humans came to possess (at least) two quite distinct forms of social status, *dominance* and *prestige*. Below, I'll layout how each of these forms of status connects to quite different psychological processes, motivations, emotions and bodily displays. However, before doing that, it's worth considering whether achieving dominance and prestige do both in fact favor greater reproductive fitness in small-scale societies. Reproductive fitness is the key currency that natural selection will seek to increase. If both forms of status are associated with greater fitness in these contexts, it's then at least plausible that both could have evolved genetically, and been sustained, over our species evolutionary history.[9]

Unfortunately, there's very little work linking prestige and dominance to measures of fitness, in part because evolutionary researchers have typically assumed that humans have only one dimension of social status. However, Chris von Rueden and his colleagues have recently studied prestige and dominance among the Tsimané in the Bolivian Amazon as part of a long-term field project. The Tsimané live in relatively independent small family groups that are now clustered in villages along rivers. They hunt, gather, and cultivate gardens scattered throughout the forest. Compared to most societies, their informal status hierarchies are relatively flat and local leaders are weak, making this a challenging population in which to test theories about status.

Chris asked a sample of Tsimané to rank the men in two villages along a number of dimensions, including their fighting ability, generosity, respect, community persuasiveness, ability to get their way, and their number of allies. Each Tsimané man could then be assigned a score based on the aggregate results from his fellow villagers. Chris argued that his measures of fighting ability and community persuasiveness provide the best proxies for dominance and prestige, respectively, in this context. He then showed that both of these proxies for social status are associated with having more babies with one's wife, having more extramarital affairs, and being more likely to remarry after a divorce, even after statistically removing the effects of age, kin group size, economic productivity, and several other factors.[10] Beyond this, the children of prestigious men die less frequently, and prestigious men are more likely to marry at younger ages (neither of these effects held for dominant men). These findings suggest that, at least in this small-scale society, being recognized as either dominant or prestigious has a positive influence on one's total reproductive output (children) or mating success, over and above the consequences that might accrue from factors associated with status, like economic productivity or hunting skills. Not surprisingly, both dominant and prestigious men tended to get their way at group meetings, but only prestigious men were respected and generous.

Key Elements of Prestige and Dominance

Table 8.1 summarizes some of the key elements of prestige and dominance and of the strategies used by individuals to attain and maintain each type of status.[11] Beginning at the top of the table, the successful use of status-seeking strategies based on dominance or prestige leads individuals to have greater *influence* on their group's behavior—on its decisions, movements, and internal dynamics. This effect, combined with the fact that both dominant and prestigious individuals receive deference from lower-status individuals, is what makes them both forms of social status. In dominance relationships, subordinates are influenced by the dominant out of fear; they submit or go along in order not to provoke the dominant. By contrast, because people seek out prestigious individuals due to their perceived success and skill, they become truly persuasive such that learners often shift their underlying opinions, beliefs, and practices to be more similar to those expressed by the prestigious individual. In addition, because lower-status people seek to pay deference to their chosen models in exchange for getting to learn from them, prestigious individuals gain influence as those with lower status seek to please them. Thus, prestigious individuals are influential both because people shift their own opinions and practices to better match

Table 8.1. Patterns of Dominance versus Prestige

Status Features	Dominance	Prestige
Influence	Based on coercion and threat	True persuasion and deferential agreement
Imitation by lower-status	No imitative bias except to satisfy dominant	Preferential, automatic, and unconscious imitation. May include affiliative imitation
Attention by lower-status	Tracking of higher-ups, avoidance of eye contact, and no staring.	Directing of attention to and gazing at higher-ups, watching and listening
Sociolinguistic behavior by higher-ups	Seizure of the floor and use of aggressive verbal intimidation (e.g., disparaging humor and criticism)	Given the "floor" and permitted long pauses. Uses self-deprecating humor
Mimicry by lower-status	No preferential mimicry	Preferential mimicry of higher-ups
Proximity management by lower-status	Avoidance of higher-ups; keeping distance to avoid random aggression	Approach to higher-ups; maintenance of proximity to higher-ups
Displays		
Lower-status	Diminutive body position, shoulder slump, crouching and gaze aversion	Attention to prestigious, open-body position
Higher-ups	Expansive body position, expanded chest, wide stance, arms wide	Similar to dominance display except muted. Less expansive use of space
Emotions		
Lower-status	Fear, shame, fear-based respect	Admiration, awe, admiring-respect
Higher-ups	Hubristic pride, arrogance	Authentic pride, tempered arrogance
Social behavior by higher-ups	Aggression, self-aggrandizement, egocentric	Prosocial, generous, and cooperative
Reproductive fitness	Higher-ups have greater fitness in small-scale societies	Higher-ups have greater fitness in small-scale societies

those of the prestigious, and because people are inclined to go along with prestigious individuals as a form of deference, even if they themselves don't agree.[12]

The effects of prestige on attention, cultural learning, and persuasion are well established (see chapter 4). To go beyond this to explore the idea that both prestige and dominance influence group behavior, I teamed up with my colleague Jess Tracy, an emotion and social psychologist, and our then junior colleague Joey Cheng (who did all the hard work). We formed small teams of strangers and asked the teams to tackle a group challenge called the Lost on the Moon task. Imagining they had just crash-landed on the Moon, each group had to rank a set of items in order of importance. The items were things like a compass, gun, signal flare, and matches. Our participants were told they would be paid according to how similar their team's rankings were to the rankings of NASA engineers. Everyone first created their own personal ranking, and then their teams were assembled to come up with the team's ranking. After this, every individual privately evaluated their fellow team members on a wide range of personal and social dimensions. From these peer-ratings, we assigned each person several scores, including assessments of their prestige and dominance.[13] Using complex statistical analyses, we showed that being either more prestigious or more dominant led to greater influence on their group's task outcome. We measured task outcomes by considering both who the participants thought were influential in determining the team's rankings (subjective assessments) and by examining whose personal rankings were most similar to their team's eventual final rankings (an objective measure).[14]

Both prestige- and dominance-based strategies were clearly distinguishable, revealing in the predicted patterns, and provided independent routes to influencing one's group—different routes to status. Dominant individuals tended to (1) act overbearing, (2) credit themselves, (3) use teasing to humiliate others, and (4) be manipulative. Meanwhile, prestigious individuals (1) were self-deprecating, (2) attributed success to the team, and (3) told jokes.

Imitation, Attention, and Mimicry

Lower-status individuals preferentially *attend* to (watch and listen to) and *imitate* prestigious individuals but not dominant ones. This attention and imitation is usually automatic and unconscious. It may also include bodily mimicry that serves two separate functions. First, mimicry can be an unconscious way of showing deference, of assenting to a person's higher prestige. This works because others are watching for cues to who is being copied, so substantial mimicry can effectively boost the prestige of the person mimicked. Second, mimicry is a tool that we use to help us get into other people's minds—to understand their thoughts and preferences. For example, when two people are having a positive conversational experience, getting to know one another, they will be unconsciously mimicking each other, in their body positions, vocal frequencies, movements, and facial expressions—a patterning known as the Chameleon effect.[15] Interestingly, however, since prestige subordinates are keener on understanding what their higher ups are thinking, wanting, and believing, they engage in relatively more mimicry—that is, subordinates unconsciously mimic prestigious individuals more than vice versa.

One study of vocal mimicry involved CNN's longtime talk-show host, Larry King. Researchers analyzed the low-frequency vocal patterns used by King and his guests to see whether King altered his vocal patterns to match the guest, or vice versa. Prior research had established that one of the ways that conversationalists mimic each other is by syncing up their low-frequency vocal patterns. But who accommodates to whom?

Twenty-five guests were analyzed, ranging from Bill Clinton to Dan Quayle (U.S. vice-president, 1989–93). As expected, when Larry was interviewing someone perceived to be highly prestigious, Larry shifted his vocal frequencies to match his guest's patterns. However, when he was interviewing those perceived to be of lower status than Larry himself, it was the guests who automatically and unconsciously shifted to match Larry's frequency. Larry most strongly accommodated to George Bush, a sitting American president, as well as to Liz Taylor, Ross Perot, Mike Wallace, and a presidential candidate, Bill Clinton. Meanwhile, Dan Quayle, Robert Strauss, and Spike Lee accommodated to Larry. Sometimes neither person shifted to match the other, such as when Larry interviewed a young Al Gore. These conversations were perceived as difficult, perhaps because both individuals saw themselves as being of higher status than their partner, so neither would defer.[16]

Part of figuring out who to learn from is attending to whom others are looking at, listening to, and emulating because, in a complex world, doing so can point us in the right direction: toward models we should be learning from. At least, this was the case for most of our species' evolutionary history. However, in the modern world, this aspect of our psychology may explain how someone can be famous for being famous—the *Paris Hilton effect*.[17] The nature of our media means that, without trying, many people end up attending to whomever the popular media is covering. An initial media exposure, accidental or by design, creates attention cues that cause people to unconsciously perceive someone as a worthy model. This means that we see others consistently watching certain celebrities and hear others talking about those celebrities because such people provide a shared point of reference for everyone who watches the same media. These attention cues can cause our prestige psychology to automatically infer that these individuals are worthy of our imitation, respect, and admiration. This nonconscious inference also causes increased emulation and mimicry in this group, as they seek physical, or at least social, proximity. This shift can then create a feedback loop, as the media continues to cover those who people want to know more about, and a celebrity is born—seemingly from nothing. This process got started because the initial media coverage caused some people to mistakenly infer that others were attending to a particular person. A parallel kind of runaway phenomena is described by Duncan Watts for the emergence of renowned paintings like the *Mona Lisa* and chart-topping popular songs.[18]

Status Displays and Emotions

Being or achieving high status has characteristic bodily displays that are predictable given the nature of the underlying relationships. In primates, as in many other species, high dominance rank is associated with large size. Even in humans, by 10 months infants expect sheer size to favor success in conflicts between two agents, even if those agents are merely rectangular shapes with faces.[19] Consequently, it's not surprising that dominants signal their status by "looking large" standing upright, expanding their chests, and spreading their limbs apart—think professional wrestlers or male baboons. Dominants also glare at others, looking for any hint of a challenge. For prestige, the displays of higher-ups appear to be muted or toned-down versions of the dominance display. These are pride displays with the aggressive elements replaced or suppressed. This makes sense, since prestigious individuals want to signal their status but do not want to inadvertently communicate aggression.[20] Figure 8.1 shows bodily displays of newly achieved prestige after victories in an Olympic judo competition. The image pairs a status display

Figure 8.1. Pride displays made by a congenitally blind (*right*) and sighted judo player (*left*) after victories in competition.

from a congenitally blind judo competitor with that of a sighted one. Can you tell which person has never actually seen a pride display by another person?[21]

The differences between dominance and prestige displays can be seen clearly in videos we took of our teams during our Lost on the Moon task. By systematically coding the video, we found that individuals pursuing a dominance-strategy (1) occupied more space, (2) used a wider posture, and (3) positioned their arms further away from their bodies than did those pursuing prestige. Meanwhile, prestigious individuals were more inclined to tilt their heads up, expand their chests, and smile. We also found that dominant, but not prestigious, individuals lowered their vocal pitch over the course of the interaction.

Focusing on emotions, psychologists have independently distinguished two forms of pride, which, though they have labelled them as *hubristic pride* and *authentic pride*, correspond closely to dominance-based pride and prestige-based pride (see table 8.1). Hubristic pride is the affective experience of seeking or achieving high status by controlling others through force or force threat, and authentic pride arises from seeking or achieving high status through the admiration of others based on one's competence, skill, success, or know-how in valued domains. Some evidence has also begun to reveal that achieving prestige or dominance can be linked to distinct hormonal responses.[22]

Displays by lower-status individuals are also distinctive. In a dominance relationship, the submissive displays of subordinates are in many ways the opposite of those made by dominant individuals. Subordinates try to "look small," shrinking their bodies, postures, and presence. They also avert their gaze from dominants, though they still keep track of them to avoid random acts of aggression. Submissive displays are associated with the emotion shame.[23] By contrast, lower-status individuals in a prestige hierarchy need to approach and engage the prestigious individual, hand around him or her, and actively and openly defer. Public displays of deference are particularly effective since they generate more prestige for their recipient. Aside from approach, attention, and an open body posture, this pattern is not highly distinctive except in contrast to the other displays described above. The relevant emotions are admiration, awe, and respect not based on fear.[24]

Why Prestigious People Are Often Generous

When asked by ABC's Christiane Amanpour about how he got involved in the Giving Pledge, billionaire Tom Steyer replied, "The invitation to me was a phone call from Warren Buffett. If he thinks it's a good idea, I start with the assumption it is a good idea." Warren Buffett, known as the "Oracle of Omaha," was ranked among the most admired and respected people in the world. His Giving Pledge asked billionaires to promise to give half of their wealth away, an amount totaling $600 billion. At the time of this 2010 interview, Buffett, working with Bill and Melinda Gates, had already signed up 40 other billionaires. The trio began by first taking the pledge themselves and giving away quite a bit of money. As of January 28, 2015, 128 billionaires had also pledged to give away half of their wealth.

Buffet and the Gateses have actually taken a page from an old playbook. During the early centuries of Christianity, an explicit campaign by people such as Saint Ambrose, the archbishop of Milan, made giving to the poor admirable. Rich Christians began to compete to see who could give the most to the poor (often through the church), inspired by paragons like Ambrose, who gave all their wealth away. Prior to this, giving to the poor was puzzling (at best) since the poor had little or nothing to give back. This move may have been crucial to the long-run success of the Church as an organization (and, no doubt, the poor appreciated it too).[25]

For the same reason, charitable organizations open their efforts to raise money by featuring donations from highly prestigious individuals, whose generosity is subsequently made known to all other potential donors. When Brooke Astor, the esteemed grande dame of New York philanthropy and a Medal of Freedom winner, gave generously to the New York Public Library, three donations immediately followed: from Bill Blass, Dorothy and Lewis Cullman, and Sandra and Fred Rose. Each mentioned the inspiration provided by Brooke's substantial donation. This copycat philanthropy is a well-established tool of charitable organizations.[26]

The psychology described above helps us understand why people might copy the generosity of particularly prestigious individuals, but why might highly prestigious individuals willingly take the lead and go first? We've seen this emerge everywhere, from Everest Base Camp to the Andaman Islands. Whereas dominants seek to manipulate others for their own ends, prestigious individuals tend to be generous and cooperative. Prestigious individuals can clearly benefit by not being aggressive, to avoid scaring away those who might pay deference. However, why would they be particularly generous or cooperative? It's not obvious from the evolutionary ideas presented so far.

The reason lies in our cultural nature. When a highly successful hunter achieves local recognition for his abilities (prestige), it means that when he actively cooperates, by pitching in during a turtle hunt or by supplying a community feast, others will copy his actions, inclinations, and motivations. Thus, by behaving altruistically, and because they are role models for others, prestigious individuals can increase the overall prosociality of their local groups or their sections of the social network. This, of course, means that any altruism is only altruism in the short-term sense. In the longer run, prestigious individuals who behave generously get to live in a social network that, by virtue of their own actions, becomes more generous and cooperative. For example, by causing others to donate, Brooke Astor gets to live in a better city with at least an excellent public library. By contrast, if low-status individuals behave altruistically, no one is likely to copy them or their motivations, so the social worlds they live in won't improve with their generosity. For this reason, I suspect that natural selection has psychologically linked the achievement of prestige with prosocial inclinations, especially generosity.

The psychological link is so tight that in many places, where not everyone will know who is prestigious, generosity actually turns into a cue of prestige. That is, cultural evolution has sharpened up this link so that attending to who is the most magnanimous is sometimes the best way to figure out who, locally at least, is the most prestigious. Anthropologists call some of these traditional communities "big man societies" because men can increase their prestige with vast generosity.[27] We don't live in such a society (well, at least I don't); however, as with the Giving Pledge, this strand of human nature does emerge in some important contexts.

In controlled laboratory settings, behavioral experiments confirm the link between prestige and generosity. In one experiment, researchers paired individuals who had just participated in a trivia contest. The trivia contest aimed to create a minor status distinction between the players, since one player had received a gold star for his or her performance (high prestige) while the other did not (low prestige). The actual assignment of the gold star was arbitrary, though players no doubt assumed that the stars marked outstanding performance in the trivia contest. Players then engaged in a series of sequential economic interactions with different players in which each had a chance to contribute money to a joint effort. If both players contributed money, both prospered and received more money. If only one contributed, the other (noncontributor) prospered while the contributor lost money.

The results reveal the power of prestige: when the gold-starred player had the opportunity to contribute money first, he or she tended to contribute to, and thus cooperate in, the joint effort, and then the following player—the low-prestige person—usually did as well. So, everyone won. However, when the low-prestige player got to contribute money first (or not), he or she tended not to contribute to the joint project (not cooperate), and then, neither did the high-prestige player. Even when the low-prestige player cooperated first, the high-prestige player still tended not to. Thus, not only did low-prestige players tend to copy the cooperative tendencies or actions of the high-prestige player, but high-prestige players responded with cooperation only when he or she knew the low-prestige player would follow him or her. Creating cooperation here, and enhancing everyone's profits, depended crucially on the high-prestige player going first.

What I find most amazing about this and related experiments is how such a relatively minor cue as one's apparent performance in a trivia contest can yield such substantial effects on cooperation. Other experimental work shows how prestige can (1) influence prices in markets such that higher-prestige individuals reap a disproportionate share of the benefits, and (2) help groups coordinate on mutually beneficial outcomes.[28] All these experiments indicate that prestige can be harnessed to foster cooperation if the organization or institution is structured with an understanding of prestige.

Prestige and the Wisdom of the Aged

In about 1943, a band of hunter-gatherers faced a severe and enduring drought in the Western Desert of Australia. With their normal water sources failing, an old man named Paralji led his band to increasingly distant water holes, only to find them dry or insufficient. After traveling far across their vast territory and checking more than two dozen water holes, Paralji faced having to lead his band to their last tribal water refuge, a place he had only been to once in his life, during his manhood initiation rite a half century earlier. When the band finally arrived, their last refuge was jammed with people from at least five other tribal groups.

Soon local food supplies at the refuge began to fail. Confronting disaster, Paralji recalled the ceremonial song cycles that his people periodically performed at rituals. The songs told of the wanderings of ancestral beings and included a sequence of places and names. Relying on these ancient lyrics to direct him, Paralji headed off into territories unknown to him, followed by several young men and their families. Combining the information in the songs with trail markings, Paralji led the group along a chain of fifty to sixty small waterholes and across 350 kilometers of desert, eventually arriving at Mandora Station, on Australia's west coast. The group had been saved by their ritual songs and by the distant memories of an old man.[29]

As chapter 4 points out, older people not only have a lifetime of their own direct experience, such as Paralji's trip to the water refuge during his initiation rite, but they also have had a lifetime of cultural learning opportunities, to memorize things like ritual songs. Once we became a sufficiently cultural species, capable of selectively focusing on and learning from certain models, older individuals often emerged as important information resources. By opening the informational floodgates between generations, cultural transmission changes the relationship between younger and older individuals. By contrast, in noncultural species, not only is the information accumulated by older individuals limited to what they can acquire through their own experience, but it's also of little consequence to others since they usually lack the psychological abilities to obtain it. Thus, in species with cultural learning, while aging individuals may be physically declining, they still possess transmittable know-how that makes them increasingly valuable to younger generations.

This accumulated knowledge may explain why the elderly are prestigious in most, if not all, traditional societies. In an extensive cross-cultural survey of the role of the aged in 69 small-scale or traditional societies, 46 societies included explicit mentions of respect, deference, reverence, homage, or obeisance to the aged, while in 5 more societies, this could be readily inferred. The remaining cases simply made no mention of how the elderly were treated, rather than suggesting the elderly were treated with reduced deference or respect. Across these societies, the aged receive many perks as part of this prestige deference. Elderly Tasmanians, for example, got to eat the best food, while aged Omaha were exempt from having to scarify themselves after someone died, and mature Crow got out of many unpleasant tasks. Meanwhile, leadership positions and governing councils were often restricted to people of a certain vintage.[30]

Crucially, many of these ethnographic accounts explain why the elderly were revered: because they possess an abundance of knowledge in important domains such as lore, magic, hunting, rituals, decision-making, and medicine. Consistent with this view, these accounts also make clear that the aged rapidly lose status and deference when their mental faculties begin to decline or they appear incompetent. Based on his extensive review, one researcher observed, "The most striking fact about respect for old age is its widespread occurrence ... practically universal in all known societies." The evolutionary reason is that older age is often a cue that someone is likely to possess knowledge or wisdom, and for this, we humans grant prestige status. It's also why most other animals don't respect their elderly.[31]

In many small-scale societies, institutions or social norms also endow the senior members of a community with dominance by giving them control of land, resources, inheritance, or marriage decisions. So the aged may sometimes simultaneously possess both dominance and prestige, just as many supervisors do in our modern institutions. Nevertheless, as discussed in the Lost on the Moon task, it's important to keep prestige and dominance conceptually separate, since the underlying cognitive and emotional patterns are distinct, as are the implications for cooperation.

If the elderly are so often prestigious across human societies, why aren't they particularly admired or respected in many Western societies? To answer this, we return to the evolutionary logic. The aged are accorded prestige and deference when more decades of experience and learning can provide a proxy measure for accumulated knowledge and wisdom. However, if a society is rapidly changing, then the knowledge accumulated by someone over decades will become outdated rather quickly. Age is only a good proxy if the world faced by the new generation is pretty similar to that faced by the oldest generation. Consider, for example, that the elderly of today grew up in a world without computers, email, Facebook, Google, smartphones, apps, or online libraries. They typed on manual typewriters, mailed handwritten letters, went to bookstores, and could only date people they met in person or through friends and family. In our rapidly changing modern societies, the accumulated knowledge of the elderly is less valuable than it might otherwise be. In fact, the faster things change, the younger and younger the best and most competent models get.

Menopause, Culture, and Killer Whales

What are the implications of the fact that once we are a cultural species, decades of accumulated individual and cultural learning make us increasingly valuable to younger generations? The longer we live, the more information we accumulate, and the potentially more valuable we are as transmitters of this wisdom, provided the world is relatively stable during one lifetime (which it probably was for most of our species' evolutionary history).

Under these conditions, natural selection should favor extending our lives in order to give us time to transmit our accumulated know-how to our children and grandchildren and to make sure they have the time and opportunity to learn what they will need. As individuals, our cultural stock is going up over the decades while our physical skills are going down, as are our abilities to produce high-quality babies. At a certain point, those lines cross, and it's time to stop reproduction and focus all of our efforts on the current children and grandchildren. However, given our declining physical abilities, one of the major ways we can help our younger relatives, especially in traditional societies, is by dispensing our accumulated wisdom. This is why humans, but not other primates, live for decades beyond when we stop reproducing, and even live past when we stop being economically productive. This longevity not only emerges in modern societies, but has also now been shown among hunter-gatherers and other small-scale societies, and likely dates back tens or even hundreds of thousands of years into the Paleolithic. By contrast, chimpanzees and other primates do not possess a long post-reproductive life. Death usually follows in relatively short order after reproduction ends.[32]

Direct evidence for this idea is just beginning to accumulate, though it's clear that the presence of nonreproductive grandmothers often increases the survival of their grandchildren.[33] The debate now centers on whether nonreproductive grandparents do this via the informational and prestige-status benefits related to cultural transmission, like Paralji, or whether it's by their contributions of labor, such as digging up tubers. I suspect both labor (e.g., child care) and information are important contributions. However, the key question remains as to why such nonreproductive individuals stick around in humans while they are largely absent in most other species, especially in other primates. My answer is, in humans, older individuals can give something that older members of other primate species cannot: information. In a cultural species, older individuals can transmit valuable know-how in addition to any helping out.

In the Fijian villages I work in, for example, grandmothers and grandfathers are crucial information sources. Older women advise their daughters and granddaughters, among others, on the fish taboos for pregnancy and breast-feeding that I discussed in chapter 7, as well as helping and counseling on issues related to birthing, nursing, infant care, weaning foods, weaving, cooking, social norms (etiquette), and medicinal plants. During work parties, older men attend but don't do much actual work. Instead, they provide administration and advice on activities related to house construction, turtle butchery, feast preparation, fish netting, gardening, and ritual performance.[34]

As a result of the selection pressures created by the opportunities for cultural transmission as we age, both male and female humans tend to cease reproduction at least two to three decades prior to death, which provides enough time to ensure that the last of their children is sufficiently well equipped. This effect is particularly relevant to females, since they carry most the expensive reproductive equipment. By shutting down their reproductive systems, the lives of women can be extended in order to provide more time for transmitting cultural information and for making sure their children and grandchildren are sufficiently prepared. In males, there's less natural selection can do to extend their lives, though men's testosterone levels and virility do decline, and most men in small-scale societies do in fact cease reproduction when their wives do.

Only our lineage crossed the barrier into a regime of cumulative cultural evolution and culture-gene coevolution. However, this idea—that the wisdom of a lifetime of experience may make the older members of a social group more valuable, and as a consequence, cause natural selection to extend their lives by halting or reducing reproduction—should still be observable in other species. To examine this, let's consider two of only a few species that live decades past when they stop having babies, killer whales and elephants.

Killer whales have big brains, long lives, and menopause. Estimates suggest that killer whales, like a few other species of toothed whales, live another twenty-five years after menopause, which is long enough to see the oldest of their grand-offspring reach sexual maturity. If menopause is a genetic adaptation to augment the length of females' lives in order to give them an opportunity to exploit the knowledge they've gained over a lifetime in cultural transmission, then this species ought to be both fairly cultural and possess a social structure in which this information can be put to use to help their relatives.

Though more research is needed, a preliminary look suggests that they have the predicted ingredients. First, there's much variation across killer whale groups in behavioral practices, foraging tactics, and communicative calls. Some groups, but not others, have developed techniques for how to take fish from fishing trawlers, and at least one group has a team technique in which one individual uses bubbles to scare salmon or herring into a clump near the surface where his buddies slap the clump with their tails, stunning the fish. Different groups also seem to possess different ecological information, for example, about the timing and location for catching particular salmon species. Second, experimental work suggests that killer whales are impressively good imitators, so cultural information can potentially flow through social networks and across generations, which probably explains many of the enduring behavioral differences observed among killer whale groups. Third, killer whales may engage in some of the most impressive teaching outside of our own species. In some places, young killer whales appear to learn from their mothers how to beach themselves to capture elephant seal and sea lion pups, and some observations suggest that killer whale mothers facilitate this learning process in various ways. For example, moms push their calves up the beach to get the prey and rescue them when they get stuck on the beach. Finally, detailed demographic studies of killer whales confirm that adult males, even those over age 30, are more likely to survive if their mother is still around. This study can't tell us what mom is doing for her adult sons, but she definitely matters.[35]

The opportunity to dispense the information accumulated over decades is possible in killer whales because the females remain in stable family groups. These matrilineal groups associate with related families, who are probably sister lineages, to form pods. So, knowledgeable grandmothers often have the chance to use their knowledge to benefit most or all of their close relatives, which may be the key selection pressure giving rise to menopause.

The story is similar for elephants.

In 1993, a severe drought hit Tanzania, resulting in the death of 20% of the African elephant calves in a population of about 200. This population contained 21 different families, each of which was led by a single matriarch. The 21 elephant families were divided into three clans, and each clan shared the same territory during the wet season (so they knew each other). Researchers studying these elephants analyzed the survival of the calves and found that families led by older matriarchs suffered fewer deaths of their calves during this drought.

Moreover, two of the three elephant clans unexpectedly left the park during the drought, presumably in search of water, and both had much higher survival rates than the one clan that stayed behind. It happens that these severe droughts occur only about once every four to five decades, and the last one hit in about 1960. After that, sadly, elephant poaching in the 1970s killed off many of the elephants who would have been old enough in 1993 to recall the 1960 drought. However, it turns out that exactly one member of each of the two clans who left the park, and thus survived more effectively, were old enough to recall life in 1960.[36] This suggests that, like Paralji in the Australian desert, they may have remembered what to do during a severe drought and led their groups to the last water refuges. In the clan who stayed behind, the oldest member was born in 1960 and so was too young to have recalled the last major drought.

More generally, aging elephant matriarchs have a big impact on their families, because those led by older matriarchs do better at identifying and avoiding predators (lions and humans), avoiding internal conflicts, and identifying the calls of their fellow elephants. For example, in one set of field experiments, researchers played lion roars from both male and female lions and from either a single lion or a trio of lions.

For elephants, male lions are much more dangerous than females, and of course, three lions are always worse than only one. All the elephants generally responded with more defensive preparations when they heard three lions instead of one. However, only the older matriarchs keenly recognized the increased danger of male lions over female lions and responded to the increased threat with elephant defensive maneuvers. This greater knowledge does in fact cash out, because older matriarchs, while not reproducing themselves, do appear to increase the reproductive success of their family group, and their knowledge is passed to their offspring and grand-offspring.[37]

My point is that under the right conditions, natural selection will favor extending the lives of individuals in order to provide them with opportunities to exploit and transmit the information they've gleaned over a lifetime. Selection also favors attending to, learning from, and respecting the senior members of one's community when they are likely to possess valuable cultural information. This goes for humans as well as less cultural species like elephants and killer whales.

Leadership and the Evolution of Human Societies

Exploring how culture-gene coevolution has shaped our species' status psychology is crucial for understanding the emergence of political institutions. In egalitarian societies, which lack hierarchical institutions, prestige lays a crucial foundation for politics and economics. As we saw above, even the smallest-scale foraging societies are disproportionately influenced by prestigious individuals, whose status is rooted in success or skill in locally valued domains like hunting or warfare. In traditional societies living in richer environments, prestigious men use their persuasive abilities, influence, and generosity to expand their sphere of influence in competition with other prestigious big men. In some places these competitions result in epic feasts at which these individuals seek to enhance their prestige by giving away more than their competitors, crushing them with their productivity, organizational skills, and generosity. These "big men," which is often the literal local translation, can accumulate substantial influence in their lifetimes, though when they die, little of this influence passes to their descendants. Similarly, understanding dominance helps illuminate the psychological underpinnings of hierarchical institutions, such as those based on hereditary chiefs or divine kings. Many modern institutions harness both forms of status as they aspire to promote individuals, based on merit, skill, success, and knowledge, into positions of dominance, where they control the costs and benefits to others (e.g., salaries, promotions, and vacations).

Effective institutions often harness or suppress aspects of our status psychology in nonintuitive ways. Take the Great Sanhedrin, the ancient Jewish court and legislature that persisted for centuries at the beginning of the Common Era. When deliberating on a capital case, its seventy judges would each share their views, beginning with the youngest and lowest-ranking member and then proceed in turn to the "wisest" and most respected member. This is an interesting norm because (1) it's nearly the opposite of how things would go if we let nature take its course, and (2) it helps guarantee that all the judges got to hear the least varnished views of the lower-ranking members, since otherwise the views of the lowest-status individuals would be tainted by both the persuasive and deferential effects of prestige and dominance. Concerns with dominance may have been further mitigated by (1) a sharing of the directorship of the Sanhedrin by two individuals, who could be removed by a vote of the judges, (2) the similar social class and background of judges, and (3) social norms that suppressed status displays.

These customs are not something that smart people often just think up, and even when they do, such practices are hard to implement. This is so because the high-status members of such deliberative bodies tend believe their views deserve special attention and want the opportunity to speak first to increase their influence on outcomes. Converging with their colleagues, low-status members are often disinclined to speak first, out of a fear of looking ill informed or of contradicting the high-status individuals, who haven't yet spoken. Thus, neither high- nor low-status individuals would necessarily be particularly supportive of a low-to-high-status speaking rule unless they understood something of status psychology and were more concerned about the institution's long-term success than their personal influence and careers. Professors in university departments, for example, regularly meet to discuss "important" issues and then vote. In my experience in departments of anthropology, psychology, and economics, the spontaneous speaking order is almost always from high to low prestige, except that often the youngest and most junior professors don't say anything at all. Similarly, though the Supreme Court of Canada uses the same speaking protocol as the Great Sanhedrin, the U.S. Supreme Court goes the opposite way, beginning with the Chief Justice and proceeding down from there.[38]

Across human societies, we see that seeking prestige, often more than wealth itself, drives much human behavior. However, prestige derives from success, skill, or knowledge in locally valued domains. While not infinitely malleable, what constitutes a *valued domain* is amazingly flexible. The differential success of societies and institutions will hinge, in part, on what domains are valued. How respectable was it to excel in reading, inventing machines, memorizing ancient texts, having children, obtaining additional wives, or growing yams?

I'll close this chapter with a lesson in leadership that reflects an intuitive grasp of prestige from the English explorer James Cook. In 1768, as Lieutenant Cook was preparing to depart for the South Pacific, scurvy continued to plague the British navy, as it had for centuries, killing many sailors. Scurvy symptoms start with spongy gums and a general malaise, which is followed by bleeding from the nose and mouth and the loss of teeth. If vitamin C is not consumed, this decline ends in death. On the suggestion of an English physician, Cook obtained a large supply of sauerkraut, which we now know will prevent scurvy. Since sauerkraut represented a rather pungent and unusual deviation from the traditional maritime fare, Cook worried that his sailors would refuse to eat it and knew that neither force nor education were likely to succeed in creating an enduring dietary shift. Instead, he ordered that plates of sauerkraut be dressed and served at the officers' mess, but not at the sailors' mess. Within a week of setting sail, inferring that the officers had a taste for sauerkraut, the rank-and-file crew began actively requesting servings of sauerkraut. Very quickly, sauerkraut became so desired that it had to be rationed. Cook finished his expedition with not a single case of scurvy, a feat theretofore unheard of among Europeans during such long ocean voyages.

CHAPTER 9

IN-LAWS, INCEST TABOOS, AND RITUALS

One evening, in the village of Teci (pronounced "tethi") on Yasawa Island (Fiji), I was drinking kava by lantern light at a crowded social gathering. Made from powdered roots mixed with water, kava is a ritually served beverage that numbs the tongue and imbues one with a peaceful feeling. In Fijian style, we were all sitting on comfortable woven mats, with the higher-status (older) men sitting toward the private end of the one-room house and lower-status people arrayed toward the other end. On this particular evening, I had achieved a small anthropological victory, as I'd managed to sit in the middle of the room with my age-mates and had not been ushered immediately to the high-status end, as a guest. Just as I was wondering when the next round of kava would be served, I looked up to see a neighbor, Kula, appear at the open doorway. He spotted me immediately (I stand out among Fijians) and noticed the open space next to me, in the otherwise crowded room. Flashing a big smile, the young man crouched according to custom as he made his way toward me. As Kula slid into position, while greeting me, his back pressed accidentally against a young woman. Almost immediately, with laughter breaking out, Kula was poked by his cousin and told that the girl behind him wanted to chat with him. Kula turned around to see who was behind him. He immediately looked terrified as he realized that, in the dimly lit room, he'd sat down beside, and inadvertently brushed against, his "sister." This behavior, though unintentional, was completely inappropriate and embarrassing. Of course, as usual, I was initially confused and not laughing, because I still hadn't quite pieced together what had just happened. With shame oozing down his body, Kula stood up and quickly exited, disappearing into the darkness. He did not return that evening.

Kula had sat down next to one of his many "classificatory sisters," who by most readers' kinship taxonomy would have been Kula's distant cousin. In these communities, as in many small-scale societies, certain types of cousins are labeled as "brothers" and "sisters," and they are supposed to be treated like one's "real" (genetic) siblings. In anthropological parlance, these classificatory siblings are your *parallel cousins*, which include your father's brother's children and your mother's sister's children, but not your mother's brother's or father's sister's children. These other cousins, your parents' opposite-sex siblings' children, are your *cross-cousins*, who are kind of like your official friends and potential lovers. Following the same logic, people also have classificatory siblings through their great-grandparents and beyond. Kula had violated, in a small way, the local incest taboo, which prohibits any direct interactions with one's opposite-sex siblings, real or classificatory. This taboo requires that opposite-sex siblings avoid interacting at all, which precludes talking or even sitting near each other. Of course, sex or marriage is out of the question, as is touching or being alone together. The logic here is that any touching or talking could blossom into sex and marriage, so it's best to nip this in the bud.

Kula had infringed on this incest taboo by sitting next to, and incidentally touching, his sister-cousin. Almost gleefully, Kula's cross-cousin highlighted his mistake by suggesting that he talk to his sister-cousin— an action that would have made things much worse (it was a joke). Cross-cousins have reciprocal and egalitarian relationships, which are reinforced and affirmed by constant joking. This joking relationship is totally unlike the relations of respect and authority between, for example, real and classificatory brothers of different ages.[1]

Kula's mishap opens a window on how traditional societies operate and organize themselves, which illuminates something of the social worlds experienced by both our ancestors and much of the world today. Even the smallest-scale human societies—unlike primate societies—are built on and organized around a set of kinship norms. While no doubt grounded in innate psychological processes that influence how we locate and treat our close genetic relatives and reciprocal partners, these social norms variously reinforce, extend, and suppress aspects of our genetically evolved psychology. Building on this idea over the next three chapters, I will show how the emergence of social norms drove a genetic evolutionary process of *self-domestication* that dramatically shaped our species' sociality. To begin, in this chapter I'll introduce you to some of the ways in which cultural evolution grabbed hold of our innate psychology and harnessed it to expand human groups and our social networks. This created new forms of social organization that intensified the cooperation and sociality in our evolutionary lineage. Along the way, we'll take a closer look at social norms related to marriage, fatherhood, incest, and rituals. In chapter 10, we'll see how intergroup competition has long shaped cultural evolution to favor the proliferation of prosocial, or group-beneficial, norms and the formation of more-complex institutions (packages of social norms). These norms and institutions have long been important selection pressures on our species' genetic evolution. Then, in chapter 11, we'll bring all these observations together and focus on the impacts of this culture-driven process of self-domestication on our psychology.

This view contrasts sharply with the canonical view of the evolution of human cooperation. For decades, evolutionary researchers, from Richard Dawkins to Steven Pinker, have argued that humans are able to organize and cooperate so effectively because our psychology has been shaped by the evolutionary forces of kin selection and reciprocal altruism (reciprocity).[2] Our kin psychology evolved genetically because it permits us to bestow help or benefits on individuals who are genealogically related to us and thus likely to share particular altruistic genes. Our reciprocity psychology emerged as natural selection readied us to take advantage of the potential for ongoing tit-for-tat exchanges of benefits (or costs) with others.[3] In what's to come, I will be substantially augmenting and amending this canonical view. We'll see that not only are kin selection and reciprocal altruism insufficient to explain cooperation in the modern world, or in other complex societies, they are insufficient to explain cooperation in small-scale societies, including nomadic hunter-gatherers. So, though humans certainly do possess innate proclivities for helping our kin and engaging in reciprocity,[4] these are, in and of themselves, too weak or narrowly delimited to explain cooperation in real human societies. For example, though motivations to help close relatives can be strong, even in small foraging bands, the average other person is a quite-distant relative, and bands contain many nonrelatives.[5] By studying a diversity of real small-scale societies, we'll see how understanding human cooperation and sociality requires exploring how our social instincts are harnessed, magnified, and recombined within an interlocking web of culturally evolved social norms.

We'll see that cooperation, even in small, nomadic hunting-and-gathering societies, hinges on the existence of culturally constructed norms that substantially augment our innate proclivities.

Social Norms and the Birth of Communities

At some level, it should be uncontroversial that culture shapes the kinds of kinship relations that I described at the outset of this chapter. Kula's behavior, when he sat down with me, was being monitored by other members of his community, who started giggling among themselves almost as soon as he sat down. The question is, why did the community care that Kula sat down next to his classificatory sister? If the pair's shared ancestor had been opposite-sex siblings instead of same-sex siblings, then Kula could have brushed her back intentionally and made sexual jokes toward her without a negative reaction from his fellow villagers. Instead, shame crashed over Kula, and an otherwise enjoyable party was ruined, because he felt compelled to vanish into the night.

The origins of this form of sociality lie in our cultural learning abilities, which became increasingly sophisticated over our species' evolutionary history. As we saw in chapter 4, simply by observing others we can acquire ideas, beliefs, values, mental models, tastes, and motivations. Growing up in Teci, as in many other places throughout the world, means gradually acquiring and internalizing the notion that sexually mature males and females who stand as classificatory siblings ought not to interact in direct ways. Cultural learning means that it's possible for people to acquire notions of how people *should* behave, both toward others and even in purely nonsocial situations. Deviations from "proper behavior" evoke negative emotions toward the deviant, even in uninvolved third-party observers. In chapter 11, we'll see that even very young children respond negatively to violations of completely arbitrary rules.

In exploring culture-gene coevolution so far, we've begun to see how effective cultural learning is for acquiring tastes for performing all manner of costly actions, including even food preferences that require overcoming our innate aversions, as we saw with chili peppers. Building on such empirical observations, evolutionary researchers have been using mathematical modeling tools to ask this question: what happens when people culturally learn from others, and then those acquired behaviors, strategies, beliefs, or motivations influence future social interactions?

The answer from *cultural evolutionary game theory* is that *social norms* spontaneously emerge. Groups of individuals who engage in social interactions and learn from each other using cues like success and prestige often end up sharing similar behaviors, strategies, expectations, or preferences, and deviations from these shared standards are penalized or sanctioned in some way. Or, in some cases, individuals come to share standards for valuing uncommon excellence, for rewarding individuals for going above and beyond the standard. Either way, the resulting behavioral patterns are stable in the sense that they tend to endure and resist efforts by one or a few individuals to change them.[6]

In both the real world and in many of these mathematical models, norm violators are sanctioned by the effects of reputation. When individuals break social norms, it often doesn't impact them immediately, though it may. Rather, observers of the violation spread the word about what happened, and this gossip has negative consequences downstream, for some later interaction. What is often underappreciated is that reputation itself is merely a type of cultural information that spreads because of many of the same psychological abilities that underpin other types of culture. Once our ancestors could learn from each other, say about which foods to eat or how to make a tool, we could also learn from each other about whom not to build a long-term relationship with for activities like hunting, sharing, mating, and raiding. Sophisticated language is not necessary, since I can convey my feelings about an incestuous norm violator to a friend in the same way I communicate my feelings about vegetarian hot dogs to my wife (using my disgust expression).

Two other interesting results arise from studies using cultural evolutionary game theory. First, it turns out that any behaviors, underpinned by certain beliefs, strategies, or motivations, that call for individuals to pay personal costs, such as not eating a tasty type of food (e.g., bacon) or not having sex with an attractive distant cousin, can be sustained by cultural evolution, for example, through reputational damage. Norms even make nonsocial behavior (e.g., masturbation) into social behaviors, because uninvolved third parties come to care about such behavior. Second, social norms will tend to remain stable even when they help neither the group nor the individual. In fact, cultural evolution can produce sticky social norms that are bad for everyone. Ethnographic examples are numerous and range from cutting off the clitorises of young girls (female genital cutting) to consuming the brains of dead relatives at funerals (which can transmit a deadly prion disease).[7]

Social norms make it possible for humans to solve—often without anyone understanding how—what would otherwise be inescapable social dilemmas. Social life is riddled with chances to exploit others, which most people don't even notice. And the more individuals interact and trust each other, the greater the opportunities there are to exploit others—to cheat or free ride on the efforts of others. Culture has several tools and some secret tricks, but two are most important. First and foremost, it brings in third parties to monitor, reward, and sanction others based on local culturally transmitted and widely shared rules. When necessary, it incentivizes third-party actions in some way, often to sanction norm violators. Second, by providing mental models of situations and relationships, it directs our attention away from opportunities to exploit others and reframes situations in ways that tap or harness our instincts in distinct, and often prosocial, ways. Behaviors like smoking, eating horse meat, or littering can go from perfectly acceptable to disgusting, once new culturally transmitted mental connections are made. This is how, over tens of thousands of years, cultural evolution forged primate troops into human communities. Now, let's take a deeper look at how social norms have shaped small-scale societies.

From Kin to Kinship

To understand how cultural evolution has shaped our kinship systems, and our forms of social organization, over our species' evolutionary history, I'll use nonhuman primates as reference points (hereafter, I'll just say "primates" instead of "nonhuman primates"). This make sense, since if you go back far enough, our ancestors were just another primate. By drawing lessons from across the primate order about kin relationships and forms of social organization, we can begin to infer what cultural evolution and culture-gene coevolution has done and what it continues to do.

Let's start with marriage. Marriage institutions are sets of social norms, including beliefs, values, and practices that regulate and reinforce our pair-bonding instincts. By firming up this somewhat flimsy bond, marriage norms can buttress spousal relationships and create affinal (in-law) relationships. They can also strengthen the paternal side of a child's kinship network. The innate psychological foundation of marriage is a long-term pair-bonding instinct, which humans appear to share with some other apes, including gorillas and gibbons, as well as with some monkeys. This instinct can be thought of as a potential strategy, which may be deployed depending on the context. We don't have to pair bond (it's not like peeing), but it's one of the things we'll be inclined to do under some circumstances. The term "pair-bonding" is often confused with notions of monogamy. It's important to realize that pair-bonding does not imply monogamous mating. Pair-bonds form between dyads, but a single individual can have multiple pair-bonds. Gorillas, for example, often form long-term pair-bonds with multiple females at the same time. In humans, both historically and cross-culturally, individuals often pair-bonded and married more than one other person at a time—85% of human societies permitted polygamous marriage in some form. Here, pair-bonding refers to enduring, or at least not ephemeral, relationships between mates.[8]

Marriage, often with its accompanying rituals and gift exchanges, involves the community in a couple's pair-bond. That is, community members are third parties who monitor (gossip about) and potentially sanction those who violate marriage norms. Widely shared standards of behavior prescribe economic, social, and sexual roles, as well as the obligations and contributions required by each spouse and their relatives. Cross-culturally, marriage norms govern such arenas as (1) whom one can marry (e.g., incest taboos), (2) how many partners one can marry (polygamy?), (3) what the inheritance rights are and who a "legitimate" heir is, (4) where the new couple will live, with the wife's parents (matrilocal) or with the husband's parents (patrilocal), and (5) what the rules are regarding sex outside the pair-bond.

By helping to guarantee that a male is in fact the genetic father of the offspring produced by his mate, pair-bonding brings males into the raising of offspring, or at least makes them more tolerant of their mates' offspring. *Paternity certainty* captures the notion that, in some species, males have to worry about whether they are the genetic father.[9] All other things being equal, the more paternity certainty a male has, the more willing he will be to invest in his mate's offspring. In many primates, including chimpanzees, females mate promiscuously, so males usually have little or no idea who their offspring are, and they don't care much.[10] Even in pair-bonding primates, male investment is pretty minimal, such as in gorillas, where males act only to protect their mates and their offspring from other males.[11]

By reinforcing pair-bonds, marriage norms can make better fathers, and failing that, they can make more fathers, as you'll see below. Most, but not all, societies have social norms that regulate the wife's sexual fidelity (i.e., no cheating), and about one-quarter also constrain the husband in some way. Both kinds of norms can increase the man's investment in his wives' children. Social norms about sexual fidelity mean not only that the husband is monitoring his wife's sexual and romantic life, but so is the rest of the community, making it much tougher for the wife to behave in ways that might lower the husband's confidence that his wife's children are indeed his children. This has a psychological impact on the husband, motivating him to invest more in his wife's offspring (because they are more likely to be his). Wives also know that if they are caught violating fidelity norms (e.g., having sex with someone else), it will influence their reputation with people well beyond their current husband and his kin.

On the husband's side, norms that constrain his sexual behavior also inhibit—not prevent—him from diverting resources away from his family in efforts to obtain extramarital sexual opportunities—that is, have affairs, pay prostitutes, etc. Again, for the same reason, a community is now monitoring him, and violations of these norms can affect his relationships well beyond that with his wife and her kin. By curbing his ability to freely divert resources in seeking sex, social norms about fidelity can help channel the husband's resources to his wife's children. Of course, in societies that permit or encourage men to have multiple wives (polygyny), men are likely to use any extra resources or wealth to obtain more wives.

By binding husbands and wives together, for longer than would otherwise be the case, and increasing paternity certainty, some marriage norms can create, or at least strengthen, links to the husband's relatives, including his parents, siblings, and even kids he might have with other wives in a polygynous society. For children, this dramatically expands their kinship networks, firming up their links to paternal grandparents, aunts, and uncles. While close genealogical relatedness often does not underpin affinal relationships (though it may), there remains a common evolutionary interest. For both the wife and husband, marriage norms create *affines* (in-laws, those whom you are related to through marriage), which come with both benefits and responsibilities, as we will see below. My sister-in-law, Ilyse, and I do not share any genetic variants through our recent common descent (we aren't related), but we both share a genetic interest in my children, who are genetically related to both of us.

To my knowledge, no evidence indicates that this shared interest has been exploited by natural selection in primates, probably because a species needs to both live in larger social groups and have enduring pair-bonds, which primates are not good at. In chapter 16, I'll return to the question of how and why pair-bonding might have emerged in our particular evolutionary lineage.

Making Dads

In building a broader kinship network, social norms and practices connect a child more tightly to his or her father's side of the family in subtle ways. In contrast to many complex societies, mobile hunter-gatherer populations usually emphasize kinship through both the mother and father and permit new couples much flexibility in where they can live after marriage. However, there's always that problem of paternity certainty for the dad's entire side. Among the Ju/'hoansi, mobile hunter-gatherers in the Kalahari Desert in southern Africa, social norms dictate that a newborn's father—or, more accurately, the mother's husband—has the privilege of naming the child. These norms also encourage him to name the child after either his mother or father, depending on the infant's sex. Ju/'hoansi believe name-sharing helps the essence of the paternal grandparents live on, and it consequently bonds both the grandparents and the father's whole side of the family to the newborn. Relatives of the grandparents often refer to the newborn using the same kinship term they use for his or her older namesake—that is, the grandfather's daughter will call the newborn baby "father."[12]

This bias to the father's side is particularly interesting since Ju/'hoansi kinship relationships are otherwise quite gender egalitarian, emphasizing equally the links to both mom's and dad's sides of the family. This biased naming practice may help create that symmetry by evening out the imbalance that paternity uncertainty leaves behind. In many modern societies, where social norms favoring the father's side have disappeared, the effect of paternity certainty emerges as *maternal* grandparents, uncles, and aunts invest more than the same *paternal* relatives do.[13] Thus, Ju/'hoansi practices link newborns directly to their father's parents and simultaneously, via the use of close kin terms like "father" and "sister," pull all of dad's relatives closer.

More broadly, in Ju/'hoansi society, sharing the same name is an important feature of social life, which has many economically important implications. Psychologically, creating namesakes may work in two interlocking ways. First, even among undergraduates and professors, experiments suggest that sharing the same, or even a similar, name increases people's liking for the other person, their perceptions of similarity, and their willingness to help that person. In one study, for example, professors were more likely to fill out a survey and mail it back if the cover letter was signed by someone with a name similar to their own name. The perception of similarity suggests that namesakes may somehow spark our kin psychology, since we already know we use other cues of similarity (appearance) to assess relatedness.[14] Second, even if this same-name trick doesn't actually spark any change in immediate feelings, it still sets the appropriate social norms—the reputational standards monitored by others—which among the Ju/'hoansi specify all kinds of important things about relationships, ranging from meat-sharing priorities to water-hole ownership. Norms related to naming or namesake relationships are common across diverse societies, and many people in small-scale societies intuitively know the power of namesakes, as my Yasawan friends with names like Josefa, Joseteki, and Joseses often remind me. My own kids are named Joshua, Jessica, and Zoey, thus matching my own first name by first initial or by rhyming.

While many evolution and economics oriented researchers have often assumed that social norms such as these are merely a superficial window dressing on our evolved psychology, the evidence suggests that such social norms run deep and profoundly shape social life. To throw these effects into stark relief, let's have a look at societies in which social norms and beliefs about marriage (1) only lightly regulate pair-bonding, (2) structurally eliminate marriage and suppress pair-bonding, thereby dispensing entirely with husbands, dads, and in-laws, and (3) encourage, or at least permit, women to obtain "secondary fathers" for their children—creating additional social fathers.

Societies with few or weak marriage norms provide us with a sense of how much "work" marriage norms are actually doing in contrast to the effect of our innate pair-bonding instincts. Consider the Aché, who before contact were mobile hunter-gatherers in the forests of Paraguay, South America. Precontact Aché did form somewhat enduring bonds between mates, and these bonds were important for linking children to their paternal relatives. However, while social norms did prohibit relationships between siblings, first cousins, and individuals in certain ritual relationships, community-wide expectations otherwise appeared to have little to say about the behavior of those involved in pair-bonds, and the formation of pair-bonds was not marked by communal rituals or public promises. Divorce was initiated unilaterally by either party and involved simply moving out, which was easy since the Aché didn't have much stuff. A sequence of pair-bonded relationships began around age 14 for women and 19 for men. Early marriages were typically interspersed with other, more fleeting romances but usually became more stable once a woman had two or three kids with the same man. By age 30, women had experienced an average of 10 marriages, and first marriages ended in divorce at a rate of 100%. Postmenopausal women reported an average of 13 marriages, and most women had children with different fathers. While enduring polygamous relationships were uncommon (4%), every woman had been in a polygamous marriage at some point. Most of these involved multiple wives with one husband, but a few went the other way. Some men had serially married and had children with three sisters. Some women reported marrying both the father and his son, at different times. And men reported having married both the mother and then her daughter.[15]

This suggests that the more demanding marriage norms found in most societies—for better or worse—operate to reinforce our otherwise flimsy pair-bonding instincts.

No Fathers

In the provinces of Yunnan and Sichuan in China, the Na and three other ethnic groups have maintained societies without husbands or fathers for at least a millennia, despite aggressive efforts by the Chinese government to introduce their own preferred marriage norms. This stable society is organized around female-headed matrilineal households. Children are conceived principally during "furtive visits," in which men slip into women's houses for sex and are gone by morning. The paternity of children is not a concern (and is often uncertain), and genetic fathers are not expected to contribute to the child's household; instead, men invest in their sister's children. There are no terms in the local language for "father," "husband," or "in-law." Of course, Na men and women do sometimes still form enduring relationships, but no norms regulate sexual exclusivity, permanence, rituals, or intra-pair obligations. Thus, social norms have organized this remarkably stable society by suppressing pair-bonding and erasing patrilineal kinship.[16]

Multiple Fathers

Even in societies with marriage, social norms and beliefs need not reinforce concerns about sexual fidelity that arise from male pair-bonding psychology but can instead promote investment in children in other ways. Many South American indigenous populations believe that a child forms in his or her mother's womb through repeated ejaculations of sperm, a belief system that anthropologists have labeled *partible paternity*.[17] In fact, people in many of these societies maintain that a single ejaculation cannot sustain a viable pregnancy, and men must "work hard" with repeated ejaculations over many months to sustain a viable fetus. Women, especially after the first fetus appears, are permitted, and sometimes even encouraged, to seek another man, or men, to have sex with in order to provide "additional fathers" for their future child. Anyone who contributes sperm to the fetus is a secondary father. In some of these societies, periodic rituals prescribe extramarital sex after successful hunts, which helps establish and formalize the creation of multiple fathers. Secondary fathers—often named at birth by the mother—are expected to contribute to the welfare of their children (e.g., by delivering meat and fish), although not as much as the primary father, the mother's husband. Frequently, the secondary father is the husband's brother.

Obtaining a second father is adaptive, at least sometimes. Detailed studies among both the Barí in Venezuela and the Aché show that kids with exactly two fathers are more likely to survive past age 15 than kids with either one father or three or more fathers.[18]

Importantly, social norms cannot just make male sexual jealousy vanish. Men don't like it when their wives seek sex with other men. However, rather than being supported by their communities in monitoring and punishing their wives for sexual deviations, they are the ones acting defiantly—violating social norms—if they show or act on their jealousy. Reputational concerns and norms are flipped around here, so now the husband has to control himself. In the eyes of the community, it's considered a good thing for an expectant mother to provide a secondary father for her child.

Marriage norms help expand human kinship systems by harnessing our pair-bonding instincts. In doing this, norms variously exploit the shared fitness interests of in-laws, the willingness of men to invest in the offspring of women they've had sex with, and the power of namesakes. They also variously sometimes suppress male sexual jealousy (in partible paternity), male parental investment (among the Na), female extramarital sexual desires (most societies), and polygynous pair-bonding (in societies with monogamous marriages). As societies expanded and became more complex, marriage norms were increasingly used to build intergroup alliances, to promote peace, and to sustain larger-scale forms of social organization. But even in the simplest human societies, these norms have long been at work shaping social life.

CHAPTER 9

From Incest Aversion to Incest Taboos

Unlike most other primates, human brothers and sisters form long and enduring social bonds. Among hunter-gather populations, brothers and sisters often live in the same bands. In many other traditional societies, either brothers or sisters continue to live in their home community while the other sex marries outside the community, but sibling bonds usually remain strong. Like other primates, the most important factors in establishing a brother-sister bond is familiarity while growing up. For opposite-sex siblings, this early familiarity breeds both deep affection and sexual aversion.[19]

An immense variety of kinship systems—sets of social norms—found across diverse small-scale societies have harnessed and extended these innate psychological tendencies to distant and even remote relatives. As mentioned, social norms identify classificatory siblings and stipulate that these individuals *should* be treated like real siblings. These norms could be "stand-alone" social rules (like putting the fork on the left), but the fact that they tap our evolved incest psychology probably makes them easier to learn, internalize, and enforce on others.[20]

However, in these societies, no one completely confuses their real (genetic) siblings and their classificatory siblings, as some famous anthropologists seem to suggest. In my own work in Fiji, for example, I've occasionally heard villagers refer to their "true" sister or brother. One time I was standing outside of a Fijian kitchen house, wondering about dinner, when I heard the wife of the house defending herself to her husband after she'd given the entire stock of the little store she'd recently set up to her older brother. "But, he's my real brother," she said defensively in Fijian, as she went on to describe how her brother came and sat so humbly at the low-status end of the house and performed the simple *kerekere* ritual of request. She felt moved and that she had to help him, even if it meant the store would close permanently. It was salient to her and her husband that it was her *real* brother, and not merely one of her many classificatory brothers, who had made the request.

Tapping this real versus classificatory sibling difference in the context of the incest taboo, my Fijian team and I wrote up two stories and had a random sample of adults from a couple of villages respond to our stories. As background, recall Kula: in these villages, social norms demand that brothers and sisters—real or classificatory—must never be alone in a house and must never talk to each other. Also, note that village houses have their three doors open all day (if someone is home), so a passerby can usually get a glance inside. In our first story, a *real* brother and sister were sitting inside a house alone, chatting. Our second version was the same except the siblings were now classificatory.

Can you guess how the villagers responded? My undergraduates usually guess wrong. Though Yasawans felt that the pair were doing something wrong in both versions of the story, it was the actions of the classificatory siblings that really got the community riled up and were turned into the subject of serious and rapidly spreading gossip. The villagers felt that while the real sibling shouldn't be breaking the rules, it was minor and nothing would happen. It seemed that my Fijian participants understood that innate incest aversion was prophylactic, so chatting was very unlikely to lead to sex. However, for classificatory siblings, the only effective prophylactic was the continuous monitoring and potential wrath of the community. Chatting alone, in most societies, is often an important step on the road to sex.

This is the difference between incest aversion and incest taboos. Incest taboos are social norms that evolved culturally to regulate sex and pair-bonding between non-close relatives by harnessing innate intuitions and emotional reactions that originally arose via genetic evolution to suppress sexual interest among close relatives, especially siblings. By harnessing innate incest aversion and labeling distant relatives as "brothers" and "sisters," cultural evolution seized a powerful lever to control human behavior, since incest taboos can strongly influence mating and marriage, and kin-based altruism can be extended through social norms. If you control mating and marriage, you get a grip on much of the larger social structure, and even aspects of people's cognition and motivation.[21]

Of course, to construct kinship systems, cultural evolution also harnesses our reciprocity psychology in various ways. One common way is through a set of social norms that define certain kinds of relatives, such as the cross-cousins who teased Kula, as governed by reciprocity-based relationships. Such relationships are egalitarian, relaxed, and often affirmed through joking. This reciprocity can be both positive and negative, since those who get teased can, and will, tease right back. Crucially, however, these are more than long-running dyadic exchange relationships because third parties are monitoring the pair to make sure they are behaving in the manner prescribed by the local norms for such relationships.[22]

The point that I'm slowly rolling out here is that human communities—whom we ally with, help, marry, and love—are forged by social norms, which variously harness, extend, and suppress our social instincts. Our species cooperation and sociality is deeply influenced by and highly dependent on culturally evolved social norms, which makes us rather unlike other animals. We acquire social rules by observing and learning from others, and we—at least to some degree—internalize them as goals in themselves. Because cultural learning influences how we judge others, it can create self-reinforcing stable patterns of social behavior—social norms.

This view suggests that, stripped of our social norms and beliefs, we aren't nearly as cooperative or as communal as we might seem. And to the degree that we are more cooperative than other mammalian species (and we are), it's because culturally evolved norms constructed social environments that, over eons, penalized and gradually weeded out aggressive, antisocial types (norm violators) while rewarding the more sociable and docile among us.[23] In chapter 11, I examine the evidence suggesting that this culture-gene coevolutionary process domesticated our species by shaping our psychology, making us uncomfortably similar to animals like dogs and horses.

As noted above, my view contrasts with that of some prominent evolutionary writers who have suggested that while the sociality and cooperation we observe in the modern world is due to modern institutions, the social behavior of small-scale societies, and especially hunter-gatherers, *directly* reflects our genetically evolved social psychology. This implies that the patterns of social interaction and cooperation among these populations should be explicable *without* reference to culturally transmitted norms, practices, or beliefs. Sociality in these populations should be easy—an automatic operation of an evolved psychology designed by natural selection to snugly fit this way of life. By contrast, if I'm right, sociality and cooperation among hunter-gathers, and everyone else, should depend on norms, practices, and beliefs that amplify or suppress our innate motivations and dispositions. We've already seen how social norms (sometimes) reinforce our pair-bonding instincts and fatherly motivations as well as extend our incest aversions. Now let's zoom in a bit closer on one kind of foraging society, that of mobile hunter-gatherers, who have been routinely used to gain insights into the lives of Paleolithic societies, before the spread of agriculture.[24]

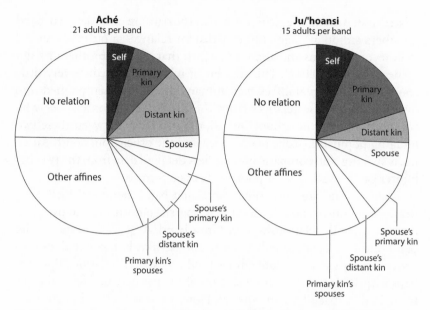

Figure 9.1. Pie charts showing the fraction of different kinds of relationships in the average band among the Aché and Ju/'hoansi, hunter-gatherers in Paraguay and Africa, respectively.

Sociality and Cooperation among Hunter-Gatherers

Mobile hunter-gatherer bands are renowned for their cooperation in activities like hunting and in their broad sharing of valuable foods, like meat. A common explanation for this cooperation has been that hunter-gatherers live in small groups of closely related individuals. If true, the argument goes, kin selection can explain much of the observed cooperation.

The problem with this explanation is that the best available evidence indicates that hunter-gatherers do not live in groups mostly composed of close kin. Based on work by Kim Hill and colleagues, figure 9.1 shows the average composition of both Ju/'hoansi and Aché bands. "Primary kin" includes siblings, half-siblings, and parents, while "distant kin" includes anyone with a blood link going back up to five generations (extending to second cousins). Together, these two categories plus oneself make up only about a quarter of the band. That means that about three-quarters of band relationships are based on something besides genetic relatedness. Among the Aché, for which the data are most detailed, band members are on average only very distant relatives, a bit more related than second cousins and about one-tenth that of a full brother or sister (genealogical $r = 0.054$). This tiny bit of relatedness predicts very little cooperation and guarantees that humans should be keenly tuned into distinguishing close relatives from the distant- and nonrelatives who compose most of the group. Overall, the similarity between these two foraging populations, one from Africa and the other from South America, is striking, and comparisons with less detailed data from thirty other hunter-gatherer societies supports this basic picture.[25]

Okay, so who are these other unrelated band members? Well, two-thirds are spouses and affines. That is, marriage norms create over half the ties in adult relationships within a band. Arguably, primates who pair-bond may create bonds that are spouse-like, but as noted, no evidence suggests that primate affines hold any special relationships. Perhaps surprising to some, the evolution of in-laws may be one of the key features that make humans special. However you look at it, bands are culturally constructed, because it's only through marriage and affines that hunter-gather bands can be said to be "mostly relatives."

A remaining quarter of the band have neither blood nor affinal ties. Yet, as in most small-scale societies, they are likely all referred to by kinship terms.[26] They aren't genetic relatives, but they are labeled as classificatory kin of some kind. Among the Ju/'hoansi, as alluded to above, many are linked using same-name relationships. For example, if we are unrelated, you might tell me to call you "mother" because your son has the same name as I do. This tells me how I must treat you and also cues everyone else into how I should behave toward you (e.g., no flirting or sexual jokes with "mom"). Through social norms, culture reinforces our kin-based and pair-bonded relationships and dramatically expands our narrow circle from genetic kin to cultural kinship.

Meat Sharing

Paleoanthropologists believe that cooperative hunting and meat sharing were crucial elements in human evolution, reaching back millions of years into our past. Among those foraging peoples studied by anthropologists, meat provides an important and highly valued contribution to the diet, and as we saw in chapter 8, hunters generally receive much prestige from their hunting success.[27] However, since even the best hunters can't reliably obtain game on a consistent basis, because streaks of bad luck, illness, and injury are inevitable, meat sharing has probably long been a problem that needed solving. By sharing meat, cooperative hunters can avoid the otherwise long stretches of time without fat or protein in their diets. For these reasons, some evolutionary researchers believe that the broad sharing of meat across foraging bands arises from innate psychological dispositions, without any cultural input. Could meat sharing among foragers be instinctual?[28]

A close look at food sharing among foragers reveals that it too is governed by social norms and fostered by what might be called "cultural-institutional" technologies.[29] For example, in addition to social norms specifying that shares of meat be delivered to certain categories of culturally constructed kin, such as the hunter's in-laws, cultural-institutional technologies like *ownership transfer* and *meat taboos* operate to make sharing psychologically easier. Let's consider these in more detail.

In many foraging groups the ownership of meat is diffused, or transferred, from the hunter to a third party who is designated to distribute the meat. Because it wasn't their sweat and skill that produced the meat, a third party may find it easier to follow the local distributional norms.[30] For example, Ju/'hoansi hunters frequently use arrowheads owned by someone else while hunting. Social norms dictate that the owner of the arrowhead becomes the owner of any kill made with that arrowhead, and this owner has the responsibility for distributing the meat. Hunters often like to use other people's arrowheads, since it relieves them of the responsibility for making fair distributions—where "fair" is defined by local standards and others will rapidly criticize any appearance of bias in the distribution. For example, elderly men and women commonly own and lend arrowheads, and anyone can receive them as gifts from one of their special *hxaro* exchange partners (see below), especially if they themselves can't make arrowheads.[31] By relieving the hunter of ownership, this institution mitigates self-interested biases and disperses the responsibility for dividing the meat to others in the band, who might not otherwise experience it.

Food taboos also influence meat distribution in many hunter-gatherer groups, and in some groups, the entire distribution is governed by a complex system of such taboos. An interesting system of taboos was observed in the early twentieth century among certain Kalahari hunter-gatherers. These taboos virtually guaranteed that large prey had to be widely distributed across the band. Here the hunter himself could only eat the ribs and one shoulder blade; the rest of the animal was taboo for him. The hunter's wife received the meat and fat around the animal's hindquarters, which she had to cook openly and share with other women (only). Taboos prohibited young males from eating anything except for the abdominal walls, kidneys, and genitals. Violations of any of these taboos were believed to result in the failure of future hunts. Such beliefs create collective interests in making sure that other members of the band don't violate the taboos: since your violations will result in hunting failures that will reduce my meat intake, I'm going to make sure you don't violate any taboos. Thus, everyone in the band (believes) they have direct and personal incentives to monitor and sanction taboo violators.[32] Such complex systems of taboos were common, having been recorded in detail among hunter-gatherers in South America, Africa, and Indonesia.[33]

Taboos on particular game species or certain parts of animals for particular categories of people are interesting because, to the learner, they appear to be facts about the world that can drive purely self-interested actions: I want to avoid illness, and certain animal parts will cause me to get sick, so I'd best not eat those. Crucially, such beliefs induce sharing in a community without anyone realizing it. However, if such beliefs are inaccurate and costly to the individual, it's sensible to ask why either individual experience or rules like "copy the successful" would not result in the eventual disappearance of such taboos. Three interrelated psychological factors work against this:

1. There is reason to suspect that we humans have an innate susceptibility to picking up meat aversions, due to the tendency of dead animals to carry dangerous pathogens.[34] Thus, we humans are primed to acquire meat taboos over other food avoidances.
2. These taboos are social norms, so violations will be monitored and judged by others. This factor is especially potent here because the punishment or misfortunes believed to result from taboo violations will (often) be felt by the whole band (e.g., hunting failures).
3. A good learner will acquire this rule while growing up and never actually violate it (meat is consumed in public), so he'll never directly experience eating the tabooed part and not having bad luck. Rare cases of taboo violations that, by coincidence, were followed by bad luck or illness will be readily remembered and passed on (psychologists call this "negativity bias"). Meanwhile, cases of violations followed by a long period when nothing bad happens will tend to be missed or forgotten, unless people keep and check accurate records.

Based on my field experience, any skeptic who questions the taboos will be met with vivid descriptions of particular cases in which the taboos were violated and then poor hunting, illnesses, or bad luck ensued.[35]

Amazingly, despite immense variation in the details of local social norms and beliefs, the consequences across mobile hunter-gatherers are similar: most or all band members obtain some meat from any large kill. Of course, this doesn't mean everyone gets an equal share. In many such societies, priority extends first to the hunter's close kin, affines, and ritual partners, and only secondarily to the full band and visitors.[36] It appears that cultural evolution has devised numerous solutions—combinations of norms—that achieve roughly the same end: a diffusion of the risks associated with repeated hunting failures across the band.[37]

Communal Rituals

As darkness fell in the Kalahari desert, Ju/'hoansi women from many bands squeezed together around a blazing dance fire and began to sing in a high chorus. Then, with soft rattles made from moth cocoons wrapped around their legs, the men assembled around the women and danced in a circle, stomping out a rhythm. Soon the women began a special high-pitched clap to complement the men's rap and rattle. Accompanied by string instruments played from the periphery, the main event began as the women started to sing loudly of the *n/um*, a powerful supernatural essence that can be either protective or dangerous. An hour or two later, the men's dance line began snaking through the women's circle to form a figure eight. As some men began to enter into trances, the dance intensified. Entranced men increasingly struck out into the darkness, shouting, as they battled spirits and hurdled invectives at their god. This ritual storm intensified and abated in cycles through the night, until dawn, when it slowly faded away.[38]

Having seen thirty-nine of these communal rituals, the ethnographer Lorna Marshall wrote, "People bind together subjectively against external forces of evil, and they bind together on an intimate social level. . . . Whatever their relationship, whatever the state of their feelings, whether they like or dislike each other, whether they are on good terms or bad terms with each other, they become a unit, singing, clapping, moving together in an extraordinary unison of stamping feet and clapping hands, swept along by the music." Similarly, Megan Biesele, an ethnographer who studied another Ju/'hoansi group fifteen years later, explained that "the dance is perhaps the central unifying force in Bushman life, binding people together in very deep ways which we do not fully understand."[39]

Psychologically powerful communal rituals like this one are common in small-scale societies and among mobile hunter-gatherers, from the central desert in Australia to the Great Basin in North America. Like Megan and Lorna, keen observers of human communities, such as the Muslim scholar Ibn Khaldûn in the fourteenth century, have argued that communal rituals have a potent psychological impact on their participants and create strong personal ties, deep trust, and a profound sense of group solidarity. Recently, however, researchers have begun to systematically measure the effect of communal rituals on social bonding and cooperation and to further break rituals down into their active ingredients. These ingredients include (1) synchronous singing and dancing or other movements (e.g., marching), (2) collaborative music making, (3) extreme physical exhaustion, (4) feelings of a common fate, (5) shared experiences of danger or terror, (6) supernatural or mystical beliefs, and (7) causal opacity or a lack of instrumentality (that is, people are not sure why the ritual must be done in a particular way, but they know it *must* be done in that way).[40]

Several recent experimental studies, for example, show that singing and/or moving in synchrony with others deepens feelings of affiliation, fosters trust, and promotes cooperation within groups. In one experiment, American university students were placed into one of four groups. Though the groups all listened to the Canadian national anthem through headphones and could read the lyrics, each group was told to do different things, some of which involved cups. The control group just listened while holding one cup above the table. In the "synchronous-singing" group, participants were told to sing along with the music, which caused them to sing in sync with each other. In the "singing and moving in sync" group, participants both sang together and moved their cups in sync with the music, which caused them to move in sync with each other. Finally, in the "asynchronous singing and moving group," participants did the same thing as in the "synchronous moving and singing" group, except that the music in their headphones started at different times, so they all moved asynchronously.[41]

After this exercise, the participants then engaged in a cooperative project in which they could contribute money to a joint investment. The more money contributed in total by the group members to their joint project, the more money everyone would take home. However, since everyone got an equal share of the money at the end, individuals could profit by not contributing money to the project, effectively free riding on the contributions of others. The results show that synchrony—both singing and moving together and just singing together—promoted greater cooperative contributions, which resulted in higher monetary payoffs for the whole group. Parallel results have been found even among four-year-old children, for whom jointly making music promotes greater prosociality.[42]

Perhaps even more enduring and powerful than synchrony are the potent social bonds forged among those who share terrifying experiences. Such experiences have been routinely created in different ways by male initiation rites in societies across the globe and throughout history, and can be found in many hunter-gatherer societies. Among the Arunta of central Australia, for example, the initiation into manhood involved four main rituals that spanned roughly fifteen years, from age 10 to 25 years. Based only in the local community, the first three of these rituals involved being thrown in the air, kidnapped at night, blindfolded, bitten, piled on by a group of men, and forced to endure periods of silence and deprivation, as well as learning much tribal lore through a series of frightening dances and song narratives performed by inhuman-looking men, who were painted and costumed. The second rite in particular, performed just after their puberty started, culminated in a ritual circumcision in which a stone knife was used to cut off the adolescent's foreskin. Then, soon after these wounds had healed, a third rite would commence and eventually climax in a ritual subincision: the boys' penises were sliced lengthwise, along the underside, slit and split like a hot dog.

The final initiation rite was performed on young men in their twenties assembled from across the tribal network. Formal invitations went out to all bands, and even to neighboring groups, to gather at a particular location for a rite that would last for months, with dances and songs performed by many of the assembled bands. People believed that these invitations could not be refused, lest they get an illness as punishment. Secluded together, often deprived, and silenced for months, the initiates experienced a long sequence of nighttime ceremonies, dance performances, and sacred narratives. In the final phase of this rite, the initiates had to repeatedly lie on a bed of glowing embers, with only a layer of leaves to protect them. Initiates had to stay on the embers, while choking on the smoke, until they were told they could get up (after roughly four to five minutes). Only after this ordeal by fire were these young men considered fully fledged men of the tribe.[43]

Lest you think that these young guys somehow took this ritual in stride and were not scared, it's worth noting that it was not uncommon for adolescents and young men to flee from the advancing front of these spreading rites by moving to distant groups who had not yet adopted the practice.[44] Nevertheless, the elderly Arunta explained that the ritual "imparts courage and wisdom" and "makes the men more kindly natured and less apt to quarrel."[45] We'll return to the spread of such rituals in Australia in chapter 10.

While systematic experimental research is just beginning on such "rites of terror," it appears that the psychological effects created by these rites establish enduring emotional bonds among the initiates, and even potentially among observers. Psychologically, these rites create a potent emotional memory that somehow binds together those who shared the experience. These bonds may be closely related to those observed among soldiers who have experienced intensive combat together, creating the "band of brothers" phenomenon.[46] Crucially, however, by incorporating these experiences into regular initiation rites, cultural evolution has engineered a way to solidify social ties within age cohorts of males drawn from across the tribal group. Older Arunta had themselves noticed the power of their rituals for social bonding, though they couldn't explain how or why the ritual worked, and they certainly knew of no one who had designed the protocol.

More broadly, communal rituals are sets of culturally evolved social norms that, though quite diverse, often exploit various aspects of our psychology in ways that foster greater solidarity, trust, affiliation, and cooperation among participants. They represent one of the institutional-cultural technologies deployed by cultural evolution to shape our sociality and cooperation across diverse societies. Even among the smallest-scale human societies, communal rituals nourish the social fibers that bind a collection of bands into a tribe.

Threads in the Fabric of Social Life among Bands

Perhaps the most important feature of social life in human hunter-gatherers, in contrast with other primates, is that individuals are socially connected into an immense network of other people scattered across numerous other groups. In many foraging societies, band membership itself is quite fluid. If an individual or family wants to leave their band, due to some acute social tensions, drought, or just to visit friends, they can tap a network of contacts who can open the doors to extended visits in other bands. By contrast, chimpanzees live in troops that patrol and defend a territory. As I'll explain in chapter 10, intruders are attacked and killed on sight, unless they are young females, who are permitted to move among troops. Let's consider how culture made us the only tribal primate.[47]

What determines the links or social ties that foragers extend *outside* their band? Kim Hill, Brian Wood, and their collaborators have recently explored this among two mobile hunter-gather groups, the Hadza and the Aché. The Hadza live in nomadic foraging bands in Tanzania's vast savannah woodlands and continue to hunt with bow and arrow and gather roots, tubers, and honey. Kim and Brian questioned people from dozens of bands in each population about their interactions with a randomly chosen set of other adults drawn from the entire society. They asked about interactions related to helping, hunting together, and many other associations. Then, to analyze their data, they asked what relationship factors predicted the existence of various kinds of interactions, such as hunting together, giving and receiving meat, sleeping in the same camp, helping, or joking around. As expected, being close kin (aunts and closer) is important, and about 5% to 10% of the other people were close kin of some kind. For example, people were overall more than twice as likely to have received food from a person when they were sick or injured if they were close relatives. Beyond genetic relatedness, affines were important too, being 50% more likely to have received food. Since affines made up 15% to 20% of the randomly selected sample, this category collectively represents a rather big contribution.[48]

Even more important than blood and affinal relationships, ritual relationships establish important social ties across bands. At Aché rituals, adult sponsors step forward to assist children through birth and puberty rites (like "godparents"). As part of the ritual, they then enter into a special named relationship with the child's parents. Each ritual relationship is associated with a particular role (e.g., cutting the umbilical cord, washing the newborn, etc.) and results in lifelong rights and obligations of mutual support governed by social norms. Being in a ritual relationship, while holding genetic relatedness constant, is strongly associated with sharing meat and information, as well as receiving help when one is sick or injured. These culturally constructed ritual relationships are much more important than close genetic relatedness, and this importance is further magnified by the fact that the Aché have twice as many ritual partners as they do close kin outside their own band.

Among Hadza hunter-gatherers, the *Epeme* taboos and ritual dances bond a select group of men in a secret covenant. Ritually, the men communally consume particular joints from large game and perform in undisturbed silence and darkness for other members of the community. Here the data again show that ritual relationships are associated with sharing meat and information as well as receiving help when one is sick. Not only are individual ritual relationships more important than genetic relatedness, but individuals have three times as many ritual partners as they do close blood relations.[49]

Taken together, ritual and affinal relationships, both of which are culturally constructed and nonexistent in primates or other animals, explain much more about the patterns of association, cooperation, helping, and sharing than blood ties.

Elsewhere, in southern Africa, the Ju/'hoansi achieve the same end as the Hadza and Aché, a vast interconnected social web that threads through many bands. They too also rely on affinal connections and communal rituals, but they also have the *hxaro* exchange relationships. *Hxaro* partnerships are special, culturally defined, relationships that come with obligations and are sustained by ongoing exchanges of goods. Since people form and inherit many such relationships, goods given as gifts between partners pulse constantly through this broad network. *Hxaro* appears to tap our innate reciprocity psychology, but this is then fueled, extended, and reinforced by social norms, all monitored by third parties.[50]

Thus, the vast tribal social networks that mobile hunter-gatherers rely on, for example, in times of drought and war, are largely constituted and nourished by social norms of various types, including those related to rituals, marriage, and exchange.

Onward

Let me close by underlining my major points. Our ability to learn from each other gives rise to sets of social norms, including practices like communal rituals, food taboos, and kinship rules, which strongly influence human social life. In shaping individual decisions, social norms are powerful for a number of reasons, but they generally

- deploy third-parties to monitor and sanction norm-violators often through reputational damage,

- shape individuals' perceptions of the costs and benefits of various actions (e.g., food taboo violations cause hunting failures), and
- harness aspects of our evolved psychology, such as the way marriage reinforces our pair-bonding psychology or rituals exploit the cooperation-inducing effects of synchrony.

Such social norms are crucial for understanding community and cooperation in all human societies, including those of mobile hunter-gatherers. Detailed studies of contemporary hunter-gatherers show how practices related to marriage, naming, exchange, and ritual influence the formation of bands and spin the broader social threads that weave bands together into tribes. Even meat sharing in hunter-gatherer bands, which is often argued to be an ancient and important feature of our species' evolutionary history, depends crucially on culturally constructed kinship ties, social norms of ownership, food taboos, and ritual practices.

Until now, I have casually described a variety of social norms that appear to promote sociality, harmony, and cooperation among the groups possessing them. But, clearly, in many cases people don't understand how or why their norms work or that their norms are even "doing" anything. And in the cases of the food taboos and communal rituals, it's probably the case that if people knew precisely what was going on—if they had the correct causal model—the practices would lose at least some of their effectiveness.

So, how can we explain the emergence of such group beneficial norms?

CHAPTER 10

INTERGROUP COMPETITION SHAPES CULTURAL EVOLUTION

In the forests of Uganda, primatologists have been studying a particularly large troop of chimpanzees at Ngogo for about two decades. As of 1999, this group of 150 or so individuals controlled a territory of 29 km² (11.2 mi²). As in other chimpanzee groups, adult males go out on "boundary patrols." During these nighttime expeditions, but unlike their other movements, males neither socialize nor feed, but silently travel in single file through regions separating their territory from that occupied by adjacent chimpanzee troops. There, moving along the border, they sometimes make targeted incursions into the territories of other troops. Over nine years, 114 of these patrols attacked and killed 21 members of other chimpanzee troops. Thirteen of these 21 murders occurred during incursions into the northeastern corner of their territory and thus targeted one particular troop. While the exact size of this other group is uncertain, this many kills implies that about three-quarters of the chimpanzees in the other group can expect to be murdered by a patrol before dying of old age at 50. In 2009, Ngogo chimpanzees, including females and infants, began regularly entering this new territory and acting as they do at the core of their own territory. It appears that their systematic raids over at least a decade drove the other troop back, allowing this large group to effectively expand its territory by 22.3%.[1]

The presence of substantial and deadly intergroup competition, and territorial expansion, in one of our closest primate relatives suggests that it may be old, even older than our species' heavy reliance on cultural learning. Cultural evolution may have emerged in a world in which intergroup competition was already prevalent.

When our genetically evolving capacities for cultural learning began to give rise to cultural evolution and social norms, it's likely that our species was already living in stable social groups. Many of these norms would have been arbitrary, such as using a particular type of stone for smashing nuts open. But, occasionally, prosocial norms might have emerged that fostered food sharing, internal harmony ("no fighting" or "no stealing others' mates"), or cooperative efforts in community defense. But how could such systems of social norms have culturally evolved to perform this function? As we saw in chapter 9, many social norms appear almost engineered to harness and extend our social instincts. Yet few if any adherents to these norms understand the "design" or implicit "function" of their institutions.

Intergroup competition provides one important process that can help explain the spread of norms that foster prosociality. Different groups culturally evolve different social norms. Having norms that increase cooperation can favor success in competition with other groups that lack these norms. Over time, intergroup competition can aggregate and assemble packages of social norms that more effectively promote such success, and these packages will include social norms related to cooperation, helping, sharing, and maintaining internal harmony.[2] Below, I discuss the most important categories of intergroup competition and highlight key lines of evidence. In chapter 11, I'll consider how a long evolutionary history of living in a social world regulated by norms, which themselves were shaped by intergroup competition, influenced our species' genetic evolution.

Once a new norm emerges in one group, intergroup competition can grab hold of it and spread it widely through a number of related processes. Consider five forms of intergroup competition:[3]

1. *War and raiding.* The first and most straightforward way that intergroup competition influences cultural evolution is through violent conflicts in which some social groups—due to institutions that foster greater cooperation or generate other technological, military, or economic advantages—drive out, eliminate, or assimilate other groups with different social norms.[4]

2. *Differential group survival without conflict.* In sufficiently harsh environments, only groups with institutions that promote coopera-

tion, sharing, and internal harmony can survive at all and spread. Groups without these norms go extinct or flee back into more amicable environments. The right institutions allow groups to enter new ecological niches, for example, by fostering survival in the Arctic through cooperation in whale hunting, or by surviving shocks, like droughts in deserts, that would exterminate or disperse less cooperative groups. Groups with superior institutions simply outlast and eventually replace those with fewer cooperation-galvanizing norms. Since humans expanded out of Africa and into harsh environments, for which they had few genetic adaptations or innate proclivities, this may have been particularly important during human evolution. For this kind of process to work, groups don't ever have to meet each other, and violence between groups need not occur.[5]

3. *Differential migration.* Since social norms can create groups with greater internal harmony, cooperation, and economic production, many individuals will be inclined to migrate into more successful groups from less successful ones. Meanwhile, few will want to move to less successful groups unless forced to. Over time, the more successful groups will expand through immigration whereas other groups will contract thorough emigration. This movement has been observed in both the differential rates of switching groups at boundaries of small-scale tribal populations and in migration patterns among nations in the modern world.[6]

4. *Differential reproduction.* Under some conditions, social norms can influence the rate at which individuals within a group produce children. Since children tend to share the norms of their group, over time the social norms of groups who produce children at faster rates will tend spread at the expense of other social norms. Some modern religions, for example, take advantage of this, with their pronatalist gods and fertility-favoring institutions.[7]

5. *Prestige-biased group transmission.* Because of our cultural learning abilities, individuals will be inclined to preferentially attend to and learn from individuals in more successful groups, including those with social norms that lead to greater economic success or better health. This causes social norms, including ideas, beliefs, practices (e.g., rituals), and motivations, to flow via cultural transmission from more successful groups to less successful groups.[8] Since individuals cannot easily distinguish what makes a group more successful, there is a substantial amount of cultural

flow that has nothing to do with success (e.g., hairstyles and music preferences).

Over time, combinations of these intergroup processes will aggregate and recombine different social norms to create increasingly prosocial institutions. To be clear, by "prosocial institutions" I mean institutions that lead to success in competition with other groups. While such institutions include those that increase group cooperation and foster internal harmony, I do NOT mean "good" or "better" in a moral sense. To underline this point, realize that intergroup competition often favors norms and beliefs that can readily result in the tribe or nation in the next valley getting labeled as "animals," "nonhumans," or "witches" and motivate efforts to exterminate them.

How Old Is Intergroup Competition?

How important was intergroup competition in shaping the culturally evolved institutions of our Paleolithic ancestors? Has it been shaping cultural evolution for long enough to have had an impact on our genetic evolution?

Several converging lines of evidence, though all admittedly indirect, suggest that intergroup competition was probably important for much of our species evolutionary history. To triangulate in on this, we'll first look at nonhuman primates, since at some point in the past we were just another primate species, so they provide a point of departure. Next, since our ancestors lived in societies very different from today's modern nation-states, we will examine small-scale societies and focus especially on hunter-gatherers. While none of these societies is in any sense representative of Paleolithic populations, they collectively provide much insight into the broad patterns and potential diversity of ancestral human societies, who faced the same problems, used similar technologies, and relied on many of the same resources. Finally, we consider these lines of evidence in light of paleoanthropological efforts to reconstruct the lifeways of human ancestors. This work draws insights from both the unearthed tools and bones of past populations and from reconstructions of ancient environments, which draw on data from ice and lake cores that help resolve long-term patterns of environmental change.

As I discuss intergroup competition, keep in mind that there are many other cultural evolutionary forces that do not favor prosocial institutions. When the forces of intergroup competition are spent or weakened, success-biased cultural learning (or purely rational self-interest) will cause individuals to seek out any "cracks" in their groups' institutions to manipulate or exploit for their own benefit or that of their kith and kin. Over time, history suggests that all prosocial institutions age and eventually collapse at the hands of self-interest, unless they are renewed by the dynamics of intergroup competition. That is, although it may take a long time, individuals and coalitions eventually figure out how to beat or manipulate the system to their own ends, and these techniques spread and slowly corrode any prosocial effects.

Let's start with intergroup conflict via warfare and raiding. As we saw at the outset of this chapter, chimpanzees have violent intergroup conflicts that can result in significant gains and losses of territory. Aggressive intergroup interactions are common in many primate species, but chimpanzees are particularly interesting because they provide a ready model for what the common ancestor of humans and chimpanzees might have been like. If modern chimpanzees do it, it's plausible that the common primate ancestor we share with chimpanzees also did it. The mortality rate we saw above at Ngogo is unusually high, but data from other sites suggest that mortality rates from chimpanzee intergroup conflicts range from 4% to 13%, which as we will see is comparable to many small-scale human societies. Aside from Ngogo, intergroup aggression has been well documented at four chimpanzee sites, and territorial expansions have been observed in two other populations besides the one at Ngogo.[9]

While chimpanzees do reveal some behavioral patterns that probably represent cultural traditions, no reliable evidence so far suggests that they have social norms, and certainly no norms that promote success in intergroup competition.[10] This suggests that intergroup competition may have preceded the emergence of cultural evolution and could have started shaping norms as soon as they began to emerge. However, even if intergroup competition did not exist initially, cultural evolution would have created the know-how to exploit clumped resources (which could be defended and controlled) and social norms that would have given rise to differences among groups—imbalances of power—that would have generated intergroup competition. Of course, it's possible that during the course of human evolution something happened to suppress intergroup competition. So we need to ask whether small-scale societies, and particularly hunter-gather groups, really do experience intergroup competition.

Conveniently, one form of intergroup competition in small-scale societies, warfare, is actually a hot topic.[11] The answer is that there's lots of conflict and lots of variability in conflict rates. And while it's certainly true that farming and herding societies fight much more than most hunter-gatherers, substantial evidence indicates that many hunter-gatherer populations experienced enduring violent conflicts between groups that inflicted high mortality rates and much loss of territory. Reviews of the evidence for warfare among hunter-gatherers show that 70% to 90% of these societies experience war or raiding as either "continuous" (with conflicts every year) or "frequent" (with conflicts at least once every five years). Estimates of the percentage of deaths directly due to violent intergroup conflicts average 15%, based on ethnographic observations, and 13% based on archeological studies of cemeteries (remember, chimp death rates ranged from 4% to 13%). These percentages are very high compared to the same rates in the United States or Europe in the twentieth century, where the numbers are all less than 1%, but low compared with many preindustrial agricultural societies.[12]

Beyond the loss of life, conflict among hunter-gatherer populations resulted in the systematic gain and loss of territory (and thus resources) over time. In the five groups for which there is ethnohistorical data, gains and losses in territory ranged from 3% to 50% per generation (25 years), with an average of 16%. We might worry about these numbers since four of the five groups derive from western North America, and thus may have been influenced in their territorial expansion by agriculturalists, at least indirectly. However, even if we take the 3%, which comes from the Warlpiri, who lived in the middle of a vast continent of hunter-gatherers in Australia, a territory of 100 square miles would more than double every 600 years. In 5000 years, which is brief given the more than 2 million years since the origins of the genus *Homo*, a territory the size of the original Washington DC (100 mi^2) would expand to the size of Indiana (36,417 mi^2).[13] In another 5000 years or so, a group expanding at this rate could displace enough other groups to cover all of Asia. In short, 3% per generation is plenty fast.

To be clear, conflicts among nomadic foragers were very different from war in more complex societies. Most conflicts involved raids or ambushes, which relied on stealth, surprise, and superior numbers to mitigate the risks. Attackers typically came out much better than the victims, at least in the short run, until the victims sought revenge with stealth raids of their own. Pitched battles did occur and sometimes involved hundreds of individuals on a side, but these were relatively rare. Groups often experienced enduring periods of hostility with neighboring groups, in which strangers were killed on sight, and maintained a no-man's land (a kind of DMZ) between them. These patterns are reminiscent of chimpanzees, except crucially, this hostility usually occurred between tribal populations, which consisted of many interlinked bands who shared customs and language, rather than between residential groups, as in chimpanzees. That is, intergroup conflict occurred on a much bigger scale in humans.

There's reason to suspect that the Paleolithic world may have been even more prone to such intergroup conflicts than these historical and ethnographic data suggest. This is because of the much more intense climatic fluctuations experienced over most of our evolutionary history compared to the last 10,000 years of relatively stable climates. Not only were ancient populations dealing with constantly shifting seasons and rising and falling seas, but they would have been hit more frequently by immense storms, floods, fires, and droughts. Two kinds of evidence indicate that these shifts would have sparked more war. First, the archeological record of California's maritime hunter-gatherer populations over a 7000-year period shows that violence is most common during periods of climatic shifts, which stress resources. Second, quantitative analyses on warfare using ethnographic data from both a global sample of diverse small-scale societies and a regional sample from East Africa indicate that unpredictable environments probably cause more warfare between groups.[14] Thus, unpredictable environments, which characterized the Paleolithic, likely intensified intergroup competition.

Intergroup competition also often shapes the cultural evolution of institutions without violence or war. To see one mechanism, let's look at a case in which a village in New Guinea decided explicitly to copy the institutions of a regionally more successful group, including practices, rituals, and beliefs.

Throughout the Highlands of New Guinea, a group's ability to raise large numbers of pigs is directly related to its economic and social success in competition with other regional groups. The ceremonial exchange of pigs allows groups to forge alliances, repay debts, obtain wives, and generate prestige through excessive displays of generosity. All this means that groups who are better able to raise pigs can expand more rapidly in numbers—by reproduction and in-migration—and thus have the potential to expand their territory. Group size is very important in intergroup warfare in small-scale societies, so larger groups are more likely to successfully expand their territory. However, the prestige more successful groups obtain may cause the rapid diffusion of the very institutions, beliefs, and practices responsible for their competitive edge as other groups adopt their strategies and beliefs.

In 1971, the anthropologist David Boyd was living in the New Guinea village of Irakia and observed intergroup competition via prestige-biased group transmission. Concerned about their low prestige and weak pig production, the senior men of Irakia convened a series of meetings to determine how to improve their situation. Numerous suggestions were proposed for raising their pig production, but after a long process of consensus building, the senior men of the village decided to follow a suggestion made by a prestigious clan leader who proposed that they "must follow the Fore" and adopt their pig-related husbandry practices, rituals, and other institutions. The Fore were a large and successful ethnic group in the region, who were renowned for their pig production.[15]

The following practices, beliefs, rules, and goals were copied from the Fore and announced at the next general meeting of the community:

1. All villagers must sing, dance, and play flutes for their pigs. This ritual causes the pigs to grow faster and bigger. At feasts, the pigs should be fed first from the oven. People are fed second.
2. Pigs should not be killed for breaking into another's garden. The pig's owner must assist the owner of the garden in repairing the fence. Disputes will be resolved following the dispute resolution procedure used among the Fore.
3. Sending pigs to other villages is taboo, except for the official festival feast.
4. Women should take better care of the pigs and feed them more food. To find extra time for this, women should spend less time gossiping.
5. Men must plant more sweet potatoes for the women to feed to the pigs and should not depart for wage labor in distant towns until the pigs have grown to a certain size.

The first two items were implemented immediately at a ritual feast. David stayed in the village long enough to verify that the villagers did adopt the other practices and that their pig production did increase in the short term, though unfortunately we don't know what happened in the long run.

Let me highlight three features of this case. First, the real causal links between many of these elements and pig production are unclear. Maybe singing does cause pigs to grow faster, but it's not obvious, and no one tried to ascertain this fact, via experimentation for example. Second, the village leadership chose to rely on copying institutions from other groups and not on designing their own institutions from scratch. This is smart, since we humans are horrible at designing institutions from scratch. And third, this transmission between groups occurred rapidly because the Irakia already had a political institution in the village, which involved a council of the senior members of each clan, who were empowered by tradition (social norms) to make community-level decisions. Lacking such decision-making institution, Fore practices would have had to spread among households and thus been much slower in spreading. Of course, such political decision-making institutions themselves are favored by intergroup competition.

More broadly, this case is not unique in any way, as much ethnography and ethnohistory from New Guinea and elsewhere indicates that the copying of institutions and rituals from more successful groups is commonplace. For example, an in-depth study of the Enga, a small-scale agricultural population in the New Guinea Highlands, reveals the effects of intergroup competition on the spread of ritually galvanized sets of norms and political beliefs (termed "cults") that promoted "identity, welfare, and unity" among local communities. These institutional packages often included the psychologically potent and terrifying initiation rites described in chapter 9. According to the ethnographers Polly Wiessner and Akii Tumu, these cults were

> readily transmitted across linguistic boundaries when (1) donors and recipients faced comparable problems, so that underlying beliefs and overt procedures were meaningful, and (2) the owners of the cults were perceived as being successful. . . . Cults were imported in order to acquire new and more effective ways to communicate with the spirit world, as well as to emulate those who appeared more successful.[16]

In some cases, less successful communities would go to more successful communities and pay them, often in pigs, to learn about their rituals and institutions in order to better ascertain the crucial details.

Elsewhere, in the Sepik region of New Guinea, villages typically break down after they exceed about 300 people, as squabbling clans fracture and move apart. However, one Arapesh community named Ilahita dramatically exceeded the size of all other villages in this region, maintaining an ethnically diverse population of 1500 people. The ability to sustain solidarity with this locally immense population led to both success and security in a region with substantial military and economic threats.

The anthropologist Donald Tuzin studied Ilahita in detail in order to determine how it managed to sustain such a large size when other communities could not. He found that in the last century Ilahita had adopted a ritually galvanized form of social organization ensconced in an encompassing mystical belief system. This package reorganized the community to create cross-cutting mutual interdependencies among subgroups that were then sacralized in rituals. The basic elements of this institutional-ritual complex, which Ilahita elaborated upon, were first copied around 1870 from a highly successful and aggressively expanding group called the Abelam. Their acquisition, retrofitting, and apparent improvement of the Abelam package permitted Ilahita to stand against this group and has since led to both military and economic success. Ilahita has also grown through the in-migration and assimilation of refugees fleeing from hostile neighbors, which represents a case of intergroup competition via differential migration.[17]

These ethnographically rich cases suggest that increasingly effective social-bonding rituals spread over time, along with rising community size and political complexity, and that this societal evolution may have been driven by the rising intensity of military and economic competition. This suggestion fits with recent cross-cultural statistical analyses of small-scale societies showing that more warfare is associated with the presence of more terrifying and costlier rites for males. In many cases, the threat of war seems to drive the spread of rituals via prestige-biased group transmission, so these two forms of intergroup competition combine synergistically to favor more cooperative cultural forms.[18]

The presence of violent conflicts, territorial losses and gains, and the wholesale copying of institutions from more successful groups show us that some of the crucial elements of intergroup competition are not only present but also common, even in the smallest-scale human societies.[19] However, these ethnographic cases don't tell us if these relatively short-term interactions matter over the longer run for cultural evolution, over centuries or millennia in ways that systematically shape institutions, forms of social organization, and ultimately social psychology. Did intergroup competition shape the social worlds that our genes and psychology faced over the long run during human evolution?

CHAPTER 10

Hunter-Gatherer Expansions

Currently, there is relatively little evidence of some groups of hunter-gatherers expanding at the expense of other groups of hunter-gatherers over centuries or millennia. Part of the reason for this is that most of the available evidence for one group systematically spreading involves farmers or herders expanding at the expense of foragers or other groups of farmers and herders. Many of these population expansions can be traced either to institutions or forms of social organization, or to technological differences, or to both. As I show in chapter 12, the size and complexity of a group's suite of tools, weapons, and other technologies are heavily influenced by the group's social institutions, so sustained technological differences cannot be neatly partitioned from institutional differences. The massive success of farmers in taking over the globe thousands of years ago has made it hard to spot the older expansions of hunter-gatherer groups. This has led some to think that sustained expansions via intergroup competition remain a peculiar condition of farming and herding societies, an affliction to which mobile hunter-gatherers are immune. However, if hunter-gather expansions are important, then we should be able to spot them when we look at the parts of the globe that farmers and herders either couldn't get to or didn't get to until late in the game. This will take us to Australia, a continent of hunter-gatherers until Europeans arrived; the Arctic; and the Great Basin of western North America. Unlike potential cases of hunter-gatherer expansions found deep in the archeological record, these more recent expansions allow us to combine linguistic, archaeological, genetic, and ethnographic evidence and thereby generate a richer picture of what happened.

The Pama-Nyungan Spread

In Australia, we saw that the Warlpiri were increasing their territory by 3% per year. However, this doesn't tell us if that rate could be sustained for, say, 5000 years. Maybe groups variously gain and lose over centuries with no *net* effect. Crucially, it turns out that the Warlpiri are part of the distinctive Pama-Nyungan language family. As shown in figure 10.1, this single language family covers seven-eighths of Australia (the white part in the figure). All of the other roughly two dozen language families of indigenous Australian languages were crowded into the remaining one-eighth of the continent, all in the north, just west of the Gulf of Carpen-

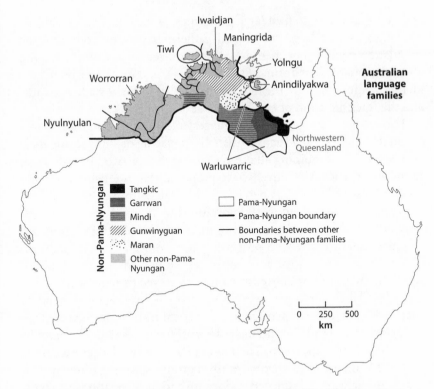

Figure 10.1. Distribution of major Australian language families. The Pama Nyungan family of languages covers most of Australia.

taria (where Burke and Wills went). This linguistic patterning, along with more detailed analyses of the Pama-Nyungan family itself, reveals a well-known linguistic signature, one that usually marks an expansion. These analyses indicate that the Pama-Nyungan expansion began in northwestern Queensland between 3000 and 5000 years ago and gradually spread over most of the continent.[20]

This linguistic picture is enriched by archeological, epidemiological, genetic, and ethnographic data. From the archaeological perspective, at about the same time as this linguistic expansion was occurring, new and different stone tools began appearing across Australia, including distinctive "backed blades." The distribution of these new tools roughly matches the distribution of Pama-Nyungan languages. New plant foods also began appearing that, like nardoo, require complex preparations, such as the grinding of various seeds. Though they are rather labor intensive, these new food sources could be stored, gradually accumulated, and eventually used to feed large gatherings. Not coincidently, evidence also suggests that large ceremonial gatherings became common, the population density increased, and people moved into new, challenging and inhospitable, environments. Studies of stone tools and their geographic sources indicate the development of substantial trade networks and the intensification of exchange.

This evidence suggests that languages, tools, rituals, and food preparation techniques spread across Australia, replacing or supplanting their alternatives. By combining all the available evidence, two linguists, Nick Evans and Patrick McConvell, have proposed that the Pama-Nyungan speakers spread because of new (1) patrilineal kinship institutions, (2) marriage rules that prescribed unions outside the local group and possibly outside the dialect group, (3) multigroup ritual gatherings or ceremonies supported by seed processing and storage capabilities, (4) intensive initiation rites for adolescents (as we saw among the Arunta), and (5) more encompassing cosmologies, conveyed through song cycles, that established sacred group identities (which saved Paralji's band in chapter 8). These kinship, marriage, and ritual institutions involved social norms that tightly bound males from different residential groups in interdependent social webs. The marriage norms of "dialect exogamy" meant that men had to seek wives from groups speaking different languages or dialects. This created an incentive to build relationships with other groups and forced local groups to remain integrated into larger populations. As discussed in chapter 9, the emotional impact of these new rites might have fostered solidarity among local bands and, especially, bound male adolescents together for life.

Both the wide-ranging affinal ties and large ceremonial gatherings created by these social norms would also have fostered better technologies and more adaptive cultural repertoires through the exchange of information about tools, weapons, skills, food sources, and medicines. Similarly, large ceremonies would have institutionalized "technology transfer" as young men learned complex skills from the most skilled members of all participating groups. Without social norms demanding such ceremonies, which required diverse groups, the population's ability to develop and sustain a large repertoire of complex technologies would have been inhibited. Recall that the Arunta believed failure to heed the ritual summons would result in illness.[21]

One key question is whether this expansion involved violent conflict, biased migration into successful groups, prestige-biased group copying, or one of the other mechanisms through which intergroup competition operates. Sparse as it is, the available evidence indicates that several of these mechanisms were at work. First, the genetic comparisons of Australian aboriginal populations suggest that speakers of Pama-Nyungan languages are often, but not always, genetically distinct from speakers of other languages. Paralleling this, Pama-Nyungan speakers tend to have high rates of the retrovirus HTLV1, while non-Pama-Nyungan have low rates. HTLV1 is principally transmitted through breast milk, which means that mothers pass it to their children with very little transmission occurring between tribal groups. This suggests that, in part, the competition occurred through the replacement of people—suggesting either *differential reproduction* or *violent conflict*, or both.

Accompanying these forms of competition, however, ethnographic and ethnohistorical data indicate that both *differential migration* and *prestige-biased group transmission* were probably also important parts of this spread, especially in light of the weak, sometimes nonexistent, relationship between language and genes (or retroviruses). While we can't be sure that some of these ancient norms, beliefs, and practices spread by being copied from group to group, we do have more recent cases that catch this process in action, as new kin-based institutions and rituals have been observed to emerge and get systematically and repeatedly copied by neighboring groups. For example, over about 60 years, a particular rite of male circumcision diffused out of the Kimberleys, to Arnhem Land, the Great Australian Bight, and eventually to Queensland.[22] Similarly, new and even more complicated sets of marriage rules spread widely in parts of Australia when two groups integrated their kinship and marriage systems. The more complex, recombinant institution that emerged probably fostered even greater integration across diverse groups than did its parent forms.

Intergroup competition by *differential survival* in a challenging environment may have also played a role, as Pama Nyungan speakers eventually entered, or perhaps reentered, the uninhabited and hostile environments of the Western Desert. Here, where earlier arrivals hadn't endured, groups bearing this new package of social norms and rituals, which permitted widely scattered bands to remain socially interconnected, could have survived more frequently when droughts or floods struck.[23] As we saw in chapter 8, Paralji saved his band from the 1943 drought in the Western desert by variously relying on the knowledge of distant waterholes he had gained during his adolescent initiation rite and on the rich

song cycles he'd gradually memorized over decades of ritual performances. Of course, Burke and Wills couldn't even find water in the Australian desert when there wasn't a drought.

The Spread of the Inuit and the Numic

The same kinds of processes of intergroup competition have also shaped
hunter-gatherer institutions in the Arctic. On the North Slope of Alaska,
around the time of the Battle of Hastings in England (around 1000 CE),
speakers of an Inuit-Inupiaq language—the Inuit—began expanding
eastward across the vast Canadian Arctic. In a few hundred years, these
hunter-gatherers would colonize Greenland and move south into Labrador, on the east coast of Canada. The territory they entered, however, was
not empty. The Dorset Eskimo, an archaeologically and probably genetically distinct population who had long inhabited these regions, rapidly
receded and vanished (mostly) in the wake of the Inuit tide. The Inuit
may also have driven out, or at least encouraged the speedy departure of,
the Norse settlers they encountered and fought in Greenland.[24]

Compared to the Dorset, the archeology shows that the Inuit (the
"Thule") had a more sophisticated technological repertoire. The Inuit
arrived equipped with advantages that included powerful compound
bows, high-quality adzes (for woodworking), kayaks, dogs, sleds, and
snow goggles. Along the coast, Inuit groups also had a full whaling
package, which included skin boats and harpoons. Interestingly, the
Dorset had had archery and dogs in the past but had lost them centuries before encountering the advancing Inuit (note, if you are skeptical
that useful tools can be "lost," stand by for chapter 12). Socially, the
Inuit probably also had the capability to rapidly organize men under a
prestigious leader for economic endeavors like whale hunting and
probably for raiding, warfare, and community defense. Their repertoire
of cultural-institutional technologies included flexible kinship norms,
special namesake relationships (like the Ju/'hoasi), rituals and other
bonding tools. For example, by providing his wife to another man for
sex, an Inuit (man) could cement an enduring and mutually beneficial
bond, even creating a special relationship between the men's children.
Together, such institutional-cultural technologies helped individuals
and communities weave and sustain vast webs of social relationships
across widely scattered populations. These networks were crucial for
maintaining trade relationships and linguistic similarity as well as for
finding marriage partners and recruiting allies for defense and raiding

(offense). As we'll see for the polar Inuit, a group's ability to sustain complex technologies depends on their sociality, on their ability to sustain broad-ranging social contacts.[25]

As with Pama-Nyungan peoples, ethnographic and ethnohistorical evidence suggests that multiple forms of intergroup competition were likely afoot during the Inuit expansion. In terms of warfare, I suspect that the Inuit gradually occupied Dorset territories and outcompeted them for local resources. When conflict did eventually break out, the Dorset tended to lose and retreat. Neither the archaeological nor the linguistic evidence tells us if success in warfare influenced the Inuit expansion. However, ethnohistorical evidence from Northern Alaska shows that raiding was an ever-present feature of Arctic life, and even pitched battles were common. Devastating surprise attacks at dawn aimed to annihilate whole communities, and sometimes did. Ambushes sometimes wiped out large trading or hunting parties. Consequently, in this hunter-gatherer social environment, strangers were viewed with great suspicion and were usually killed, a fact that placed a premium on having lots of personal contacts in other communities.[26]

Differential extinction could also have played a role. With their superior technology and more diverse set of foraging strategies, the Inuit likely endured and adapted to changing environments and ecological shocks more effectively than the Dorset, and thus would have reproduced faster. Prior to the arrival of the Inuit, the Dorset seemed to have periodically gone extinct in particular regions. If the Inuit's institutional advantages meant that they went locally extinct less frequently than the Dorset, Inuit norms and practices would have spread and eventually come to dominate—even if the Inuit and Dorset never confronted each other violently.[27]

Some cultural transmission did occur between the Inuit and the Dorset; for example, archaeological evidence suggests that late Dorset populations were acquiring Inuit house designs. Researchers have also located a few isolated Arctic populations who are not genetically related to the Inuit yet have clearly adopted many Inuit practices. These may be descendants of the Dorset.[28]

As in Australia, intergroup competition among foraging populations would have favored the spread of social institutions that permitted widely scattered small groups to sustain broad and enduring social relationships and to pull together local teams to engage in cooperative activities, like whaling, community defense, and raiding. Groups less effective at this, because of their social norms, would have lost to more capable groups. In this environment, intergroup competition would not have fostered trust and fairness toward strangers, but rather would have highlighted the need to sustain a tight social network of trusted allies, friends, and kin.

The story was very much the same in North America's Great Basin. The Great Basin is a vast watershed between the Rockies and the Sierra Nevada Mountain Range. Between about 200 CE and 600 CE, Numic-speaking hunter-gatherers expanded out of eastern California in a fan shape across the Great Basin. Three Numic groups—the Paiutes, Shoshoni, and Utes—gradually replaced the pre-Numic hunter-gatherers living there, as well as driving off some encroaching agricultural peoples at the fringes. Like their Australian counterparts, they were fueled by a combination of new forms of social organization and advanced technologies. Their flexible fission-fusion social organization and rituals permitted them not only to aggregate periodically for hunting, sharing information, making marriages, and raiding, but also to seasonally scatter as independent families for hunting, gathering, and defense (it's very hard to effectively attack mobile nuclear families). Unlike the foragers they replaced, these groups relied heavily on intensive plant-processing techniques and stored foods (and fancy twined water containers) that permitted them to both sustain higher population densities and to better withstand environmental shocks—like droughts.

After 1650 CE, when Numic groups first entered the historical record, they rapidly became renowned for their bravery and feared for their raiding parties and surprise attacks. During this period, Numic groups drove out the non-Numic inhabitants of both Warner and Surprise Valleys and seized the territory for themselves.[29] One Numic-speaking group, the Comanche, would eventually enter the Great Plains, adopt horses, and expand dramatically. The Comanche rapidly drove out other indigenous groups, most of which were farmers, and permanently pushed back the Spanish. These mobile hunting bands would come to dominate a vast territory and would only be driven back by another rapidly expanding group, the United States.[30]

Ancient Expansions

These cases of intergroup competition, in which one group of hunter-gatherers expands at the expense of another group or groups of hunter-gatherers, are rich in part because we know what these societies looked like at the earliest European contact, and we have some sense of their institutions, languages, and lifeways. However, archaeological evidence suggests that these kinds of expansions, on large and small scales, go deep into our species' evolutionary history. Over a million years ago, our genus expanded out of Africa into a vast range of Eurasian environments that were experiencing rapid climatic and ecological shifts. To the degree that survival in these evolutionarily new and harsh environments depended on cooperation or social networks to sustain technologies (like fire, bows and arrows, fishing, and clothing), differential extinction would have favored any culturally transmitted behaviors that fostered either.[31]

Around 60,000 years ago, groups of *Homo sapiens* expanded out of Africa (our lineage), this time at the expense of other members of our genus and species. As they did for the Pama-Nyungan speakers in more recent millennia in Australia, "backed blades" marked the expansion of these populations into Neanderthal-occupied Europe after 50,000 years ago. Like the Inuit, they may have advanced with the help of superior technology, specifically bows and arrows. These African varieties did interbreed to some degree with the other human lineages they encountered, but the Africans eventually replaced them culturally and dominated them genetically. This should now sound familiar. The archaeology can't tell us much about the processes behind these expansions, but the evidence of violence is consistent with some degree of warfare and raiding (as with chimpanzees). The recurrent presence of cannibalism among Paleolithic humans, which involved the consumption of fully grown adults, suggests violent intergroup conflicts.[32] As is typical in these expansions, the European variants—that is, the Neanderthals—also appear to have been copying the newcomers from Africa, suggesting prestige-biased group cultural transmission.

In more recent millennia, especially since the origins of plant and animal domestication some 12,000 years ago, the intensity of intergroup competition has dramatically escalated, driving the rise of increasingly large and complex societies. At the global level, Jared Diamond has argued that intergroup competition is crucial for explaining the expansion of particular agricultural groups around the globe and that the elevated intensity of this competition in Europe, and Eurasia more broadly, helps explain why it was Europeans who conquered the world after 1500 CE, and not the Aztecs or Warlpiri.[33]

Overall, this combination of evidence suggests that intergroup competition, in a variety of forms that include nonviolent competition, has been shaping cultural evolution and the social worlds we live in for eons, well back into our species' evolutionary history. If this evidence provides even a roughly correct view, then intergroup competition, through its influence on the social norms, reputational systems, punishments, and institutions experienced by individuals, will have shaped our genetic evolution. Let's now turn to this process, which is a form of self-domestication.[34]

CHAPTER 11

SELF-DOMESTICATION

Upon entering the Developmental and Comparative Psychology Laboratory at the Max Planck Institute in Germany, three-year-old participants engage in a series of tasks. First, they meet a hand puppet named Max and an experimenter for a warm-up, which allows them to get comfortable. As part of the warm-up, the experimenter uses some familiar objects in a typical way, such as using a colored pencil to draw. Then, the child has a chance to use the same objects. After the child's turn, Max has a chance with the objects. Sometimes Max uses the objects incorrectly; for example, by using the wrong end of a pencil to draw. Most kids immediately point out Max's mistake, and the few who don't will highlight these mistakes once asked about Max's actions. For the children, this establishes that Max sometimes makes mistakes and that it's okay to point these out.

In the next phase of the experiment, the child and Max are sitting at a table, and Max decides to take a nap. At a nearby table, off to the side, an adult is performing a multistep procedure using several unfamiliar objects. For example, one of these "target tasks" involves a Styrofoam board with a gutter, a wooden block, and a black suction head. The adult at the other table—the model—puts the wooden block on the board and uses the suction head to push the block across the board, into the gutter. Without looking at or addressing the child, the model either (a) acts like he knows what he's doing and is familiar with the task, or (b) acts like it's all new to him, and he's making it up as he goes along.

Then, after the model finishes, the original experimenter returns and brings the unfamiliar objects over to the child, saying, "Now you can have it." The child can do whatever he or she wants with the objects, while the researchers covertly record any imitation of the model by the child.

Finally, Max wakes up, and it's his turn to have a go at the objects. He uses the unfamiliar objects in perfectly sensible ways, but in ways that are *different from how the model used them*. This is the key moment in the experiment. The researchers carefully record the child's reaction to Max as he uses the objects in divergent ways.

Most children immediately protested against Max's "aberrant" actions (see figure 11.1), both when they had seen the confident model and when they'd seen the model who wasn't quite sure what he was doing. However, the kids protested much more when they'd seen the confident model. Many of these protests were in normative form, like "No! It does not go like this!" or "You must use this!" Other times the children just gave commands such as, "No, don't put it there!" The kids who most accurately imitated the model themselves were more likely to react in protest. Yet, even those who had only weakly imitated the model reacted negatively to Max's deviations from what the model had done. It was as if they had inferred a social norm without having themselves mastered the techniques necessary to live up to the local standards.

The psychologist Mike Tomasello and his collaborators have performed many experiments like these, all of which tell the same story.[1] By observing others, young children spontaneously infer context-specific rules for social life and assume these rules are norms—rules that others should obey. Deviations and deviants make children angry and motivate them to instill proper behavior in others. What's striking about these findings is that children can and will do all this without any direct teaching or pedagogical cues (like pointing or eye contact) from adults— though no doubt these must help convey the rules in many circumstances. The children's peculiar motivations to reprimand Max's actions are not imitated from adults in the experiment because no adult ever reprimands Max but instead are spontaneously applied by the child to violations of inferred rules. This experiment illustrates some of the essential features that distinguish human social life in all societies from that of other species:

- We live in a world governed by social rules, even if not everyone knows the rules.

Figure 11.1. An experimental subject wagging his finger at Max the puppet, who is violating the rules for this context.

- Many of these rules are arbitrary, or seem arbitrary (e.g., fish taboos in Fiji).
- Others care whether we follow these rules, and react negatively to violations.
- We infer that others care about whether we follow these rules.

As in the small-scale societies seen in earlier chapters, the social world faced by our Paleolithic ancestors would have been increasingly shaped by the emergence of an immense variety of norms, and by the selective spread of specific norms packaged in institutions, that fostered success in intergroup competition. From the gene's-eye view, survival and reproduction would have increasingly depended on the abilities of one's bearer (the individual) to acquire and navigate a social landscape governed by culturally transmitted local rules—those appropriate to whatever group a particular gene happened to find itself in. Typically, in small-scale societies, as in many communities, the sanctioning of norm violators begins with gossip and public criticism, often through joking by specific relatives (as with Kula), and then intensifies to damage marital prospects and reduce access to trading and exchange partners. If violators are still not brought into line, matters may escalate to ostracism or physical violence (e.g., beatings) and occasionally culminate in coordinated group executions.[2] In parallel with how wolves were domesticated into dogs by killing those that wouldn't obey and refused to be trained, human communities domesticated their members.[3]

In research in the villages of Yasawa Island, my team and I have studied how norms are maintained. When someone, for example, repeatedly fails to contribute to village feasts or community labor, or violates food or incest taboos, the person's reputation suffers. A Yasawan's reputation is like a shield that protects them from exploitation or harm by others, often from those who harbor old jealousies or past grievances. Violating norms, especially repeatedly, causes this reputational shield to drop and creates an opening for others to exploit the norm violator with relative impunity. Norm violators have their property (e.g., plates, matches, tools) stolen and destroyed while they are away fishing or visiting relatives in other villages; or they have their crops stolen and gardens burned at night. Despite the small size of these communities, the perpetrators of these actions often remain anonymous and get direct benefits in the form of stolen food and tools as well as the advantages of bringing down a competitor or dispensing revenge for past grievances. Despite their selfish motivations, these actions sustain social norms, including cooperative ones, because—crucially—perpetrators can only get away with such actions when they target a norm violator, a person with his reputational shield down. Were they to do this to someone with a good reputation, the perpetrator would himself become a norm violator and damage his or her reputation, thereby opening themselves up to gossip, thefts and property damage. This system, which Yasawans themselves can't explicitly lay out, thereby *harnesses* past grievances, jealousies, and plain old self-interest to sustain social norms, including cooperative norms like contributing to village feasts.[4] Thus, individuals who fail to learn the correct local norms, can't control themselves, or repeatedly make mistaken violations are eventually driven from the village after having been relentlessly targeted for exploitation.

Over our evolutionary history, the sanctions for norm violations and the rewards for norm compliance have driven a process of self-domestication that has endowed our species with a *norm psychology* that has several components. First, to more effectively acquire the local norms, humans intuitively assume that the social world is rule governed, even if they don't yet know the rules. The violation of these rules could and should have negative consequences. This outcome means that the behavior of others can be interpreted as being influenced by social rules. This also means that, at a young age, we readily develop cognitive abilities and motivations for spotting norm violations and avoiding or exploiting norm violators, as well as for monitoring and maintaining our own reputations.[5] Second, when we learn norms we, at least partially, *internalize* them as goals in themselves. This internalization helps us navigate the social world more effectively and avoid temptations to break the rules to obtain immediate benefits. In some situations, internalizations may provide a quick and efficient heuristic that saves the cost of running the mental calculations that consider all the potential short- and long-term benefits and probabilistic penalties of an action; instead we simply follow the rule and abide by the norm. This means that our automatic and unreflective responses come to match the normatively required ones. Other times, internalized preferences may merely provide an additional motivation that goes into our calculations.[6]

The experiments involving Max are cool because we get a look at kids' reactions to novel arbitrary rules for specific contexts. These rules aren't about cooperation or helping others; they are just context-specific rules. Nevertheless, kids automatically infer that they are social norms and get mad when they are violated. More important, this same pattern emerged when psychologists focused on studying altruism in children during the 1960s and 1970s. In the classic experimental setup, a school-child is brought alone to a testing area to get acquainted with an experimenter. The child is then introduced to a bowling game and shown a variety of attractive prizes that he or she can obtain with tokens won in the bowling game. The child is also shown a charity jar for "poor children" and can voluntarily put some of his or her winnings from the game into it. This jar often has a March of Dimes poster, or some facsimile, posted behind it. A model, who could be a young adult or another child, demonstrates the game by playing for 10 to 20 rounds. On preset rounds, the model wins some tokens and donates some of these token to the charity jar. Children experience one of three situations: (1) a generous person who puts lots of tokens in the charity jar, (2) a stingy person who puts only a few tokens in the jar, or (3) no demonstration. After the demonstration is complete, the child is left alone to play the bowling game and to donate to charity, if he or she wants.

The results of many versions of this experiment demonstrate four key findings. First, children spontaneously imitated the model, becoming either more generous or more selfish depending on which model they saw. Those who saw the generous model gave more than children in the "no model" condition, while children who saw the stingy person gave less than children in the "no model" condition. Second, beyond merely the effects on their donations, children also imitated other aspects of the model's behavior, including the model's verbal statements. Children even repeated verbal statements from the model when they seemed to contradict their own and the model's actual behavior. That is, they say how important it is to give to poor children, but then not give much. Third, the effect of exposure to a model—generous or stingy—endures for weeks or months in retests. But the effect does not extend to quite different contexts, those that don't resemble a bowling game.[7] Finally, children readily imitate standards for self-reward or self-punishment and readily impose those standards on others. When children are assigned to help a younger novice with the bowling game, they will often demonstrate generosity to the novice and then impose the standard they've acquired on the novice by scolding her or him, if he or she doesn't spontaneously adopt it.[8]

Overall, children are not culturally learning to be altruistic in some general or dispositional sense; they are acquiring norms about proper behavior in the bowling game context, and those behaviors include proper donation sizes. Because they have inferred that social norms exist, they impose these behaviors on other children in the same way that the puppet Max was reprimanded for his "mistakes."[9]

How Altruism Is like a Chili Pepper

It's clear that when people encounter a new situation, they try to figure out which norms, among those they've already acquired, might apply to the situation and are also prepared to acquire new norms specific to this unfamiliar context. With this in mind, we can now look at findings from economic games. In classic experimental social dilemmas, two or more strangers interact anonymously and make decisions that influence both their own payoffs and those of the other players. All decisions in these experiments are real in the sense that these decisions are implemented and determine how much money people take home. Many valuable insights about social norms and psychology come from the use of economic games. Properly interpreted, economic games are valuable tools for measuring social behavior and for teasing apart the complex packages of motivations, understandings, and beliefs that jointly influence decisions. Well-known economic games include the Prisoner's Dilemma, Ultimatum Game, and Dictator Game. To understand these experiments, imagine this situation:

You enter an experimental economics laboratory at Big City University. It's filled with college-age strangers seated at computer terminals. You are told to sit down at an open terminal, which has partitions that prevent others from seeing your screen. After some preliminaries, the computer screen informs you that your ID has been randomly assigned to interact with another person in the room, but neither you nor this person will ever know the other's identity. You each make one decision and the game is over. If you get any money from this decision, it will be added to your "show-up fee" (which is $20). You will receive all money in cash at the end as you exit.

In this interaction, you have been randomly assigned to the role of the "proposer" and the other person is the "responder." As the proposer, it's your job to divide $100 between you and the other person by making an offer between $0 and $100 (in increments of $1) to the responder, who knows everything, except your identity. The responder then has two choices: they can either accept your offer or reject it. If the responder accepts, he or she will receive your offer, and you get the remainder. If the responder rejects, you both get nothing (no money). This means you'll go home with only your show-up fee.

This is the Ultimatum Game. Using game theory, we can determine what a person would do if they were interested only in maximizing their take-home pay. To figure this out, put yourself into the shoes of the responder. If the proposer offers you any money more than zero, you face a choice between zero (if you reject) and some positive amount of money (if you accept). If, for example, the proposer offers you $1, you can leave with $1 more by accepting it. Thus, if you are a money-maximizing responder, you should accept any positive offer. Proposers, realizing this, should offer only $1, which would be accepted. If humans were money-maximizers, Ultimatum Game experiments should reveal many low offers and few rejections of nonzero offers. Not surprisingly, it turns out that this never happens in any human society. By contrast, experiments with primates show little or no evidence of motivations besides narrow self-interest *in dealing with strangers*. Chimpanzees, for example, never reject in the Ultimatum Game.[10]

In Western societies, most people offer half ($50 out of the $100), and enough people reject offers below 50% that it doesn't pay to give less than half because the risk of having your offer rejected is too great. Interestingly, among people over about the age of 25, this willingness to offer 50% is mostly not driven by concerns about being rejected. To explore this, we can turn the Ultimatum Game into a Dictator Game by removing the possibility of rejection. In the Dictator Game, the proposer gives some portion of the $100 to the other player and the proposer keeps whatever is left. If people were strictly self-interested, the proposer would give nothing to the other player and take home the entire $100. But, rather than giving $0, most Western adults continue to give half. This outcome suggests that people have an internalized equality norm toward strangers, applicable in this context, that is calibrated to allow them to effortlessly and without constant strategic recalculation navigate through a world with punishers, like the responders in Ultimatum Games who would sanction them for keeping too much for themselves. But people continue to adhere to the norm even when no punishment is possible and no consequences outside the game are plausible.[11]

My collaborators and I have systematically performed economic games across diverse societies and among chimpanzees. In humans, the evidence is clear that such games often tap social norms that people bring into the laboratory from their social lives outside, and consequently game play varies dramatically across societies. In modern industrialized societies, these experiments often measure social norms that regulate impersonal exchange and other social interactions, and evolved culturally to facilitate mutually beneficial interactions in large-scale societies with lots of strangers and anonymous interactions. The strength of these impersonal norms, though unusual, is a key feature of many modern societies. By contrast, the smallest-scale human societies tend not to offer very much nor reject low offers because they lack social norms for monetary exchanges with strangers or anonymous others.[12]

However, when games are played repeatedly in the laboratory, participants begin to develop "lab-specific" social norms as they adapt to a new context. These social norms include motivations, beliefs, and expectations, including concerns about what other people will think of those who violate whatever norms govern the game.

It's Automatic

Internalized social norms help guide us through complex social environments, allowing people to automatically—without conscious reflection or complex mental calculations of the reputational consequences—do the "right thing" (i.e., comply with local norms). This can be seen in how people respond in the Public Goods Game. The structure of this game captures the logic of real-life situations, like recycling, giving blood, paying taxes, and defending the community, in which the group does best if everyone cooperates but the individual does best if he or she acts selfishly while everyone else cooperates. In this classic cooperative dilemma, individuals are placed into groups with three strangers for a single interaction. Each person gets $4 to start. Without knowing what others will do, they have to contribute between 0 and $4 to a common project. Whatever enters the project is doubled and then distributed equally among all four group members, regardless of what they contributed.

To highlight the cooperative dilemma, consider that the group gets the highest payoff if everyone contributes all four of their dollars to the common project (4 × $4 = $16). This money doubles to $32 and is distributed equally so that everyone goes home with $8 (twice what each started with). However, every individual does best if they keep their $4 and free ride on those who contribute to the common project. For example, if three people contribute $4 and one free rider contributes nothing to the common project, then the three contributors go home with $6 each, and the free rider goes home with $10—his initial $4 plus the $6 he got from the common project. If three people free ride and only one person contributes his entire $4, then the free riders go home with $6 each while the contributor gets only $2. Thus, those aiming to maximize their payoff should contribute zero. However, most educated Westerners agree that—if asked—players *should* contribute all the money to the common project. Among the typical experimental subjects (undergraduates), the average contributions are commonly between 40% and 60%, with many people contributing either 100% (cooperators) or 0% (free riders).[13]

To examine whether high contributions in the Public Goods Game, and prosocial choices in other such games, result from automatic norm following, David Rand and his colleagues examined the relationship between the time people spent making their contribution decisions and

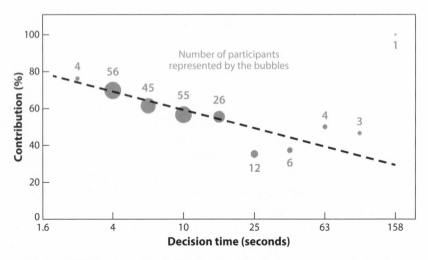

Figure 11.2. Plot showing that the longer people took to decide what to do, the less they cooperative they were.

the size of their contributions. Figure 11.2 shows one of Dave's findings: the more rapidly participants made their decision, the higher their contribution was to the common pool—that is, quick, gut responses were more cooperative.[14]

Such findings are provocative, but it could be that cooperative people happen to also be people who respond quickly to questions. To address this, Dave ran participants through the same experiment again, but this time rather than letting them take however long they wanted, people were randomly put into one of three different treatments. They were alternatively (1) forced to answer in less than 10 seconds, (2) unconstrained as before, or (3) forced to delay their decision for 10 seconds and asked to reflect on it. Figure 11.3 shows the results: under time pressure, participants were more cooperative. When forced to delay and reflect, participants became less cooperative than when they were unconstrained.

Rand and his team showed the same effects across many different experiments, including experiments in which they unconsciously cued participants to either "reflect" or "go with their gut." Going with one's gut leads to more cooperation, if you have the appropriate norms.

Above, we saw young children immediately express anger when Max violated a social norm that the children had inferred. This fits well with much work on rejections in the Ultimatum Game. Not only do people

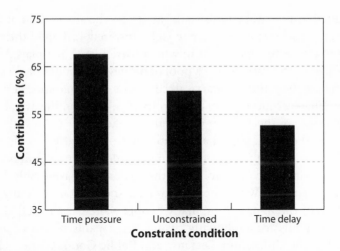

Figure 11 3. Graph showing average percentage contributed in three treatments. Under time pressure people cooperated more.

from some societies get angry when they receive a low offer, but also the participants are quicker when deciding to reject low offers. By contrast, deciding to accept a low offer—the rational and self-interested thing to do—seems to take much careful consideration. When placed under time pressure for their responses, individuals from these societies reject more of the unfair offers. In one experiment, researchers used drugs to reduce people's impulse control (serotonin depletion). Loss of their impulse control resulted in more rejections of low offers, but not of 50/50 offers. Negative emotional reactions are our automatic and unreflective response to norm violations and norm violators.[15]

The power of norms in economic games first impressed me in 1995 when I was administering the Ultimatum Game among the Matsigenka, in the Peruvian Amazon. Lacking strong social norms specifying equality toward strangers in monetary exchanges, these people were happy to be offered any money in the game, didn't expect proposers to offer half, and weren't inclined to punish proposers for low offers. Nearly twenty years of subsequent research across two dozen diverse societies showed that these sentiments are common in the smallest-scale human societies.

Converging evidence for the importance of norm psychology in economic decisions comes from the economist Erik Kimbrough. Late one night, while returning from a pub in Amsterdam, Erik noticed that people waited at the "walk/don't walk" lights on the street, even at broad intersections when no cars were in sight. Inspired, Erik used this observation to create an experiment in which participants first played a very simple game. They were given a pool of money that would begin slowly draining as their little avatar walked down a virtual street on their screen. Whatever money remained in their pool when the avatar reach the other side of their screen was theirs to take home. Following an explicit rule, the avatar would automatically stop at red lights along the virtual street and wait, while the money would continue to drain from the participant's account. To make the avatar go, players only had to press a key, any key. While they could make their avatar go at any time, regardless of the light's color, many people waited at all the lights for green. After this "rule-following" game, participants played economic games, like the Ultimatum, Dictator, and Public Goods Games. The results confirmed Erik's suspicions: the amount of time people waited at the lights was associated with making more equal offers in the Dictator Game, contributing more in the Public Goods Game, and punishing low offers more frequently in the Ultimatum Game. Observance of a costly nonsocial rule, like waiting at traffic lights, appears to be underpinned by the same psychological machinery as complying with, and punishing, social norms in behavioral games.[16]

As both Adam Smith and Friedrich Hayek argued long before Erik and me, it's our automatic norm following—not our self-interest or our cool rational calculation of future consequences—that often makes us do the "right thing" and allows our societies to work. This means that how well a society functions depends on its package of social norms.

And in the Brain

The effects of internalizing norms can be seen in our brains when economic games are combined with tools from neuroscience. When people cooperate, give to charity, or punish norm violators in locally prescribed ways, the "rewards circuits" in their brains fire up. Some of these are the same circuits that fire when people are rewarded with money or food, yet in these costly social contexts, the circuits are firing despite the fact that individuals are actually losing money.[17] Neurologically speaking, people "like" to comply with norms and punish norm violators.

Using these brain-imaging tools, it's instructive to consider what people's brains do when we decide to break a social norm. Consider lying. Neurologically, lying requires most people, though presumably not law-

yers or car salesmen (just kidding), to override their automatic or unreflective reactions by engaging those brain regions responsible for cognitive control and abstract reasoning. That is, violating a social norm
requires mental effort and "higher" cognition.[18] Most Westerners, for
example, have to override an internalized norm to lie to strangers in
many contexts. Note, of course, that intentionally not telling the truth
isn't always a norm violation, such as with "white lies." And in many
places it's considered totally fine—if not encouraged—to lie to strangers
or foreigners to benefit oneself or one's family (no "over-ride" needed).

Why would natural selection have built us to be norm internalizers?
Broadly speaking, internalizing motivations helps us to more effectively
and efficiently navigate our social world, a world in which some of the
most frequent and dangerous pitfalls involve violating norms. Such motivations may help us avoid short-term temptations, reduce cognitive or
attentional loads, or more persuasively communicate our true social
commitments to others. The logic here parallels that which we encountered in chapter 7, where I explained how cultural learning could overcome an innate aversion to chili peppers and other spices in order to
reduce the dangers of meat-borne pathogens. Reinterpreting the pain as
pleasure helps individuals navigate the ecological landscape by solving
an adaptive problem (meat-borne pathogens) without us even being
aware of it. Analogously, internalizing norms as tastes helps us more easily and intuitively navigate the social landscape.

Why Spotting Potential Norm Violations Is Easy

In addition to the internalization of social norms, culture-gene coevolution has honed our cognitive abilities, motivations and emotions in various ways, including ways that permit us to effectively manage our reputations. On the cognitive side, both children and adults are more skilled
at solving logic problems when the latter are contextualized as norm
violations. This helps us avoid committing norm violations ourselves
and pick out other norm violators, whom we might be required to
punish, avoid, or ostracize (and even be rewarded for doing so). As we
saw in Fiji, spotting norm violators results in opportunities to justifiably
steal their crops or take revenge for past grievances.

To understand these abilities, consider this experiment with three-
and four-ear-olds: the children hear one of two stories and then have to
solve a logic problem. In both stories, they are told about some mice
that go out to play in the evening. Some of these mice tend to squeak

while playing, which attracts the neighborhood cat, who comes and tries to catch them. In one version, the children hear a *descriptive claim*, which states that all squeaky mice stay in the house in the evening. In the other version, they are told about a *social norm* prescribing that all squeaky mice *must* stay in the house. Now for the test. The children are placed in front of the mouse house, with ten yellow rubber mice inside the house. They are also shown that "squeaky" and "quiet" mice can only be distinguished by squeezing the mice and listening for a squeak. Then, evening arrives at the mouse house, and four mice leave the house to play in the backyard. Depending on which version of the story they heard, children were tasked either with (1) checking to see if the descriptive claim was true or (2) locating norm violators. The answer is the same in both cases: you have to check all the mice in the backyard, not in the house. Checking the mice in the house tells you little, since quiet mice might be in the house in either case, and you don't know how many of each kind of mice there are. When checking for norm violations, most three- and four-year-olds decided to check the mice in the backyard. However, when verifying the descriptive statement, most of the children did not think to check the backyard mice.[19] This suggests that setting up the task to cue norm psychology made the children better at solving the logic problem.

This self-domestication process has also tinkered with our feelings and emotional displays to better navigate a world governed by social norms. Primate emotions related to shame and pride have been retrofitted to apply to social norms. Shame in humans evolved (genetically) from a primate "proto-shame," the package of feelings and bodily displays that we see in primates when individuals demonstrate or signal their subordinate status to a dominant group member. The shame and proto-shame display in both humans and primates involves slumped shoulders, downcast gaze, crouching, and a diminutive body posture— the idea seems to be to look small and unimposing. However, as the anthropologist Dan Fessler has persuasively argued, shame in humans emerges when someone violates a social norm or delivers a substandard performance as well as in status hierarchies (see chapter 8). Norm violators display shame to their communities for communicative reasons that parallel those that drive subordinates to display shame in the presence of more dominant animals. In both cases, the shame display reaffirms their acceptance of the local social order. In the context of norm violations, the ashamed is effectively saying to the community, "Yes, I know I violated a norm, and should be admonished for it; but please don't be too harsh on me."[20]

The same kind of coevolutionary process may have provided some of the basic mental tools for assigning reputations to individuals as well as certain default settings and motivations for judging things like harm, fairness, and status. These psychological adaptations evolved genetically in response to the broad spreading—by intergroup competition—of social norms that (1) suppressed the harming of community or in-group members, (2) prescribed equitable treatment to peers, and (3) established enduring status relationships. One of my UBC colleagues, the developmental psychologist Kiley Hamlin, has shown that before the end of their first year of life, babies make rather nuanced social distinctions consistent with these predictions. By using puppet shows as simple morality plays, Kiley's efforts reveal that infants prefer puppets who help others but don't generally like those puppets who hinder or otherwise hurt others. Crucially, however, babies rapidly develop a key nuance: by eight months of age, babies prefer puppets who *hurt* previously antisocial persons (those spotted harming other puppets) over those who help antisocial types. Hurting others is fine according to babies, as long as those others are known to harm others or are members of other groups. Similarly, toddlers actively punish, by taking treats away from guys who help antisocial others; instead, they prefer puppets who hurt antisocial guys. This work shows that, early in development, babies already possess some of the key reputational and motivational elements that sustain social norms in small-scale societies and seem prepared to apply that reputational logic to simple circumstances of helping and hurting.[21]

In short, to survive in a world governed by social rules enforced by third parties and reputations, we became norm learners with prosocial biases, norm adherers internalizing key motivations, norm-violation spotters, and reputation managers. This makes us rather unlike any other species.

Norms-Created Ethnic Stereotyping

When a group of chimpanzees bumps into a lone individual from a neighboring group, hostility erupts immediately with a volley of aggressive hoots and barks. If the group is large enough, they will likely attack and kill the unlucky traveler. Human societies, even the smallest-scale ones, are quite different in this respect from chimpanzee populations because local groups, whether those are bands, villages, or single households, are enmeshed in larger tribes, or at least diffuse tribal networks. Tribal members, or co-ethnics, share a dialect or language and often many other obvious markers of membership, such as dress, greetings, gestures, rituals, and hairstyles. Less obvious is that co-ethnics tend to share a set of social norms, beliefs, and worldviews that govern their lives and allow them to anticipate each other's behavior, to coordinate, and to cooperate.[22]

It's not that humans are so nice or generally friendly to strangers. In many human societies past and present, lone travelers could easily find themselves fleeing for their lives because they have encountered a large party of strangers—as I discussed among the Inuit. Among small-scale societies, this often happens at language boundaries, or when local communities are at war. However, it's also the case that within a tribal network, which often involves many more individuals than anyone can know personally, people could often hunt, gather, farm, travel, and look for mates in relative security. Strangers, especially if they are wearing the relevant symbols and make the appropriate greetings, can be approached, and shared networks of relationships can be determined.[23]

Tribes or ethnolinguistic groups, and the psychology that permits us to navigate the social world they create, likely arose through a culture-gene coevolutionary process. Here's the idea: cultural evolution gave rise to a variety of different social norms, so different groups became increasingly characterized by different practices and expectations about such things as marriage, exchange, sharing, and rituals. Then, natural selection acting on genes responded to this world governed by social norms by endowing individuals with the cognitive abilities and motivations to help them better navigate and adaptively learn. The success of persons growing up in this emerging landscape of social norms depended—at least in part—on their ability to acquire the appropriate social norms for their own group and to preferentially target their interactions toward those most likely to share their norms. If learners acquire social norms that don't fit with others in their group, the learners will end up violating the local norms, getting a bad reputation, getting punished, etc. Even if they acquire the appropriate local norms but then interact with people from other groups, who have different norms, they can end up getting sanctioned, wasting time, or miscoordinating. For example, a boy and girl from different ethnic groups might fall in love and carry on a romance for years only to find out that a marriage is impossible, since his family demands a dowry but her family is looking for a bride price (a payment in exchange for their daughter). Both sides expect to be paid by the other, and this is serious business.

However, social norms are tricky because they are often hidden from view until it's too late. Many of our norms are so profoundly part of how we view the world that it's hard to imagine anyone could believe otherwise. For example, you might marry a lovely man from the Horn of Africa, only to find out years later that he had your eight-year-old daughter ritually circumcised while visiting his family. His decision might not fit your preferred customs, though it seemed a matter of course for your husband and his mother, who is disgusted by the idea of clitoris-bearing women. In this part of Africa, as well as in the Middle East, female genital cutting is a longstanding tradition and is associated with purity and fertility. They can't understand why you are so upset.

To deal with the non-obvious nature of social norms, natural selection took advantage of the fact that the cultural transmission pathways of social norms are often the same as those for other more observable markers, like language, dialect, or tattooing practices. Such markers can then be used as cues to both (1) figure out who to learn from and (2) whether a potential partner is likely to share one's norms.[24] The best markers are those that are difficult to fake. The reason why difficult-to-fake markers, or complex combinations of simple markers (e.g., dress, gestures, *and* manners), are best is because an easy-to-fake marker, like a distinctive hat, can be simply put on in order to trick or manipulate another person. For example, a Gentile physician living in Manhattan might place a mezuzah outside her office in hopes of attracting or retaining more (Jewish) patients. A mezuzah is a tiny piece of parchment with specific Hebrew verses inscribed on it, often stored in a decorative case. It's typically attached to the doorframe at the entryway of houses. I suspect that many Jews readily notice these tiny boxes (my wife does), while they remain invisible to most non-Jews. In contrast to something like a mezuzah, language and dialect are better markers because they are not easy to get right unless one grows up in a certain place or within a certain social group. This suggests that language or dialect might be a priority cue for determining who to learn from and interact with and for making guesses about the probable actions of the speaker.

We've already seen (in chapter 4) the evidence that infants and young children preferentially learn tool use and food preferences from those who share their language or dialect. The developmental psychologist Katie Kinzler and her colleagues have also shown that young children preferentially seek interaction with those who share their language, especially when it's spoken in their dialect. This holds for diverse populations, with similar experimental results emerging from children in Boston, Paris, and South Africa.[25] At 5 to 6 months of age, infants preferentially watch those who share mom's accent. By 10 months, infants preferentially accept toys from those who speak with their mother's accent.[26] Later, preschoolers tend to pick those who share their language or dialect as "friends."

The importance of language as an ethnic marker came across most strikingly while my wife, Natalie, was conducting her PhD dissertation research among Chaldeans in Michigan. Chaldean immigrants from Northern Iraq have been gradually clustering in metropolitan Detroit over the last century. By the late 1990s, this ethnic group had come to dominate the small-grocery-store business sector of the city. By forming tight social networks, hiring mostly relatives or fellow Chaldeans, and preferentially using Chaldean doctors, lawyers, and other professionals, this group has consistently prospered in an often-challenging economic environment (it's Detroit). Being considered "Chaldean" by the community was and is crucial, since it gives one access to jobs, handshake contracts with other Chaldean businesspersons, broad social networks, and substantial marriage possibilities. Speaking Chaldean, the language spoken by Jesus, as any Chaldean will remind you, is very important for establishing one's Chaldean identity. It was so important that many in the second and third generations would take Chaldean language classes. Even some first-generation immigrants from urban areas in Iraq, like Mosul, would also take the language classes, since Iraqi city dwellers of Chaldean descent sometimes would only learn Arabic. Of course, speaking Arabic was absolutely not a Chaldean marker, since Detroit is full of Muslim Arab immigrants that Chaldeans want to distinguish themselves from. Of course, practicing Chaldean Christianity was also an important cue of Chaldean identity.[27]

Ethnic markers go well beyond language and dialect, however. For thousands of years and across the globe, many populations have shaped their skulls, including Europeans until recently. By using a variety of techniques on infants, such as strapping boards to the head, people have created distinctive and beautiful (to them) cranial forms, including flat, round, and conically shaped heads.[28] The shapes often marked distinct ethnic groups or classes. Because cranial reformation must begin in infancy and requires serious investments by one's family, it's nearly impossible to fake this cue (see figure 11.4).

Figure 11.4. Device traditionally used on infants for head flattening among Chinookan-speaking populations in the Pacific Northwest, USA.

Thus, the world that cultural evolution often creates is one in which different groups possess different social norms and where norm boundaries are often marked by language, dialect, dress, or other markers (e.g., head shape). This social environment would have favored the reliable development of cognitive tools for navigating such a world. In this world, knowing a person's dialect would have allowed one to predict with some confidence many other aspects of his or her preferences, motivations, and beliefs, because dialects get transmitted along the same learning pathways as social norms, beliefs, and worldviews. The situation may also have favored an evolved psychology for recognizing the groups in the world, determining their markers, and making generalizations about their members using category-based induction (as discussed in chapter 5). That is, if you learn something about one member of a group—for example, he doesn't eat pigs—you tend to assume that this applies to all members. Of course, the downside of such tendencies and abilities is that they sometimes yield incorrect inferences and tend to throw the whole social landscape of groups and their behaviors into a starker relief than reality (sometimes) supports. Cognitive scientists call these abilities our folksociological capacities.[29]

We can see how deeply norms are intertwined with our folksociology by returning to the experiments with Max the puppet. The child subjects now encounter Max along with Henri. Max speaks native-accented German, but Henri speaks French-accented German. Young German children protested much more when Max—their co-ethnic as cued by accent—plays the game differently from the model than when Henri did. Co-ethnics are favored because they presumably share similar norms, but that also means they are subject to more monitoring and punishment if they violate those norms. This appears to hold cross-culturally: in experiments like the Ultimatum Game, people from places as diverse as Mongolia and New Guinea willingly pay a cost to preferentially punish their co-ethnics over their non-co-ethnics for norm violations.[30]

This approach to how and why we think about tribes and ethnicity has broader implications. First, intergroup competition will tend to favor the spread of any tricks for expanding what members of a group perceive as their tribe. Both religions and nations have culturally evolved to increasingly harness and exploit this piece of our psychology, as they create quasi-tribes. Second, this approach means that the in-group versus out-group view taken by psychologists misses a key point: not all groups are equally salient or thought about in the same way. Civil wars, for example, strongly trace to ethnically or religiously marked differences, and not to class, income, or political ideology.[31] This is so because our minds are prepared to carve the social world into ethnic groups, but not into classes or ideologies.[32]

Finally, the psychological machinery that underpins how we think about "race" actually evolved to parse ethnicity, not race. You might be confused by this distinction since race and ethnicity are so often mixed up. Ethnic-group membership is assigned based on *culturally transmitted* markers, like language or dialect. By contrast, racial groups are marked and assigned according to *perceived* morphological traits, like skin color or hair form, which are *genetically transmitted*. Our folksociological abilities evolved to pick out ethnic groups or tribes. However, cues such as skin color or hair form can pose as ethnic markers in the modern world because members of different ethnic groups sometimes also share markers like skin color or hair form, and racial cues can automatically and unconsciously "trick" our psychology into thinking that different ethnic groups exist. And this by-product can be harnessed and reified by cultural evolution to create linguistically labeled racial categories and racism.

Underlining this point is the fact that racial cues do not have cognitive priority over ethnic cues: when children or adults encounter a situation in which accent or language indicate "same ethnicity" but skin color indicates "different race," the ethnolinguistic markers trump the racial markers. That is, children pick as a friend someone of a different race who speaks their dialect over someone of the same race who speaks a different dialect.[33] Even weaker cues like dress can sometimes trump racial cues. The tendency of children and adults to preferentially learn and interact with those who share their racial markers (mistaken for ethnic cues) probably contributes to the maintenance of cultural differences between racially marked populations, even in the same neighborhood.

My point is that because of culture-gene coevolution, humans reliably develop the psychological equipment to map and navigate a world of immense cultural diversity. However, in mapping the social world around us using both our own observation and culturally acquired categories (like race, see chapter 7), our folksociological system, like our visual system, errs on the side of providing us with only the essential landmarks and main avenues around us, while ignoring lots of detail. Thus, the dynamically shifting gradations and clines of cultural variation are often rendered as a snapshot in stark relief.

Why Kin-Based Altruism and Reciprocity Are So Strong in Humans

Efforts to apply evolutionary theory to humans have long emphasized the importance of kinship and reciprocal altruism (reciprocity), as I noted in chapter 9. There's no doubt these are important. However, what's interesting is how potent kinship and reciprocity are in humans compared to other species. Genetic relatedness certainly matters in primate social life, but it doesn't matter nearly as much as it does in humans. Humans help more relatives more often than other mammals, who miss both helping opportunities and often whole classes of relatives (like paternal half-siblings). For kin-based altruism to emerge, individuals have to be able to identify *when* and *whom* to help; yet natural selection misses many situations in which relatives could help each other but don't because relatives are often hard to spot and it's not always easy to know when they need help.

The cultural evolution of social norms can strengthen the power of kin-based altruism by creating social norms that point to specific situations and relatives who need help. Social norms bring the community in as monitors, to make sure people don't overlook their responsibilities to their relatives. Brothers will be naturally inclined to help each other, but brothers monitored by a community possessing norms about brotherly responsibilities will be even more inclined to help each other. Thus, sanctions for norm violations can then strengthen natural selection's potency in shaping our nepotistic instincts, especially those that bind fathers to their offspring.

For reciprocity, the effects of social norms may have been even more striking, given that reciprocity is relatively rare outside of humans and especially outside of primates. Social norms can galvanize reciprocity in ways that make it more enduring and applicable to more domains. Governed by local norms, third parties can help monitor and decide whether someone has defected (failed to reciprocate) and can assist in sanctioning nonreciprocators. For example, in many small-scale societies men exchange sisters as wives. You permit me to marry your sister now, and I promise that when my sister is old enough, she will marry you. After my marriage, in the meantime, circumstances may change. Perhaps you get permanently injured or my sister disappears with another man. As a consequence, I might be inclined not to pay you back, even if it ends the relationship between our families. However, in many places, I'd not only be violating our personal agreement but also social norms about sister exchange. Failure to meet my obligations will impact my reputation broadly. In one well-studied case among the Gebusi, in New Guinea, my failure to meet my sister exchange obligations would increase the chances that I would, at some future date, be found guilty of witchcraft and then executed by the community. In this culturally constructed world, if I defect on our reciprocity relationship, I risk not only the end of our relationship but possibly my own end.[34] In such a world, natural selection will favor potent motivations for reciprocity.

Cultural evolution created a social world that magnified natural selection's ability to shape our instincts for kin-based altruism and dyadic reciprocity.

War, External Threats, and Norm Adherence

In Nepal from 1996 to 2006, Maoist rebels battled the Nepalese Armed Police Force and then the Royal Nepalese Army. The conflict killed over 13,000 people, destroyed property, and displaced hundreds of thousands from their homes. Violence in small rural communities was random and unpredictable. Sometimes violence was used to intimidate locals, coerce support, or gather information. Other times it was used for revenge or as an excuse to settle old political scores. To study the impact of this war on people's social motivations, the political scientist Michael Gilligan and his colleagues deployed a battery of behavioral games, including both the Public Goods and the Dictator Games, in six pairs of communities. Each pair of communities was selected to match on a range of geographical and demographic dimensions. The key difference between each community in the pair was that one had experienced high levels of fatalities during the war whereas the other had not experienced any war-related fatalities.[35]

People from communities that had experienced more war-related violence, even if their own households had not experienced any violence, property loss, or displacement, were more likely to cooperate with their fellow villagers in the Public Goods Game. They also gave more in the Dictator Game, but this appears to be mostly due to the violence experienced specifically by their household. The sums of money involved were not trivial, with most people taking home somewhere between one-half and a full day's wage from these games.

The effects of the strengthened social norms and a more tightly bonded community appears to have resulted in the formation of more, and more active, community organizations. *None* of the communities *unaffected* by the war established new local organizations, such as farming cooperatives or women's groups. By contrast, 40% of those communities affected by the war had subsequently established new organizations. Even if the communities affected by violence didn't establish any new organizations, the ones they already had or those started by outsiders were more active than those in unaffected villages. By strengthening prosocial group norms, the experience of war resulted in more, and more energized, community organizations.

Why would war have these prosocial effects?

During hundreds of thousands of years, intergroup competition spread an immense diversity of social norms that galvanized groups to defend their communities; created risk-sharing networks to deal with environmental shocks like drought, floods, and famines; and fostered the sharing of food, water, and other resources. This meant that, over time, the survival of individuals and their groups increasingly depended on adhering to those group-beneficial social norms, especially when war loomed, famine struck, or droughts persisted. In this world, culture-gene coevolution may have favored a psychological response to inter-group competition, including threats that demanded group solidarity for survival. Under such threats, or in environments where such threats are common, intergroup competition favors cultural practices that monitor individuals more closely and sanction norm violators particu-larly harshly, thereby suppressing the increased temptation to break the norms (e.g., not sharing food during a famine). Under threat, increased sanctions in the form of ostracism, injury, and execution may have fa-vored an automatic and unconscious innate response to cling more tightly to our social norms and groups, including their beliefs, values, and world views. This means that cues of intergroup competition should promote greater solidarity and identification with one's group, as well as stronger norm adherence. Stronger norm adherence means both more compliance with norms and stronger negative reactions to norm violations.[36]

Though historians have long speculated that war influences our pro-social motivations, several recent studies, including my opening account of the Nepal work, have now rigorously documented these effects by studying the devastating quasi-natural experiments that can still be found around the globe. We are far from nailing this down, but it's clear that war creates enduring psychological effects in a manner consistent with what we'd expect in a cultural species that evolved in a world torn by intergroup conflict.

Now, let's head to the Republic of Georgia, in the Caucuses, and Si-erra Leone, in West Africa.

The economists Michael Bauer, Julie Chytilová, and Alessandra Cas-sar (and later me) wondered whether the experience of war might affect children more than adults. This is a good question since many social norms are acquired and internalized during middle childhood and early adulthood. The team also wondered whether war creates some generalized prosociality, or if it might be galvanizing in-group solidarity. In other words, do the effects of war bias people toward those in their own community and against those outside one's social sphere? To explore this, the team conducted experiments with children between the ages of 3 and 12 in the Republic of Georgia six months after Russia had attacked in 2008, and with adults in Sierra Leone in West Africa ten years after the horrendous civil war there. It should be noted that many of the adults in Sierra Leone were adolescents, or even children, during the war. In each place, the effect of war on these populations was essentially random, providing a kind of natural experiment. Using interview data, our team divided up participants into three categories according to how much the war had impacted them. The three categories were (1) those most affected by the war (e.g., they had relatives killed and were displaced from their homes), (2) those somewhat effected by the war (e.g., injured relatives), and (3) those least effected.[37]

To facilitate studying children, the team used simple experiments in which the children just had to choose between two options. For example, one experimental game called the Costly Sharing Game gives participants the choice between (a) keeping two for themselves and giving zero to another person, or (b) keeping one for themselves and giving one to the other person (a 50/50 split). They also varied the identity of the other person, making them either an in-group member or an out-group member. For the children in Georgia, the in-group person was someone from their school class, while the out-group player was from a distant Georgian school. In Sierra Leone, the in-group person was from their village, while the out-group person was from a distant village in Sierra Leone.

The results reveal that the experience of war has its maximum impact on sociality during a developmental window that opens during middle childhood, at roughly age 7, and remains open into early adulthood (roughly age 20). If war is experienced during this age range, it sharpens people's motivations to adhere to their egalitarian norms, but only for their in-groups. That is, those more exposed to war increased their egalitarian choices, choosing, for example, the even split in the Costly Sharing Game, but only for members of their in-group. Crucially, the effect endures at least a decade after the conflict. By contrast, the experience of war seemed to have no effect on the treatment of distant strangers, though of course it bears emphasis that these distant strangers were not members of the attacking group.

Outside of the developmental window (ages 7 to 20), the results were different. Those beyond their early twenties did also show an increase in their in-group egalitarianism, but the increase was quite small. So, the window doesn't shut, it just substantially narrows. Meanwhile, those younger than age 7 showed no effect from the war in these experiments.

These wars in Asia, Europe, and Africa are not isolated cases or unusual conflicts. Research on the effects of war in Burundi, Uganda, and Israel, which were studied using both behavioral games and survey data on things like voting and community engagement, tell the same story.[38] Together these findings suggest that the experience of World War II in their developmental window may have forged America's Greatest Generation, permanently elevating their national commitment and public spirit.[39]

Overall, when disaster threatens and uncertainty reigns, people cling more tightly to their community's social norms, including their rituals and supernatural beliefs, because it's these social norms that have long allowed human communities to adhere, cooperate, and survive.

Across centuries and millennia, cultural evolution, often propelled by intergroup competition, created social environments replete with social norms, which influenced diverse domains ranging from marriage, ritual, and kinship to exchange, community defense, and valued domains of prestige. Over tens and hundreds of thousands of years, the diverse social environments produced by this process became important selection pressures driving human genetic evolution and shaping our sociality. The greater sociality generated by this process interacts with our cultural nature, on our ability to learn from others, to generate greater technological sophistication and larger bodies of adaptive know-how. This process gives rise to our collective brains.

CHAPTER 12

OUR COLLECTIVE BRAINS

Surrounded by a sea of ice above the seventy-fifth parallel, the Polar Inuit live in an isolated region of northwestern Greenland, at the farthest reaches of the Inuit's massive expansion across the Arctic (see chapter 10). They are the northernmost human population that has ever existed. Sometime in the 1820s an epidemic hit this population of hunters and selectively killed off many of its oldest and most knowledgeable members. With the sudden disappearance of the know-how carried by these individuals, the group collectively lost its ability to make some of its most crucial and complex tools, including leisters (figure 3.1), bows and arrows, the heat-trapping long entry ways for snow houses, and most important, kayaks. With the loss of kayaks, the Polar Inuit became effectively marooned, unable to maintain contact with other Inuit populations from which they could relearn this lost know-how. As noted by the Arctic explorers Elisha Kane and Isaac Hayes, who encountered the Polar Inuit while searching for Sir John Franklin (see chapter 3), these technological losses had a dramatic impact, leaving the group unable to hunt caribou (no bows) or harvest the plentiful Arctic char from local streams (no leisters).

The population declined until 1862, when another group of Inuit from around Baffin Island ran across them while traveling along the Greenland coast. The subsequent cultural reconnection led the Polar Inuit to rapidly reacquire what they had lost, copying everything, including the style of Baffin Island kayaks. Decades later, with their population again increasing, and with ongoing contact with other Inuit in the rest of Greenland, the style of Polar Inuit kayaks gradually shifted back from the large beamy kayaks learned from the Baffin Islanders to the small sleek kayaks of western Greenland.

Though crucial to survival in the Arctic, the lost technologies were not things that the Polar Inuit could easily recreate. Even having seen these technologies in operation as children, and with their population crashing, neither the older generation nor an entirely new generation responded to Mother Necessity by devising kayaks, leisters, compound bows, or long tunnel entrances. These sophisticated technologies had evolved culturally over generations, and this process of cumulative cultural evolution had imbued these technologies with nuances that implicitly depended on subtle, or even counterintuitive, engineering principles. And, lest there be any doubt that they really needed these lost technologies, it bears emphasis that they immediately readopted all the missing know-how once they had been reconnected to the broader Inuit collective brain—which began when the Baffin Islanders happened by.[1]

This simple historical case gives us a glimpse into one of the secrets to our success—and our Achilles heel. Once individuals evolve to learn from one another with sufficient accuracy (fidelity), social groups of individuals develop what might be called *collective brains*. The power of these collective brains to develop increasingly effective tools and technologies, as well as other forms of nonmaterial culture (e.g., know-how), depends in part on the size of the group of individuals engaged and on their social interconnectedness. It's our collective brains operating over generations, and not the innate inventive power or creative abilities of individual brains, that explain our species' fancy technologies and massive ecological success. Even individuals facing a life-and-death situation with weeks or months to prepare weren't nearly smart enough to figure out how to make even the basic tools for survival, as we learned from Burke and Wills, Franklin's men, and the Narváez expedition. Our collective brains arise from a number of synergies created by the sharing of information among individuals.

Here's the idea, broken down into its simplest form. We've established that from a very young age, humans focus on and learn from more skilled, competent, successful, and prestigious members of their communities and broader social networks. This means that the new and improved techniques, skills, or methods that emerge will often begin spreading through the population, as the less successful or younger members copy them. Improvements may arise through intentional invention as well as from lucky errors and the novel recombinations of elements copied from different people. The biases in human cultural learning mean that as transmission proceeds across the group, and over generations, the numerous errors, recombinations, and intentional modifications that don't lead to greater success will be filtered out, while those that do will tend to collect and spread.[2]

The size of the group and the social interconnectedness among these individuals plays a crucial role in this process. The most obvious way that the size of a group can matter is that more minds can generate more lucky errors, novel recombinations, chance insights, and intentional improvements. To see this in starkest terms, consider how group size influences the chance of coming up with an invention, say using feathers to fletch an arrow. Suppose any one individual operating alone will only figure out—by luck or effort—"arrow fletching" once in one thousand lifetimes. The chance that at least one person in a group of 10 people will figure out fletching in their lifetimes is then 1%. So, on average, a group of 10 persons will take 100 generations to come up with this invention (2,500 years). In a group of 100 people, at least 1 person will devise it 10% of the time in one lifetime. Consequently, on average, it will take 11 generations for the group to figure this out (275 years). For 1000 people, there's a 63% chance they will get it in 1 generation, and on average, they will figure it out in 1.6 generations (40 years). If you can unite 10,000 minds, you will have fletching in 1 generation (well, technically, a 99.995% chance). So, bigger groups have the potential for more rapid cumulative cultural evolution, especially since these effects further compound when you consider that many inventions require combining several elements, so their rate of emergence depends on the slowest element.

This, of course, assumes that the members of this group are sufficiently socially interconnected to other members of the group so that their improvements can rapidly spread through the group. The bigger the group, the more implausible this assumption is. To see the importance of this sociality, imagine that every person is a social island who keeps any insights he has secret from all others. What happens?

Well, not much. Some individuals will make slightly better tools, but then they will die and their improvements will go with them. No fancy tools will emerge. And the size of the group doesn't matter. This is the case for most animals.

Thus, along with group size, the degree of social interconnectedness is very powerful in generating cumulative cultural evolution, even more powerful than individual smarts. Consider two very large prehuman populations, the *Geniuses* and the *Butterflies*. Suppose the Geniuses will devise an invention once in 10 lifetimes. The Butterflies are much dumber, only devising the same invention once in 1000 lifetimes. So, this means that the Geniuses are 100 times smarter than the Butterflies. However, the Geniuses are not very social and have only 1 friend they can learn from. The Butterflies have 10 friends, making them 10 times more social. Now, everyone in both populations tries to obtain an invention, both by figuring it out for themselves and by learning from friends. Suppose learning from friends is difficult: if a friend has it, a learner only learns it half the time. After everyone has done their own individual learning and tried to learn from their friends, do you think the innovation will be more common among the Geniuses or the Butterflies?

Well, among the Geniuses a bit fewer than 1 out of 5 individuals (18%) will end up with the invention. Half of those Geniuses will have figured it out all by themselves. Meanwhile, 99.9% of Butterflies will have the innovation, but only 0.1% will have figured it out by themselves. Keep in mind that the Geniuses were 100 times smarter than the Butterflies whereas the Butterflies were only 10 times more social. Bottom line: if you want to have cool technology, it's better to be social than smart.

Now, suppose the invention mentioned above is the bow and arrow, and the Butterflies and the Geniuses come into conflict over territory. Who would win, the smarter group or the more social one? It's unclear, but the Butterflies would have a good chance since they would all be armed with bows and arrows, but only 18% of the Geniuses would.[3]

This cultural evolutionary process may also shape the *learnability* of the know-how and skills related to tools, technology, and practices. Over generations, the details, techniques, and protocols that go into producing complex technologies should—all other things being equal—tend to simplify in ways that make them easier to learn and more intuitive. This propensity suggests that larger and more interconnected populations not only will have a larger variety of more sophisticated tools and technologies, but also that they will have more learnable techniques for producing them.

As our species self-domesticated and we became increasingly social, our collective brains would have expanded, making possible greater technological sophistication and larger bodies of know-how. However, remember that our species' ability to live in large groups, well beyond residential populations, still depends heavily on social norms. Under the influence of intergroup competition, social norms that effectively maintain internal harmony in ever larger groups would have spread. Institutions that expand social groups and help forge broad alliances, through kin ties, naming practices, affinal ties, spousal exchange, and rituals, would have expanded our collective brains, making them better able to culturally evolve and sustain more-complex bodies of cultural knowhow, including more sophisticated tools and weapons.

To the Laboratory!

My graduate student Michael Muthukrishna and I wondered if we could capture the effect of social interconnectedness on the accumulation of skills in a controlled way in our psychology laboratory. We set up a "transmission chain" in which a first "generation" of naïve undergraduates was tasked with trying to recreate a complex "target image" using an unfamiliar image-editing program. They could then write up to two pages of instructions and tips for the next generation of novices, who would receive this written information as well as the image created by their teacher or teachers from the previous generation. We repeated this process for ten laboratory generations. Each participant's skills were scored by measuring the similarity between the image they created and the target image. Participants were paid in cash, according to both their own performance and that of their students. We recruited 100 undergraduates and assigned them to one of two treatment groups. In the 5-Model treatment, novices received images and instructions from five individuals from the previous generation. By contrast, in the 1-Model treatment, novices received the image and instructions from only one teacher from the previous generation. Keep in mind that both groups were the same size, but those in the 5-Model treatment were five times *more socially interconnected*.

Figure 12.1 shows the results. The average skill of each generation did not increase when people could learn from only one teacher. However, when they could learn from all five individuals in the previous generation, the average skill increased dramatically, from just above 20% to over 85%. In the final generation, the *least skilled person* in the 5-Model treatment was superior to the *most skilled* person in the 1-Model treatment. Social interconnectedness made everyone more skilled by Generation 10.[4]

In the 5-Model treatment, participants were not just copying the most skilled teacher in the prior generation, they were actually integrat-

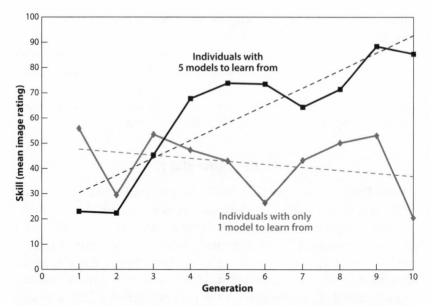

Figure 12.1. The effects of greater sociality on the change in skill over ten laboratory generations. Average skill levels in image editing for those with access to five models across 10 laboratory generations (*black line*). Average skill for those with access to only one model across ten generations (*gray line*).

ing insights from the top four teachers (of five), but putting relatively more weight on what the top teacher did. This is important, since by acquiring distinct elements from different people, learners can create "innovations" without "inventions"; that is, by recombining things copied from different models, novelties can emerge without individuals themselves thinking up a new technique on their own. This process turns out to be crucial for understanding innovation.

Similar research led by Maxime Derex in France reveals convergent findings. In a computer task, participants had to virtually construct either a *simple* arrowhead or a *complex* fishnet, which they could then use to "go fishing." Fishing yielded a catch, measured in points, which could be converted to cash after the experiment. After each round of fishing, participants could construct an improved arrowhead or fishnet. To accomplish this, they could look at other members of their group, see how many fish they caught, and potentially copy their techniques. They could also rely on their own wits, causal mental models, and trial-and-error learning. The task went on for fifteen rounds in groups ranging in size from 2 to 16 men. During the fifteen rounds, the arrowheads made by the larger groups (with 8 or 16 men) got much better while those made by the smaller group remain equally efficient or got slightly worse. By the end, the performance of the arrowheads of the largest group (with 16 people) was 50% better than those in the smallest groups (with 2 people).[5]

Size and Interconnectedness in the World

Of course, while such laboratory experiments permit us to isolate the causal effect of sociality, it's crucial to see if these processes are actually important and operating in the real world. To explore this, the anthropologists Michelle Kline and Rob Boyd looked around the globe for a natural experiment, one that would permit them to get good measures of things like population size and technological complexity. Continental societies were worrisome since these populations are all interconnected to some substantial degree, and technologies are known to spread across ethnic and linguistic boundaries. However, in the Pacific, islands or island clusters provide a natural way to partition populations into discrete chunks. Unlike with continents, hundreds of miles of ocean between islands guarantees at least some degree of isolation.

To standardize the type of technology compared across islands, the duo focused on fishing toolkits, which are a crucial part of subsistence technologies through much of Oceania. Using detailed ethnographies of these populations, the researchers obtained estimates for the size and complexity of these toolkits as well as of their population sizes at the time of European contact. Since these islands were far from true isolates, Kline and Boyd also assessed each group's relative degree of contact with other human populations.

Sure enough, islands or island clusters with larger populations and more contact with other islands had both a greater number of different fishing-tool types and more-complex fishing technologies. Figure 12.2 shows the relationship between population size and the number of tool types. People on islands with bigger populations had more tools at their disposal, and those tools tended to be more sophisticated.[6]

Another team, led by the evolutionary anthropologist Mark Collard, found the same kind of strong positive relationship when they examined forty nonindustrialized societies of farmers and herders from around the globe. Once again, larger populations had more-complex technologies and a greater number of different types of tools.[7]

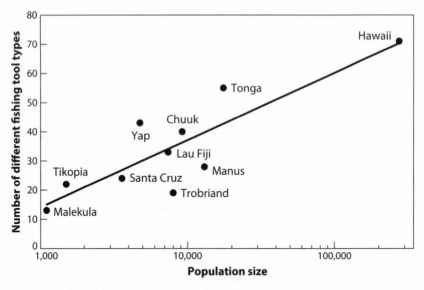

Figure 12.2. The relationship between population size and the number of different fishing-tool types in each population. The horizontal axis is a log scale.

These effects can even be observed in orangutans. While orangutans have little or no cumulative culture, they do possess some social learning abilities that result in local, population-specific traditions. For example, some orangutan groups routinely use leaves to scoop up water from the ground or use sticks to extract seeds from fruit. Data from several orangutan populations show that groups with greater interaction among individuals tend to possess more learned food-obtaining techniques. Of course, young orangutans are lucky if they can access two or three models, and in some groups, mom is the only one around to learn from.[8]

The point is, larger and more interconnected populations generate more sophisticated tools, techniques, weapons, and know-how because they have larger collective brains.

The Tasmanian Effect

This approach to understanding cumulative culture—to explaining the sophistication of technology and the size of a group's body of know-how—has two other less intuitive implications. First, if a population suddenly shrinks or gets socially disconnected, it can actually lose adaptive cultural information, resulting in a loss of technical skills and the disappearance of complex technologies. Second, a population's size and social interconnectedness sets a maximum on the size of a group's collective brain.

The loss of adaptive cultural information can result from two different processes. The first is what we saw happen to the Polar Inuit: a random shock (an epidemic) happened to strike the most knowledgeable members of the community, wiping out a chunk of their cultural know-how. Because it removed the kayak (their transportation) from their cultural repertoire and they were already geographically very isolated, they went into a slow downward spiral as their populations dwindled due to the technological losses. This kind of phenomena may be common. In his 1912 essay "The Disappearance of Useful Arts," the famed anthropologist and psychologist W.H.R. Rivers recorded many such puzzling disappearances. In one case, all of the master canoe builders from the Torres Islands, a geographically isolated island cluster in the northernmost part of the Vanuatu archipelago, died out. As a consequence, the inhabitants became isolated for a time, since no one had the skill to construct something better than a crude bamboo catamaran, which was not sufficiently seaworthy to travel outside the local island cluster and was of no use in fishing.[9]

The other process is more subtle and I suspect more important. If someone is copying the techniques and practices of a highly skilled and knowledgeable expert, they will often end up with a level of skill or knowledge that is less than that of the expert they are copying. Copies are imperfect versions of the original, usually. To see this, imagine a master archer who always scores 100 points on the archery test. He is tasked with training 100 novices. Suppose that by the time the master dies, all the novices have learned 95% of his skills. That is, they get 95 points on the archery test. One of them is then picked to teach next set of novices, and so on over 20 generations. How good will the twentieth generation be at archery?

Not good. They will get only 35 points on the test.

The reason is that some information was lost every generation, because copies are usually worse than the originals. Cumulative cultural evolution has to fight against this force and is best able to do so in larger populations that are highly socially interconnected. The key is that most individuals end up imperfect, worse than the models they are learning from. However, some few individuals, whether by luck, fierce practice, or intentional invention, end up better than their teachers. To see this, let's rerun the archer thought experiment and assume now that 90% of the novices end up with 80% of the master's skills. Ten percent, however, end up with 105% of his skills. The best among the novices is again picked to teach the next generation. How will the twentieth generation look now?

Excellent. After 20 generations, everyone will score over 200 on the archery test, twice the original master's score. Interestingly, the average score of the novices won't exceed the master's original score of 100 points until the fifth generation. This occurs because although 90% of the learners are still ending up worse than their teachers, the rankings of the teachers are gradually improving, and they eventually pull the learners along with them.

Larger populations can overcome the inherent loss of information in cultural transmission because if more individuals are trying to learn something, there's a better chance that someone will end up with knowledge or skills that are at least as good as, or better than, those of the model they are learning from. Interconnectedness is important because it means more individuals have a chance to access the most skilled or successful models, and thereby have a chance to exceed them, and so can recombine elements learned from different highly skilled or successful models to create novel recombinations.

These ideas first struck me when I was reading about the aboriginal inhabitants of the island of Tasmania. Roughly four-fifths the size of Ireland and comparable to Sri Lanka, Tasmania lies about 200 kilometers south of Victoria, Australia. When the earliest European explorers made contact with the Tasmanians in the late eighteenth century, they discovered a population of hunter-gatherers equipped with the simplest toolkit of any society ever encountered (by Europeans). To hunt and fight, men used only a one-piece spear, rocks, and throwing clubs. For watercraft, the Tasmanians relied on leaky reed rafts and lacked paddles. To ford rivers, women would swim the raft across, towing their husbands and offspring. In the cool maritime climate, Tasmanians slung wallaby skins over their shoulders and applied grease to their exposed skin. Curiously, the Tasmanians did not catch or eat any fish, despite fish being plentiful around the island. They drank from skulls and may even have lost the ability to make fire. In all, the Tasmanian toolkit consisted of only about twenty-four items.[10]

To put this simplicity into perspective, let's consider the Pama-Nyungan-speaking Aborigines who were contemporary with the Tasmanians in the eighteenth century and lived just across the Bass Strait in Victoria. These Aborigines possessed the entire Tasmanian toolkit plus hundreds of additional specialized tools, including a fine array of bone tools, leisters, spear throwers, boomerangs, mounted adzes (for wood working), many multipart tools, a variety of nets for birds, fish, and wallabies, sewn-bark canoes with paddles, string bags, ground-edge axes, and wooden bowls for drinking. For clothing, rather than draped wallaby skins and grease, the Aborigines wrapped themselves in snuggly fitting possum-skin cloaks, sewn and tailored with bone awls and needles. For fishing, the aboriginal populations used shellfish hooks, nets, traps, and fishing spears. Somehow, the Tasmanians ended up with a much simpler toolkit than did their contemporary cousins just across the Bass Strait.[11]

The Tasmanian toolkit is simple even when compared to that of many ancient Paleolithic societies. The archeological record from many parts of the world, going back tens and even hundreds of thousands of years, reveals the emergence of more-complex toolkits than those possessed by the Tasmanians at the time of European contact. Tasmanian stone tools are much cruder than many of the tools found in Europe after about 40,000 years ago and are consistent with the stone tools made by many Neanderthals and even by more ancient members of our genus. The Tasmanians lacked bone tools, yet elsewhere finely crafted bone harpoon points date to at least 89,000 years ago. Similarly, stone points for spears date to a half a million years ago, well before the emergence of our species. And even if we put this particularly old find aside, it's generally accepted that hafted tools date back before the origins of our species, 200,000 years ago. While Tasmanians were limited to one-piece wallaby skins for clothing and leaky rafts for water transport, the mere presence of humans in the Arctic 30,000 years ago and in Australia at least 45,000 years ago attests to the existence of both warm, fitted clothing and reliable watercraft that can cross open oceans. Consistent with this, humans in Indonesia were catching tuna, sharks, and wrasses in the open ocean 42,000 years ago.[12]

The puzzle deepens when we realize that Tasmania was connected to the rest of Australia until about 12,000 years ago. As the seas rose, the Bass Strait flooded and transformed Tasmania from an Australian peninsula to an island. Until this isolation, the archaeological remains left by Tasmanians cannot be distinguished in terms of complexity from those found in Australia. With their isolation, Tasmanians began to lose complex tools. The number of bone tools gradually dwindled until about 3,500 years ago, when they vanished entirely. As evidenced by fish bones, at least some ancient Tasmanian groups probably relied heavily on fish. One archaeological site that was occupied consistently between about 8000 and 5000 years ago indicates a fairly heavy reliance on fish, second in importance only to seals. But, gradually, fish dwindle and disappear from the record. By the time Captain Cook's men offered freshly caught fish from the bountiful waters around the island in 1777, the Tasmanians reacted with disgust; yet, they gladly took and ate the bread Cook offered.[13]

Of course, life did not stand still on the north side of the Bass Strait. As we saw in chapter 10, during the millennia when the Tasmanians were losing their fishing technologies, bone tools, and fitted clothing, the Pama-Nyungan expansion was underway. From the north of Australia, this expansion brought to Victoria not only fancier stone tools, dogs, and seed grinders but also new social institutions and communal rituals that more effectively integrated local groups into larger regional populations—creating greater social interconnectedness among of bigger overall population. This would have permitted cumulative cultural evolution to generate more-sophisticated tools and more-complex toolkits. But the Pama-Nyungan expansion never crossed the Bass Strait to Tasmania.

By isolating Tasmanians for eight to ten millennia, the rising seas cut them off from the vast social networks of Australia, suddenly shrinking their collective brains. A gradual loss of their most complex and difficult to learn skills and technologies ensued. Their isolation also prevented them from acquiring the technological and institutional innovations that would have expanded their collective brain, by fostering greater interconnectedness and fancier tools, weapons, and know-how.

Tasmania in the Lab

Michael Muthukrishna and I also wondered if we could create a Tasmanian-like phenomenon in our laboratory, under controlled conditions. Using a design very similar to that described above for the transmission of image-editing skills, we now trained a first "generation" of experts at tying a complex system of knots used in rock climbing. While wearing a head-cam, each participant also made a video of how to tie the knots they learned. Novices in the next generation received the videos and their teachers' skill scores for the knot system. In the 1-Model treatment, each novice received only the video and score from one person from the previous generation. In the 5-Model treatment, they received the videos and scores from all five people in the previous generation. As

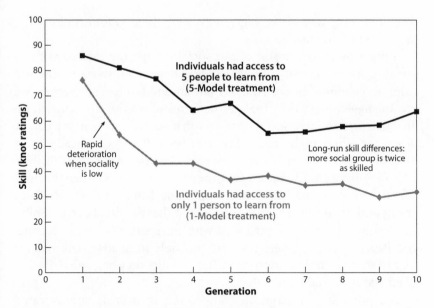

Figure 12.3. Plot shows the deterioration of skill for the 1-Model and 5-Model treatments across 10 laboratory generations. The gray line reveals the more rapid deterioration in skill for the less socially interconnected group.

with the earlier experiment, participants were paid for both their own scores and those of their students in the next generation.

Figure 12.3 shows the results. Remember, the first generation is composed entirely of trained experts, so we should expect skill levels to drop over generations. The question is whether social interconnectedness influences the rate of decline and final level of skill in the group. When learners could access only one model, knot-tying skills plummeted rapidly relative to those who could access five models. By roughly the fifth generation, the two populations appear to be approaching a steady state, with the 5-Model population sustaining a much higher average skill level than those in the 1-Model population. As before, everyone in the tenth generation of the 5-Model group is more skilled than the most skilled person in the tenth generation of the 1-Model group.

Maxime Derex's team also found parallel results using their experimental setup, as described above. As noted, the simple arrowheads improved over the 15 rounds in the larger groups. Meanwhile, the more complex fishing nets held fast at their initial performance in the big groups with 8 or 16 members, but deteriorated in the smallest groups, those with only 2 or 4 members. By round 15, this deterioration meant that the fishing nets of the big groups were three to four times better than those of the small groups.

Among humans, the size of our social groups, the intensity of our sociality, and the density of our networks depend on social norms and cultural technologies like rituals. Even among hunter-gatherers, as we saw in chapter 9, it's the ritual partners and affinal ties that enmesh individuals within large social networks that stretch across many bands. When the ethnographers Kim Hill and Brian Wood specifically asked Aché and Hadza whether they'd ever watched other members of their ethnolinguistic groups make a tool, the effects of ritual partnerships and affinal ties dominated, being more important than even close blood ties or physical proximity. These results suggest that the rituals, ceremonies, and kinship systems that expanded with the speakers of Inuit, Numic, and Pama-Nyungan languages were probably intimately tied to their sophisticated technical know-how. The sociality they engendered nourished and sustained their collective brains.[14]

The point is, if intergroup competition is favoring sophisticated tools and weapons, it has to favor the social norms and institutions that can sustain a larger collective brain—technology and sociality have to coevolve.

Children versus Chimps and Monkeys

Small groups of three- and four-year-old children faced off against similarly sized groups of chimpanzees and capuchin monkeys. The challenge was a puzzle box that was meant to provide an opportunity for cumulative cultural learning. In one version of the experiment, if individuals could perform a series of actions on the box in a specific order, they could receive increasingly desirable rewards with each correct action. The box had to be slid, pressed, and twisted in various ways. Capuchins and chimpanzees got a carrot for the first correct action, followed by an apple, and finally the coveted grape. The kids got better and better stickers. It's a nice task for our purposes since it has the feel of the food-processing techniques that I've emphasized had a real impact on human survival and evolution (also, see chapter 16).[15]

The interspecies competition on this task, unlike many of the cognitive tasks I described in earlier chapters, was actually no contest. The kids ruled. Forty-three percent of the children got to the final stage (stage 3), whereas only one chimpanzee did and none of the capuchins. Overall, for each step of cumulative complexity above the first, the data suggest that 1000 hours of access to the box yielded 1.7, 6.1, and 388.6 additional steps in complexity for the capuchins, chimpanzees, and humans, respectively. This pattern is impressive, especially given the familiarity of the nonhumans with these kinds of reward-containing puzzle boxes. Clearly, cumulative culture is our thing.

What led to the success of the children? The kids who did better (1) imitated others more often (matched others actions), (2) received instructions more often (were taught), and (3) received more gifts from others (were social). So, imitation, teaching, and sociality are what mattered. Meanwhile, the other species floundered precisely because they lack these social and cultural abilities. Chimpanzees and capuchins revealed zero instances of teaching or altruistic giving, whereas the preschoolers performed 215 acts of giving and 23 unambiguous instances of teaching. For imitation, chimpanzees did copy some stage 1 behaviors, but none for stages 2 or 3. These experimental findings fit nicely with observations of these primates in the wild, where researchers have found no clear examples of teaching or other tools that require cumulative cultural evolution.[16]

Innately Dumber than Neanderthals?

In class, I show my undergraduates unlabeled pictures of four different stone toolkits from (1) eighteenth-century Tasmanians, (2) seventeenth-century Australian Aborigines, (3) Neanderthals, and (4) late Paleolithic modern-looking humans (30,000 years ago). I ask them to assess the cognitive abilities of the toolmakers by looking at the tools. Remember, there are no names or dates, only images of stone tools. My students always rate the Tasmanians and Neanderthals as less cognitively sophisticated than both the seventeenth-century Australian Aborigines and the late Paleolithic toolmakers. Of course, there's no reason to suspect any innate cognitive differences between Tasmanians and Aborigines, who only became separate populations after the Bass Strait flooded. Unfortunately, no one raises their hand to suggest that, actually, it's not possible to infer innate cognitive abilities from tool complexity because of the importance of sociality in generating tool complexity (that's the correct answer). To sharpen this point, consider whether the Polar Inuit of 1820 were smarter than the Polar Inuit of 1860? In 1820, they could make kayaks, fancy fishing spears, and compound bows. In 1860, they could not.

Similarly, when many paleoanthropologists look at the differences in the sophistication of the toolkits produced by different species of ancestral humans, they sometimes make direct associations to the innate cognitive endowments of those species. Neanderthals provide an interesting case because they coexisted in Eurasia with our own African lineage and must have interacted in some fashion because many modern human populations carry some Neanderthal DNA. Because most Neanderthal groups possess a toolkit substantially less complex than the more modern-looking African intruders (our ancestors), the assumption has often been that Neanderthals suffered some innate cognitive deficits relative to the African immigrants. This has been argued consistently, despite the fact that Neanderthal brains were as big, or bigger, than our brains.[17]

In primates, the strongest predictor of cognitive abilities across species is overall brain size.[18] Consequently, it's not implausible that we were dumber than the bigger-brained Neanderthals. Researchers, however, have typically proceeded under the assumption that our lineage must have been innately smarter, so the real question appeared to involve figuring out what subtle differences in our brains or our specific cognitive abilities (e.g., language) would account for our relative success (our global domination vs. their extinction).

Let me toss out an alternative: individually, the African immigrants (our ancestors) who were contemporary with Neanderthals were a touch innately dumber. However, they had larger *collective* brains capable of generating greater cumulative cultural evolution. These larger collective brains resulted from bigger and more tightly interconnected social groups as well as from individuals who experienced longer adult lives (on average). Longer adult lives means more time for learning, both from diverse other people (to create novel recombinations) and from individual experience, as well as for retransmitting this wisdom to others (as I discussed in chapter 8). More specifically, Neanderthals, who had to adapt to the scattered resources of ice-age Europe and deal with dramatically changing ecological conditions, lived in small, widely scattered groups and periodically suffered shocks that reduced their population size. Adults often died young, usually from injuries suffered during hunting by, for example, engaging and surrounding dangerous prey like elephants, horses, and red deer with only thrusting spears. Meanwhile, the invading African immigrants lived in larger and more interconnected groups, perhaps due to some particular set of institutions that emerged in the richer, warmer, or otherwise more fertile climates to the south (e.g., coastal marine environments).[19] Many more of them lived past forty, probably due in part to their projectile weapons (making hunting much safer) and other superior cultural adaptations, like high-quality warm clothing. Under the right ecological conditions, social norms can evolve to create institutions that permit a larger collective brain, which can in turn extend life-span and further energize that collective brain. So Neanderthals may have needed those individually larger brains to compensate for their ecologically imposed smaller collective brains.[20] The extra edge created by more individual brainpower in Neanderthals would have been dwarfed by the power of social interconnectedness and longer lives on the collective brain sizes of the Africans.

Because of their larger collective brains, this African population may have been able to invade and replace Neanderthals in Europe for the same reasons that the Inuit were able to replace the Dorset. A different environment of origin had fostered different social institutions that generated a more sophisticated and adaptable technological suite. Curiously, despite the extinction of the Dorset and the pre-Numic people of the Great Basin, no one has suggested innate intellectual inferiority. Thus, it's plausible that the Neanderthal replacement may have been just an earlier chapter in a cultural evolutionary story that has repeated itself time and time again over the history of our genus.[21]

To be clear, my argument is that cultural evolution has been driving genetic evolution for hundreds of thousands of years or longer, so I do think that we should expect that genetic differences existed between, for example, members of our species at 150,000 years ago and at 50,000 years ago. We were self-domesticating during that period. My points here regarding groups like the Neanderthals is only that you can't read innate intelligence out of the material record without knowing about the population size and interconnectedness, and that cultural evolution driven by intergroup competition was occurring then, as now.

This view of cultural evolution underlines another point about technology and our species' evolutionary history. It's not a straight line. Researchers look for the earliest evidence of things like fire, spears, hafted weapons, fishhooks, jewelry, trade, bone tools, etc. and then all too often assume that, once invented, these persist forever, or at least until they are replaced by something better. The expected picture seems to be one of continuous, if often halting, technical progress. However, all the evidence I've presented so far indicates that the flourishing of technology and nonmaterial know-how will depend on the interaction of ecological conditions, environmental fluctuations, disease shocks, and social institutions. Groups can and will lose know-how and never get it back.

Consider something as precious as bows and arrows. Current evidence suggests that bows and arrows may have emerged by 60–70,000 years ago in Africa, though some argue for a much earlier date. Meanwhile, at the time of European contact, there were no bows or arrows anywhere in Australia. An entire continent of hunter-gatherers relied on spears, throwing clubs, and boomerangs. We've already seen that the Dorset lost their archery expertise, which cost them when the Inuit arrived. Bows and arrows were also lost in Oceania, perhaps several times. Like ocean-going boats and pottery, it turns out that the know-how to make and use bows and arrows is frequently lost by human groups. As previously noted, some foraging groups have even lost the know-how for starting a fire.[22]

Because humans in the Paleolithic faced more extreme and more rapidly shifting environments than at any time in recorded history, we should expect technological evolution to appear erratic, with bursts of flourishing and then sudden crashes as populations were decimated, dispersed, or otherwise disconnected. This view is consistent with the data. However, the scattered and fragmentary nature of the paleoanthropological records still leaves a fair amount of uncertainty.[23]

Since it is far more important for a cultural species to be social than innately smart in producing complex technologies, when we see technological and other cultural flourishing we ought to also consider the realm of the social—the institutions, marriage practices, and rituals—in trying to understand the origins of that cultural flourishing. By contrast, genetic explanations for cultural flourishing seek to link the complex products of cumulative cultural evolution—like languages—to genetic differences. However, the emergence of complex languages, far from being an ultimate explanation for cultural flourishing, represent merely one of the products of cumulative cultural evolution and is subject to many of the same forces as toolkits. We'll turn to this in chapter 13.

Tools and Norms Make Us Smarter

Collective brains make us individually smarter even in the absence of any genetic evolution. Larger and more interconnected groups generate more tools, expanded bodies of know-how, and fancier techniques. These larger repertoires, once learned, equip individuals to solve (by themselves) more problems, including problems they couldn't otherwise solve. In the longer run, these culturally constructed worlds filled with concepts, tools, techniques, procedures, and heuristics will favor genes for more-sophisticated cognitive abilities. However, let's put genes aside and assume them fixed. Cultural evolution will still make us smarter.[24]

To see this effect in its most basic form, let's start with chimps. Thibaud Gruber, Richard Wrangham, and their colleagues setup an artificial "honey well" among different chimpanzee populations and watched what the apes would do. The honey well was designed such that the chimps could access the honey, which they love, only by using a dipping tool (like a stick or leaf). Different populations of chimpanzees had been selected because some of them were known to engage in "fluid-dipping" to obtain water from logs. The question was: would knowledge of the fluid-dipping technique influence the apes' chances of figuring out how to obtain the delicious honey in this somewhat new context?

Yes. Only chimpanzees from groups in which the practice of fluid-dipping was common figured out how to access the honey. This remained true even when the researchers supplied a stick for the apes. Even when the chimpanzees found the stick already stuck into the honey well (a nice hint), still only the chimpanzees from fluid-dipping communities figured out the puzzle.[25] So, the ability of chimpanzees to solve a novel problem depends on their culture.

Now imagine an ape who grows up in a world already filled with wheels, pulleys, springs, screws, projectiles, elastically stored energy (e.g., bows, spring traps), levers, poisons, compressed air (blow guns), rafts, leisters (figure 3.1), and heating (fire and cooking), just to name a few things. This ape has a head full of options—of prefab solutions—to apply to any new problem. These tools, or really concepts, may seem easy or obvious to the reader, but they are hard for us to (innately) figure out. Consider the wheel. The wheel was invented relatively late in human history, long after agriculture and dense populations had emerged, and only in Eurasia. It was never invented in the Americas, Australia, or New Guinea, despite the availability of llamas (in South America) and dogs (everywhere) to pull carts and the general need for wheelbarrows, pulleys, and mills. Similarly, it appears that nothing using elastically stored energy was ever invented and maintained in Australia. That means no bows, stringed instruments, or spring traps. Tools using compressed air also appear to have been absent, which eliminates blowguns, flutes, and horns.[26]

Many of the products of cumulative cultural evolution give us not only ready concepts to apply to somewhat new problems, and concepts to recombine (bows are *projectiles + elastically stored energy*), but actually give us cognitive tools or mental abilities that we would otherwise not have. Arabic numbers, Roman letters, the Indian zero, the Gregorian calendar, cylindrical projection maps, basic color terms, clocks, fractions, and right versus left are just a few of the cognitive tools that have shaped your mind and mine. These have evolved culturally to "fit" the mental constraints imposed by the genes that build our brains by harnessing, enhancing, and recombining our innate capacities to develop new and unexpected abilities.

To see what I mean, consider the abacus, a tool that has not shaped the minds of most Westerners. Skilled abacus users can solve math problems, like 345 + 675 + 853, faster than calculator users. That's impressive, but it's really just one tool versus another. However, here's the thing: skilled abacus users can also beat calculator users even when they are disarmed, without their abaci. That is, they can make purely mental calculations faster than calculator users; this is because the abacus evolved from counting boards, which date back to before the Babylonians, through a long process of selective cultural evolution. Analyses of the abacus suggest that it is optimized to the constraints and affordances of our visual perception and memory, making it easier for us to store numbers in visual working memory. Once fully trained, people, including young children, can do superfast mental arithmetic. They accomplish this by visualizing the abacus and often actually manipulate their fingers nonconsciously *as if* they were rapidly sliding the abacus beads around. Language appears to play little or no role in the calculation, except that at the end, people have to report their answers in Arabic numerals or their verbal equivalents.

The approach abacus users take to mental arithmetic contrasts with that taken by typical Western undergraduates who, when facing the same challenge, lean heavily on their linguistic tools (number words). Thus, the abacus provides a mental prosthesis that, by harnessing aspects of our visual memory, can deliver mental powers that seem almost unimaginable to those unfamiliar with this simple but elegant tool.[27] As we turn to take a look at language, it's worth noting, however, that in doing mental arithmetic both Western undergraduates and abacus users had to rely on the products of cumulative cultural evolution.

CHAPTER 13

COMMUNICATIVE TOOLS
WITH RULES

The origins and evolution of language have fascinated people for centuries.[1] Linguistic skills emerge nearly miraculously in young children as they effortlessly acquire a sophisticated communicative repertoire without instruction. More striking, even skilled orators cannot readily explain the underlying, often complex, grammatical rules or hierarchical structures that lie beneath the serial flow of sounds that issue forth from their mouths. They just do it. How can we explain both the ease with which children acquire languages as well as the intricate complexities of languages, which somehow emerge without the conscious understanding of language users themselves?

Converging lines of research from several fields now point to an answer: languages arise from long-term cumulative cultural evolution. Like other aspects of culture, including sophisticated technologies, rituals, and institutions, our repertoires of communicative tools—including spoken languages—have evolved via cultural transmission over generations to improve the efficiency and quality of communication, and to adapt to the details of local communication contexts, including physical environments and social norms (like taboos). Languages, then, are cultural adaptations for communication.

These communication systems had to adapt (culturally) to our brains, exploiting features of our ape cognition, and at the same time, created new selection pressures on our genes to make us better communicators.

These genetic evolutionary pressures were powerful, shaping both our anatomy and psychology: they pushed down our larynx to widen our vocal range, freed up our tongues and improved their dexterity, whitened the area around our irises (the sclera) to reveal our gaze direction, and endowed us with innate capacities for vocal mimicry and with motivations for using communicative cues, like pointing and eye contact.

The problem with many efforts to understand the evolution of language is that the lenses used are often focused too narrowly. By placing language within the context of our species' overall repertoire of communicative abilities and then seating this within culture-gene coevolution, we can begin to see the synergistic relationships between tools, practices, norms, communication, and language. Languages are a subset of culture that are composed of communicative tools (words) with rules (grammar) for using those tools. Many of the challenges to developing an account of the origins and evolution of language diminish once we (1) understand how cultural evolution can construct complex but highly learnable repertoires without inventors or intentional designs, and (2) realize that language was merely one element in a broad and synergistic process of culture-gene coevolution.

The idea is that, among our ancestors, cultural evolution accumulated, integrated, and honed many useful communicative elements over long stretches of time into increasingly complex repertoires—as it did with our technologies and institutions. Since none of this leaves a material record for paleoanthropologists to dig up, we can't directly observe this accumulation. However, if I'm right, the communication systems of living and historically known societies will reveal various signatures of cultural evolution, which we can spot if we look closely. We'll first see that communication systems are adapted to the challenges of local environments (they are cultural adaptations), which include both the social and physical environments. Then, we'll see that the conditions of cultural evolution can influence the size and complexity of communicative repertoires. Just as with technologies, the size and interconnectedness of speech communities influence their vocabulary sizes, phoneme inventories, and grammatical tools.

If you are not a linguist, this all probably sounds reasonable, or at least not obviously silly. Why shouldn't languages operate like every other domain of culture? Almost no one would deny that the tools used by Tasmanians were less complex and fewer in number than the tools used by Victorian Aborigines or by modern Australians in Melbourne. Few would deny that societies possess quite different kinds of institutions and that these institutions could influence economic success. Most would agree that, at least in principle, it's possible that the institutions possessed by Pama-Nyungan or Inuit speakers could have facilitated their expansions. Linguists and linguistic anthropologists, by contrast, have often assumed that all languages are more or less equal, along all the dimensions that we might care about—equally learnable, efficient, and expressive.[2] Accompanying this assumption is the view that languages are not strongly shaped by nonlinguistic factors—that they are somehow hermetically sealed off from the rest of the world. Together these two assumptions have inhibited cultural evolutionary approaches to language. Recently, however, the cracks in these intellectual barricades have begun to multiply, as researchers use newly available language datasets to study languages as products of cultural evolution. Let's take a closer look.

Communicative Repertoires Culturally Adapt

There seems to be little argument about whether languages are learned from other people—that is, culturally transmitted. Even before birth, fetuses begin acquiring elements of the sounds and rhythms of the languages spoken by their mothers. Infants soon begin carefully watching the tongue and mouth movements of those around them and engage in vocal mimicry, providing themselves with immediate feedback for adaptive learning. Infant brains assemble specialized phoneme (sound) filters that allow them to effectively identify the sounds relevant in the local languages and to ignore the irrelevant ones. Around the end of their first year, children interpret pointing as a request to direct their attention at something and begin communicating with a combination of pointing and emotional reactions long before they talk. Infants and young children rapidly and unconsciously acquire elements of these communicative systems using many of the same cultural learning biases—based on cues of competence, reliability, and ethnicity (dialect)—that are used for learning about tools, practices, and social norms (see chapter 4). Learners watch their models using words or gestures *to do something*, like help a friend, request information, or obtain a particular object. They infer what their models are trying to do and imitate those goals, desires, and behaviors; meanings can be inferred from the rich context found in the routines of social life. Children's play often involves acting out scenes from adult social life, which uses language and other communicative elements in context and alongside noncommunicative practices.[3] From the perspective of the learner, communicative systems are just one important aspect of culture—a relevant local skill governed by social norms—that has to be acquired, like fire starting, ritual practices, manioc detoxification, and taboo adherence. Of course, learning the local communicative repertoire gradually opens the floodgates to acquiring many more of the other locally relevant practices, beliefs, motivations, and ways of thinking.

Operating over eons, this cultural evolutionary process has equipped human populations with powerful toolkits for communication. These toolkits involve complex repertoires of elements, including gestures, body positions, facial expressions, and vocalizations, which include clicks, whistles, grunts, growls, hisses, and squeaks. Like our toolkits for food processing, these systems have adapted to the local environments of their populations and are influenced by the same kinds of factors that influence other aspects of culture. Let me start with some of the non-speech elements of our repertoires, since these remind us that (1) spoken language represents only one component in our communication system, albeit an immensely important one in modern repertoires and (2) we humans can link meaning, by inferring the intentions of communicators, to an immense range of behaviors, not just to speech. Such systems may also provide insights into some of the earliest communicative systems used by our ancestors.

Face-to-face spoken conversations are interspersed with lots of communicative gestures, as well as facial expressions. People in different societies variously point with their index fingers, middle fingers, chins, or pursed lips. Be careful, though; at least two of these are considered rude in some place or another. We shake hands to greet others, or sometimes refuse to shake. If you are in Fiji, make sure to do a long shake (basically, hold hands for a bit). Elsewhere, people bow deeply, bow shallowly, or refuse to bow. To say "OK" or "good," I give the thumbs up or I circle my index finger and thumb in the OK sign. But, do NOT use the OK sign in Iran or Brazil, as President Richard Nixon mistakenly did (it's like giving the finger in the United States). We nod our heads to say "yes," but in some places that means "no" (e.g., Bulgaria). Some peoples subtly raise their eyebrows for "yes." Even ultra-verbal academics frequently use air quotes, an iconic gesture derived from the use of scare quotes in writing, to imply some disagreement with their terminology or its implications. I've seen many a humanities scholar, with a latte in one hand and a book in the other, struggle to communicate, unable to deploy air quotes to shield themselves from any undesirable implications of their words.

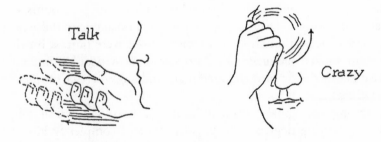

Figure 13.1. Plains Indian Standard Sign Language. The signs for "talk" and "crazy."

Sign and Whistle Languages

Gestures, however, are not merely a communicative sideshow, interwoven with spoken language. Fully fledged sign languages were common among hunter-gatherers, both in Australia and North America, and simpler gesture systems can be found across the globe, from African foraging bands to Christian monasteries. In North America, many groups of Plains Indians maintained both their own tribal sign languages, for hunting, warfare, and oratory at large ceremonies, and were conversant in Plains Indian Standard Sign Language. Plains Standard provided an important means of intertribal communication in a region of great (spoken) linguistic diversity. Using broad and open gestures, this language appears to be well adapted to communication over open plains, which favors a visual system over an auditory one, and for communication between individuals who might not want to approach each other too closely, as during intertribal communication. Probably because it was acquired after middle childhood, Plains Standard remained highly iconic, meaning that the signs contained elements of what they represented, and relied on much standardized pantomime (see figure 13.1). The sign for "bird," for example, involves flapping one's hands in a stylized way. For "eagle," you first make the sign for "bird," then the sign for "tail," and then for "black." So, "bird-tail-black" would be translated as "eagle."

The grammatical system of Plains Standard employed space instead of any temporal or sequential ordering. The equivalent of noun phrases (e.g., "this black dog") for subjects and objects can be constructed in space, at different locations, and then connected by the relevant verb sign. The hierarchical structure of spoken languages, which so impresses linguists, is relatively easy to construct if one can sketch "air pictures," as they were called, and essentially draw the links. By most accounts, Plains Standard possessed levels of expressiveness comparable to those of spoken languages, although signed conversations were quite a bit slower than in spoken language. As European-descent peoples entered the Great Plains, they copied and used many Indians signs (see the *Boy Scout Handbook*).[4]

Among Aboriginal Australian hunter-gatherers, gestural signing systems of varying degrees of expressiveness and complexity have been documented across the continent, but particularly in the central desert. Some of these societies maintained fully fledged sign languages with the full expressiveness of their spoken counterparts. Elsewhere, the local sign systems possessed only a few dozen or a few hundred signs, which could be effective in particular contexts, like hunting, raiding, or during ritually imposed taboos on speaking, but didn't have the full expressiveness of spoken languages.

Interestingly, the most developed sign languages tend to occur in populations with ritually imposed taboos on speaking that endure for many months or years. Often, as among the Warlpiri, these taboos are associated with a period of mourning after the death of a close male relative. The length of the taboo period varies substantially with the closeness of the relationship, status of the deceased, and the unexpectedness of the death. For the closest female relatives, the taboo is maintained until the deceased spirit is avenged, which often takes about a year. Unlike Plains Standard, these sign systems are not particularly iconic and tend to follow grammatical rules derived from the local spoken tongue, although the sign vocabularies are heavily shared among groups. In the northern central desert, the closer groups live to one another, the higher the percentage of shared hand signs. This holds independent of the similarity of their spoken languages and suggests that these signs were diffusing independently from the spoken languages.[5]

For our purposes, hunter-gatherer signing systems illustrate three important points. First, they appear adapted to their local context—to the functions they serve. Plains Standard, with its large iconic gestures, has evolved to foster communication among adults over distances or at large ceremonies. By contrast, the subtle gestures of Australian sign languages appear to be, in part, solutions to communication when ritually imposed taboos prohibit speech. In places where, for example, women cannot speak for months after a relative dies, sign languages achieve the same level of expressiveness as the spoken language. Ergo, nonlinguistic social norms probably drove the development of these communicative systems. Second, the complexity or expressiveness of Australian signing systems vary from group to group across the continent. Some groups have a limited number of signs (a limited vocabulary) and rather undeveloped grammatical systems, whereas others have the full deal. There's no argument that these various signing systems are equally complex or expressive. They aren't. Finally, the nature of the Australian systems suggests that they are learned by children, and thus the systems are not particularly iconic (unlike Plains Standard)—most hand signs lack any direct similarity to what they represent in the world. The reason is that iconicity aids adult learning, but when children learn languages they strip out the iconicity.

Remember, cultural evolution is smarter than we are, and so it has figured out many ways to facilitate communication. Now, let's consider whistle languages. Among remote populations from Turkey to Mexico, people can have a full conversation by whistling to each other, sometimes over long distances. Many of these whistled languages appear acoustically adapted for communication in steep mountainous terrain, where households might be relatively close, perhaps across a gorge, but an hour or more away in travel time. Some of these languages use finger whistles and can communicate easily over 2 to 3 kilometers and up to 10 kilometers under the right conditions. Other whistle languages use lip whistles for casual, close conversations. So, cultural evolution can and has built various languages out of whistles, because whistles were appropriate for the local context. Beyond whistles, anthropologists have found an incredible diversity of systems for communicating over distances using drums, horns, flutes, and gongs, among other techniques.[6]

Sonority

A successful communicator is one who can most effectively be understood, given the local social, environmental, or ecological conditions. As young or naïve learners focus on and learn from more successful communicators—who are using more effective communication tools—cumulative cultural evolution will gradually assemble sign or whistled repertoires, over time, in the same way that it hones kayaks, spears, and boomerangs. Given this, there's no reason to suspect that such cultural evolutionary processes somehow apply only to whistled or gestural sign languages, and not to typical spoken languages. Thus, spoken languages should—under the right circumstances—show some response to the local acoustic environments and to nonlinguistic social norms, just as whistled and sign languages do. While researchers have done little work on such topics, there's some preliminary evidence.

Spoken languages vary in their *sonority*. The sonority of our voices decreases as the airflow used for speech is obstructed and is highest for open vowels, like the /a/ and lowest for so-called voiceless stops like the /t/ in *tin*. Pronounce each of these sounds and note the difference in the constriction of your airflow. Both vowels and consonants vary in sonority, but vowels generally have much higher sonority than consonants. This means that more sonorous languages tend to have more vowels (e.g., Hawaiian), while less sonorous ones pack the consonants together (e.g., Russian). For the same energy and effort, more sonorous speech sounds can be heard at greater distances and over more ambient noise than less sonorous ones.

If languages adapt culturally, then we can predict that in situations in which people do relatively more talking over greater interpersonal distances with more ambient noise and sound dispersion, languages will be more sonorous. Many environmental variables might influence this, but Robert Monroe, John Fought, and their colleagues reasoned that climate, and specifically temperature, might have a big effect. The idea is simple: in warmer climates, people work, play, cook, and relax outdoors. Compared to life indoors, living outside means that communicators more frequently face the challenges of distance, noise and poor acoustics. Their team generated measures of sonority from word lists for dozens of languages and then looked at the relationship between sonority and measures of climatic temperature, like the number of months per year when it's below 10°C (50°F).

It turns out that if all you know is climatic temperature, then you can account for about one-third of the variation in the sonority of languages. Languages in warmer climates tend to use more vowels than those in colder climates and rely more heavily on the most sonorous vowel, /a/. For consonants, languages in warmer climates rely more heavily on the most sonorant consonants, like /n/, /l/, and /r/. By contrast, languages in colder climates lean more heavily on the least sonorous vowels, as the /i/ in *deep*.[7]

This simple idea can have much nuance added to it. For example, not all warm climates are equally conducive to sonorous speech. In regions with dense forest cover, the advantages of high sonority might be less pronounced, or as the anthropologists Mel and Carol Ember have argued, very cold and windy climates may select against linguistic practices that involve opening one's mouth widely, due to the increased heat loss. To this they added the idea that social norms about sexual restrictiveness might also influence sonority. Adding both of these nuances to the basic climatic temperature analysis, they managed to account for four-fifths of the variation in the sonority of languages (or at least of the few dozen languages they had data for). This is impressive, though it's important not to overstate the strength of these findings. This research program is just getting started. Nevertheless, something is clearly afoot here: the basic relationship between climatic temperature and sonority has been tested in several ways and stood up to retests by different researchers.[8]

Given the way cultural learning shapes our tools, diet, and toolkits to local environments, it shouldn't be surprising if cultural evolution similarly honed our phonemic toolkits.

Biologists have long been interested in how animal communication systems have adapted to the acoustics of a species' local habitat. Unlike language, much of nonhuman communication involves the deployment of a relatively fixed set of signals. However, research with monkeys, birds, and prairie dogs suggests that natural selection has calibrated these communication systems to diverse acoustic environments to make them more effective. Here, we have a parallel case in which selective cultural learning has—well outside of conscious awareness—adapted our communication systems to fit their local acoustic environments. Cultural evolution can solve many of the same adaptive problems as genetic evolution, only faster and without speciation.[9]

Culturally Evolving Complexity, Efficiency, and Learnability

Shifting around the sounds used by languages is one thing, but can cultural evolution actually aggregate, assemble, and integrate linguistic elements—tools and rules—in the same way it does for food-getting technologies or archery packages? Can cumulative cultural evolution endow languages with greater expressiveness or efficiency? The evidence below suggests that the answer is yes; but just as tool sophistication and toolkit sizes are limited by the systematic mistakes and misunderstandings inherent in the transmission of large and complex repertoires, so it is with languages. To survive and spread, languages too need to be learnable, especially by young children, and sometimes by adults in rapidly expanding groups. Thus, like other aspects of culture, we should expect vocabularies, phonemic inventories (repertoires of distinctive sounds), and grammatical toolkits to be influenced by factors such as population size and interconnectedness, as well as by the details of social networks (class stratification and ethnicity), technologies (e.g., writing), and institutions (e.g., formal schools).

Vocabulary Size

Let's begin with vocabularies. Words are tools for communicating, and having the right words can increase the ease, speed, and quality of communication. Although precise measures of the vocabulary sizes of different languages are difficult to establish precisely, rough estimates are sufficient because the differences are so big. Many of the languages carefully documented by linguists and anthropologists working in small-scale societies contain between 3,000 to 5,000 words—that's the entire dictionary. Meanwhile, slim dictionaries for European languages often contain about 50,000 words. A solid college-level English dictionary contains about 100,000 words, and the full-size *Oxford English Dictionary* contains more than 300,000 entries. Of course, this is totally unfair since the dictionaries of nonliterate societies reside entirely in the heads of speakers (until an anthropologist arrives), but European dictionaries reside in books and online databases, and few or no actual speakers know all the words. However, the average American 17-year-old knows 40–60,000 words while their professors know about 73,000 words. These estimated in-the-head vocabularies contain 8 to 14 times more words than are available to the speakers of many small-scale languages, assuming they know all the words in their languages.[10]

These estimates no doubt exaggerate the effective differences in vocabularies, since many of the words missing from small-scale languages are words for pieces of technology, actions, or concepts— like satellites, sit-ups, and emoticons—that don't exist in small-scale societies. To get a perspective on this, let's consider two kinds of words that don't fall into this category: color terms and integer numbers.

Languages vary in their number of basic color terms. At one end of the spectrum, English has 11 basic terms: "black," "white," "red," "yellow," "green," "blue," "purple," "orange," "pink," "brown," and "gray." By contrast, many languages have no basic color terms, or only two, which are always "black" and "white" (or, more accurately, "light" and "dark"). In languages with only 2 or 3 color terms, a brilliant "blue" sky can be described as "black" and compared to the color of dark dirty water.[11] If a language has 3 basic color terms they are "black," "white," and "red," which usually partitions the entire color spectrum. Numerous languages have 5 color terms, which can be glossed as "white," "black," "red," "yellow," and "green-blue" (grue). Of course, there are always ways around a lack of particular color terms, for example, by referring to some object with the desired color. But, this involves some linguistic work and is thus often less efficient.[12]

Ancient literary works tell the same story as the cross-cultural research. The Old Testament, the works of Homer, and ancient Vedic poems are remarkably vague about color and at times devoid of colorful descriptions. Neither the sky nor the sea is "blue." The worlds of these works are largely cast in black, white, and red. Terms for green, blue, and yellow—as basic descriptive terms—don't emerge until later in the literary histories of these societies. It's amazing that these rich, poetic cultural treasures are nearly colorless.[13]

Thus, color terms provide a narrow but clear window into language evolution. The color spectrum is actually continuous, so the borders between colors are arbitrary—the physics of light does not actually provide discrete categories like "red" and "yellow." This could create a prickly problem for a learner to solve: what region of the visual spectrum does the label "red" actually apply to? How can I as a learner converge on the particular region of color space that you mean by "red"? The answer may lie in the fact that while the actual color spectrum is uniform, our visual system perceives it as containing contours. Think of these contours as perceptual clumps or bumps in our cognitive color landscape that stand out. These contours might be genetically adaptive, reflecting things we have particularly needed to distinguish over our evolutionary history; or they might reflect simply quirks in the system created by the physiological mechanisms needed for perceiving color (or, some of both effects). Nevertheless, these contours help learners triangulate in on the local meaning of "red" or "blue." Over time, color terms evolve culturally so as to maximize the perceptual similarity within a term;[14] that is, examples or shades of "red" are *perceptually* more similar to each other than surrounding shades, even if these "reds" aren't actually any closer based on objective measures of hue and saturation. To be clear, these innate physiological-perceptual constraints are nonlinguistic in origin but get exploited by cultural evolution to help us discretely partition the otherwise continuous color spectrum using words. As with so many cultural adaptations, the system looks engineered for communication about color and fitted to the constraints of our visual perception; but, of course, the only engineer was cultural evolution.[15]

As an English speaker, you have 11 basic color terms at your disposal because cultural evolution has figured out how to make them learnable. Meanwhile, Korean speakers have 14 color terms. However, most societies over human evolutionary history probably only had two or three color terms. Learning additional color terms influences our cognitive abilities and increases our gray matter, improving our discrimination and memory for different color shades *across* terminological boundaries at the expense of shades within the boundaries of one term. Of course, if you have more nonoverlapping terms, you have more terminological boundaries.[16]

Integer numbers provide a potentially even more potent example. Since you are reading this book, you have very likely received via cultural inheritance a linguistically encoded counting system that permits you to count without bound using only ten numerals, along with some rules to structure them. This system permits you to easily distinguish a pile of 27 cherries from a pile of 28 cherries. However, this ability has not been available for most of human history. To our knowledge, all societies have number words for 1, 2, and many, and numerous societies count "1," "2," "3," "many"—but that's it; they can't count any higher. They can't, for example, use their counting system to distinguish those two piles of cherries. This fits with research in cognitive science suggesting that 1, 2, and 3 have corresponding innate mental representations.

Beyond this, some small-scale societies augment their innate system using culturally learned body-part counting systems that count objects or people by assigning them to specific body parts in sequence, like counting on one's fingers (but also using elbows, knees, noses, eyes, etc.). In New Guinea, these systems top out at values as low as 12 but variously go as high as 22, 28, 47, or 74 in different societies. In some of these societies, there are ways to go through the cycle two or three times to obtain even higher values. Of course, populations have worked out ways around not having many integers, but groups with only limited numbers of integers usually rapidly adopt the more extensive counting systems of other languages, when they get the chance.[17]

In children, full comprehension of both integers and color terms develops relatively late, at least compared to other aspects of language. Interestingly, Western children now master their basic colors at a younger age than they did in generations past, suggesting that the cultural system is evolving to be better at transmitting this knowledge.[18] Acquiring these products of cumulative cultural evolution—integers and color terms—shapes our brains and influences our cognitive abilities. So, they provide specific cases of how expanding vocabularies influences our "smarts."

More broadly, words are useful for thinking, so possessing a large vocabulary likely improves some kinds of problem solving. The active vocabularies of American adults, which is one component of IQ, have been increasing for at least a half century, and probably longer. While broader education at the university level can account for some of this growth, most of this expansion has occurred because adults have unintentionally continued to grow their vocabularies after their college years. This may have occurred because of shifts in the demands of many jobs in the United States during the twentieth century. The upshot is that American adults score higher on IQ tests than their parents or grandparents did at the same age, in part, because they have learned to use and understand more words.[19]

Finally, there's no reason why vocabularies couldn't have started expanding as soon as our ancestors were sufficiently good cultural learners. Nonhuman primates can already create mental categories or concepts for objects (banana), events (storm), and relational categories (rival), so cultural learning only needs to provide a means to connect these categories to particular gestures or sounds (that is, make words). Once connections are made, cultural evolution can cause these category-word relationships to become widely shared in a social group over generations.[20]

My point: words are like other aspects of culture, and consequently we should expect the vocabulary sizes of languages to expand with a population's collective brain. As both color terms and integers illustrate, such additions are not merely incidental, but may give us new cognitive abilities and raise our IQs.

Sound Inventories

Along with words, the inventories of phonemes used by societies are also influenced by the sociality of populations. Of course, the sounds of spoken languages are constrained, at least in the short run, to the anatomy of our mouths, tongues, and air passages, but within the range of possible sounds, the variation is impressive. Rotokas is spoken by about 4,000 people on an island off the coast of New Guinea. The sound inventory of Rotokas consists of 11 sounds composed of 5 vowels and 6 consonants. Hawaiian too has 5 vowels but combines them with 8 consonants. Among the Pirahã in Amazonia, the men's phonemic inventory parallels the Rotokas, with 11 sounds. Pirahã women, however, don't distinguish two of the consonants that men do, so it appears that they may have the smallest known inventory, at 10 distinct sounds. By contrast, English has roughly 24 consonants and at least 12 vowel sounds, depending on the dialect. In southern Africa, !Xóõ speakers use roughly 47 non-click consonants and 78 click sounds and have a sound inventory of over 140 sounds. That's at least 10 times larger than Hawaiian, Pirahã, and Rotokas.[21]

So, languages vary massively in their number of phonemes. But, why do they vary?

Well, mostly we don't know, and detailed studies motivated by cultural evolutionary thinking are only just beginning. Nevertheless, one interesting factor may be that some people learn their languages in large communities with lots of strangers, where many interactions occur among people who share little in common. By contrast, in small-scale societies, children learn in relatively small and quite homogenous communities. When learners do not share much in common and have to learn from very different voices with substantial variation in pronunciations across individuals, learners are better able to triangulate in on sound contrasts, and getting those contrasts is more important to understanding meanings. In small communities, minor sound differences can be missed in the learning process, in part because the people who are communicating tend to share so much background knowledge and context. This effect may create a tendency for larger populations in highly integrated market societies to maintain more phonemes than smaller, more isolated speech communities.

Consistent with this, there is now evidence from several studies using different linguistic databases that show that *languages with more speakers tend to have more sounds—that is, larger inventories of phonemes.*[22]

Of course, the size of the phoneme inventory should interact with other aspects of language. If a language has more phonemes, it should be able to sustain efficient communication or expressiveness using shorter words. Since the selective processes of cultural evolution will favor greater communicative efficiency, languages that can support more total phonemes—due to larger and more diverse speech communities—can shorten their words. Indeed, several analyses also support this. Languages with more phonemes in their inventories tend to have shorter words.[23]

Overall, this growing body of evidence shows that larger and more interconnected populations tend to have more phonemes and shorter words. And while the analysis is not yet complete, there is also some reason to suggest that the words of larger populations will be better optimized for communicative efficiency than those of smaller populations.

Routes to Grammatical Complexity

Here's a simple story; see if you can understand it.

Girl fruit pick   turn   mammoth see
Girl run   tree reach   climb   mammoth tree shake
Girl yell yell   father run   spear throw
Mammoth yell   fall
Father stone take   meat cut   girl give
Girl eat   finish   sleep

This story, modified from Guy Deutscher's *The Unfolding of Language* (p. 210), contains 23 different words arranged to *violate* English grammatical rules. Yet you probably more or less understood the story. Deutscher used some basic principles that are probably rooted in our ape cognition to construct the story in an understandable way without relying on English grammar. First, he used proximity: words for things that are closer in space are closer in speech. They "go together." This doesn't have to be the case in many languages, but it's our initial inference. Second, he used temporal sequencing: the events in the story follow the events in reality. Third, he relied on nonlinguistic causal structures: research with speakers from diverse languages suggests that actors (subjects) are first in thought. After actors, humans are next inclined to think of objects (patients) and, finally, actions (verbs). You can understand "girl fruit pick" more easily than "fruit girl pick" or "pick fruit girl," even though none are the subject-verb-object that English normally requires.[24]

So, you can achieve some degree of communication—even telling a story—with just some shared vocabulary. In fact, not only do you not need grammar, but you can do it even when the organization of the words violates the rules of your own language's grammar. This means that cultural evolution could gradually increase the complexity of a simple protolanguage (involving, say, 23 words) by incrementally adding grammatical rules and tools to its stockpile of words. We now know something of how this happens. Detailed historical and comparative studies show where specific grammatical elements come from and how they usually evolve. They typically come from first hijacking concrete words, then gradually bleaching them of their original meaning, and often then honing them down to size, presumably to increase communicative efficiency. This process is called *grammaticalization*.[25] Here is a taste of grammaticalization:

1. Adjectives: Nouns and verbs, which are built on an ancient cognitive distinction between *things* and *actions* found in nonhuman primates and presumably in our ancient precultural ancestors, get deployed to build adjectives and adverbs. For example, a word for a rock-solid thing like "concrete" (cement, water, and gravel) can evolve to also operate as an adjective, meaning "not abstract," as in "Give me a concrete example." Someday, when new building materials replace concrete, this word may come to only mean "not abstract."

2. Tense: Verbs are sometimes gradually turned into markers of tense, for the past and future. The English word "gonna" provides a nice example of a word in transition to a grammatical element. Consider the sentence, "I'm gonna stay here and not move." Clearly, "gonna" is from the verb phrase "going to," but I just used it to express my intention to "stay." "Gonna" has not fully transitioned yet, but it appears to be doing what the word "will" did centuries ago. "Will" is derived from a verb meaning "want," which is a very common source for the future tense (e.g., in Swahili). Similarly, past tenses often evolve by modifying and then gradually trimming down the verb for "finish" or "done." In the unwritten Yasawan dialect of Fijian I've been studying, this can be seen in a not-yet-trimmed-down version, where the literal translation of "I have eaten" is "I am eating done."

3. Prepositions: Adverbs often get re-tasked to create prepositions, like "after" or "behind." For example, the body-part word for "back" or "buttocks" often gets first modified into an adverb like "behind" ("The sad events lay behind them"). Then, the adverb evolves into a preposition, as in "I put my laptop behind the door." In English, curiously, the euphuistic use of "behind" (as in "Get off your behind") to mean "buttocks" is turning this preposition into a body-part noun (which is rare).

4. Number and plural pronouns: Pluralization is often achieved by adding "all" to nouns and pronouns and then trimming it down. For example, in the dialect of English spoken in the southern United States, a second-person plural pronoun has emerged in the form of "y'all." This new pronoun, which many languages have but

no other dialects of English, arose as "you" crashed into an "all" that was added to remove any ambiguity about whether "you" was plural. Incidentally, "y'all" is not ungrammatical; it's an example of how languages evolve. If it feels wrong to you, that's because it violates one of your internalized linguistic social norms.

5. Subordinating conjunctions: Tools for marking subordinate clauses are often built from those for marking relative clauses, which are often built from verbal pointers (demonstratives like "this" and "that"). Verbal pointers in phrases like "that picture" or "this hat" evolve to mark relative clauses, as in "Zoey likes the picture *that* Jessica painted." These relative markers evolve into subordinating conjunctions, as in "I hope *that* Zoey likes the picture that Jessica painted."

Practically speaking, learners have to acquire the relevant social norms for how they can and cannot arrange words to communicate, but these rules are gradually altered as successful communicators, who have made adjustments to increase efficiency or expressiveness, are preferentially attended to and learned from. This can even be seen in relatively brief experiments in which people first had to learn an unfamiliar grammar and then had to use it to communicate. In one experiment, after only three days of interaction, the patterns of a just-learned grammar shifted in ways that increased communicative efficiency. Amazingly, the observed grammatical modifications parallel the actual distributions of these grammatical elements in real languages.[26]

All of this suggests that grammatical tools and rules should evolve like other aspects of culture. At some point, there must have been communicative repertoires that lacked some of the shared norms that languages have for referring to the past or future, marking number, or for embedding clauses. If true, we might find existing or historically known languages with smaller toolkits of grammatical tools or less efficient grammatical rules, just as we do for technologies. However, it may be that such tools, once they emerge in one language, rapidly spread to other languages, as good counting systems do. This would prioritize looking at more-isolated groups or languages, that may not have had time to adopt the new tools, or by looking back in time at past languages using ancient written sources.

Supporting this expectation, subordinating conjunctions, like "after," "before," and "because of," may have evolved only recently, in historical times, and are probably no more a feature of *human* languages than composite bows are a feature of human technological repertoires. The tools of subordination seem less well developed in the earliest versions of Sumerian, Akkadian, Hittite, and Greek. This makes these languages slow, ponderous, and repetitious to read.[27] Subordination tools also seem to be missing in some living languages, including languages found in Australia (e.g., Warlpiri), the Arctic (Inuit), New Guinea (Iatmul) and the Amazon (Pirahã).[28] People use phrases that are subordinate to some degree, but this subordination is provided by a combination of concatenation and context, not by specialized grammatical tools like coordinating conjunctions. To see what I mean, here's subordination using only concatenation and context in action, in English: "I'm taking Sharon's son to the circus next week . . . she's the young woman we met last week . . . we were outside the Starbucks."

Philosophers have long been impressed by "language's" ability to embed phrases within larger hierarchical structures, often using grammatical tools for subordination. However, these findings suggest that much of the hierarchical structuring and fancy subordination that we see in most modern languages is the product of long-term cumulative cultural evolution. This is not to say that we humans don't have some souped-up innate abilities for dealing with hierarchical structures, which may also be useful for making tools or understanding social relationships, but merely that the elegant bits of grammar that permit us to fully harness these abilities were built by cultural evolution.

Finally, grammatical complexity, like other aspects of culture, is influenced by social institutions and adapts to the demands of the communicative contexts of a society. For example, larger populations with more heterogeneous speech communities and more adult language learners tend to have morphologically simpler grammars. In morphologically complex grammars, words have many different, often subtle, grammatical elements that must be inserted, at the front, back, or in the middle. These elements augment the meaning of the word, conveying tense, mood, number, gender, etc. Morphologically simpler languages do much less of this, but can (often) communicate the same information with additional words and word ordering. One interesting explanation for this is that larger populations have more adult language learners, and adult language learners are bad at learning morphologically complex languages.[29]

In summary, larger speech communities seem to possess bigger inventories of words, phonemes, and grammatical tools. That is, the languages possessed by these communities are more complex. Historical evidence suggests that inventories of words and grammatical tools have expanded over centuries and millennia. Psychological evidence indicates that the contents of vocabularies and the rules of grammar, such as those related to color terms and integers, may alter our cognitive abilities, including both memory and perception. A combination of evidence indicates that modern world languages may in fact be rather unusual compared to most of the languages spoken over our species' evolutionary history.

Culturally Evolved to Be Learnable

As our communicative repertoires gradually expanded, there would have been both selective pressure on our genes to make our ancestors better at learning culture, including the elements of these emerging communicative systems, and pressure on the repertoire of communicative elements to be *more learnable*, especially by children. Here I'll briefly emphasize the role of the second partner in this duet—culture. Like other kinds of tools, which have evolved culturally to fit our hands, shoulders, and physical abilities; languages have culturally evolved to fit our brains, becoming more learnable by adapting themselves to the contours and constraints of our psychology (remember the color terms). Children are good at learning languages—in large part—because difficult-to-learn elements don't get learned, and ergo, don't get transmitted to the next generation. There are possible languages that are really hard for children to learn, but those lost out in competition to more learnable languages.[30]

Languages that more effectively exploit innate aspects of our psychology would be more learnable and less likely to be replaced by alternatives in the future. Computer simulations of the process of cultural transmission across generations among imperfect learners with limited memories favors the emergence of many of the key traits that distinguish languages from the communication systems of other species. As the size of vocabularies increased, the cultural evolutionary pressures on languages to become rule bound and regular would rise.[31] A learner might be able to acquire, for example, an idiosyncratic past-tense form for each verb in a language with a small lexicon, of say 50 words. But, they won't be able to do it for a language with 5,000 words. Thus, languages without such rules (syntax) didn't make it. They got beaten out by those languages with more-regular rules, like adding "-ed" for the past tense, that children could learn to apply. This will be the case both because it's tough to remember all those specific cases, and because as vocabulary size increases many more specialized verbs will only be heard rarely. Irregular forms can persist (like "ate" instead of "eated"), but only on frequently used words, where the constant repetition eases such memory constraints.

Direct evidence for these processes in human evolutionary history is nonexistent. However, complementing the computer simulations mentioned above is indirect evidence from laboratory experiments like those described for the accumulation of technical skills in chapter 12. These experiments show that when people observe the performance of others using an artificial language and then make inferences from those observations, both structure (syntax-like forms) and compositionality (discrete words) emerge over many transmission events as people learn from each other over many laboratory generations. No single individual does much at all, and no one is trying to achieve this as a goal. It's an unconscious emergent product of cultural transmission over generations.

Research with children speaking real languages further confirms that regularity makes languages more learnable. Children speaking *more*-regular languages understand sentences, like "The horse kicked the cow," *better* than children who speak *less* regular languages. Specifically, Turkish- and English-speaking children outperformed children speaking the quite irregular Serbo-Croatian language. Of course, these performance differences vanish in later years. Nevertheless, such research suggests that not all languages are equally easy for kids (or adults) to learn.[32]

My point is that cultural evolution is a key reason why existing languages are so easily learned by children, and some of the recurrent features of languages, like syntax, are likely the result of cultural evolution working to keep languages learnable, especially as vocabularies expand.

The Synergies of Manual Skills, Norms, Gestures, and Speech

During our species' evolutionary history, coevolution was relevant in two important ways. First, increasingly complex communicative repertoires were coevolving with increasingly complex tools, practices, and institutions. This coevolutionary interaction was synergistic, since two or more of these domains of culture will create selection pressures for genes that improve our psychological abilities to acquire, store, organize, and retransmit valuable cultural information. This interaction would have sharpened our abilities to culturally learn from others—improving our skills at inferring others' goals and intentions (to better learn from them), inferring underlying rules or norms, and learning complex sequences with hierarchical structure. Second, as noted earlier, these culturally accumulating communicative repertoires put selective pressures on genes for communicating: they pushed down our larynx to widen our vocal range, drove axonal connections from our neocortex down deep into our spinal cords to improve the dexterity of our hands and tongues, whitened the sclera of our eyes to reveal our gaze direction to cue others, and endowed us with reliably developing cognitive skills for vocal mimicry and communicative cueing, like pointing and eye gaze.

It's impossible to know in what behavioral domain cultural evolution first began to accumulate: in toolmaking, food processing, or gestural communication. It's also plausible that cumulative cultural evolution in some ancestral groups began with gestural repertoires while other groups, due to different ecological constraints, started to accumulate techniques for tool manufacture. The key point is that whatever came first would have likely catalyzed developments in the other domains. For example, human brains have axons that project from our neocortex directly to the ventral horn, where the motor neurons are, and then down deep into our spinal cords.[33] This anatomical change helps to explain our species' impressive manual dexterity and specifically our ability to learn tasks requiring manual dexterity. This change would have been favored by the cultural emergence of techniques requiring greater dexterity, like making nets, clothing, blades, or fire, or culturally transmitted skills and tactics, like using projectiles such as rocks, clubs, or spears for hunting or defense. Plausibly, either an emerging gestural communication system or increasingly complex toolmaking skills could have created genetic selection pressures favoring more direct axonal connections into the ventral horn and spinal cord. If gestural communication emerged first, it's plausible that toolmaking skills would have benefited from enhanced dexterity. If it were tools driving the boat, then enhanced gestural communication would have gotten the free ride. Better communicative skills also might have improved the quality of the information transmitted about making and using tools.

This process helps explain our species' impressive tongue dexterity and capacity for vocal mimicry, which are unique among primates. Vocal mimicry is rare in nature, although many bird species who are also excellent vocal mimics have an analogous brain setup.[34] Our neocortical projections, which also innervate the muscles of our jaws, tongues, faces, and vocal chords, could have initially been driven by gestural communication, toolmaking, or any number of other skills. However, once they spread, they could have opened the door to learning vocalizations as a by-product. Vocalizations would have first developed as an addendum to an existing repertoire of communicative gestures, which could have included facial expressions, pointing, and bodily positions. In this scenario, the genetic selection pressures for speech would have been created by a non-speech-based communicative repertoire that gradually favored adding speech elements as a way of freeing the hands and suppressing the need to necessarily look at each other. The cross talk between the tongue and limbs is highlighted at weddings, when inexperienced dancers stick their tongues out while focusing on challenging dance moves.[35]

Whatever the precise evolutionary sequence was for the emergence of gestures, vocalizations, social norms, and tool-use is not crucial since cultural evolution would have created substantial genetic evolutionary pressures for using verbal gestures—speech—to communicate, to free our hands for using tools, throwing weapons, cooking, communicating at night, and maintaining balance while running down prey. Comparative work with other apes shows that our larynx dropped, which lengthened our vocal tracts and expanded our potential phonetic repertoire. Our tongues were freed up to move both side-to-side and up-and-down. We also lost the soft, inflatable air sacs that apes have around their larynx. These changes, however, came with a significant cost. Other mammals, and human infants, can breathe and swallow at the same time and thus can't easily choke on food.[36] Older humans can and do choke.

These genetic evolutionary selection pressures were for communication in general—not only for spoken language—so our ancestors may have maintained gestural languages alongside spoken languages, as many modern hunter-gathers did. In the increasingly cooperative worlds constructed by social norms and reputations, it became relatively more important to communicate and teach others rather than deceive or exploit them. Such pressures explain why humans are peculiar in having our rather small irises set against a white background—the sclera—in our eyes. Anyone watching us can infer where we are looking or whom we are looking at. Experiments with apes and infants show that apes watch the orientation of the heads of others while human infants watch the eyes. This behavior makes it harder for people to hide what they are interested in (a cost), but it is quite useful for actively transmitting cultural information or communicating more generally. Motivated to convey cultural information, diverse populations around the globe make use of a somewhat variable set of "pedagogical" cues to facilitate learning. Commonly used cues include eye contact, head movements, and pointing. With infants, people in diverse societies speak in "motherese" (baby talk). With its slow and exaggerated intonations, motherese speeds language learning by providing input that is easier for infants to process.[37]

Recognizing that languages, or at least simpler communicative systems, were coevolving with tools, practices, and social norms helps explain how some of the features of languages might have evolved and why. For example, linguists are often impressed by the hierarchical structure, sequential ordering, and recursion found in modern languages.[38] In a world with only language evolving, it's not obvious where the cognitive abilities that underlie these patterns might have come from. If a genetic mutant with new mental abilities for creating hierarchical linguistic structures emerged in a society having only a simple protolanguage, she wouldn't have anyone with whom to use her advanced abilities. No one lacking her cognitive abilities could understand her or copy her complex patterns.

However, complex tools are also assembled sequentially, hierarchically, and sometimes recursively, just like (most) modern languages. To make a spear, you have to obtain the wood, straighten it, and balance it for throwing. Then, or maybe first, you have to obtain the flint, bone, or obsidian to make the spear tip. This material has to be knapped or carved to make a spear point. Then you need some sinew or resin to attach the spearhead to the shaft. Thus, there are componential parts (objects), actions (knapping), and rules (orderings) for assembling the parts. All of this has to be done with the goal of obtaining a good spear, just as sentences are assembled with the goal of communicating. Similarly, food-processing techniques to detoxify seeds, such as nardoo or acorns, involve sequential protocols with components and subcomponents (grinding, leaching, and cooking), which must be done in order. Beyond sequences and hierarchies, some technical skills, like weaving and knitting, demand recursive applications of the same techniques in a manner that could potentially continue without bound.[39]

Here however, unlike language, there is no need for someone else to have the same abilities to get the benefits. It's directly beneficial to the individual to be able to organize and make a more complex tool. And the existence of these more complex tools will put an even greater selective pressure on those who cannot make them. These cognitive abilities can get redeployed for organizing and transmitting communicative repertoires, or perhaps in transmitting norms and practices about kinship systems. Consistent with this, when humans compete with primates in sequential learning tasks involving manual actions, we achieve superiority only when the task involves a hierarchical structure.[40]

This idea has been explored in computer simulations in which neural networks had to retain a "genetically" evolved ability to learn complex sequences in a nonlinguistic task (like nardoo processing) while at the same time facing the challenge of learning a grammar. The neural networks could evolve genetically to be better learners of the grammar, and the grammars could evolve culturally to be more easily learned by these networks. The results are threefold:

1. Genetically evolving to learn nonlinguistic sequences improved grammar learning—so, synergies exist between tools and language.
2. But having to remain a good nonlinguistic sequence learner (a toolmaker) inhibited genetic evolution's ability to improve grammar learning—so, no genes specifically *for* grammar learning are favored.
3. Nevertheless, neural networks got much better at learning grammars because *the grammars* evolved culturally to be readily learnable by the existing neural networks.[41] In short, cultural evolution made the grammars more learnable.

Genes and Brains for Learning Complex Sequences

Researchers may have isolated a gene that was favored during human evolution to improve our procedural or sequence learning abilities. The gene, *FOXP2*, which is located on human chromosome 7, codes for a protein that influences brain development, particularly in regions related to procedural and motor learning. For 170 million years, this gene appears to have remained relatively fixed across many species; however, since we split with chimpanzees only 5–10 million years ago, there have been two changes in our copy of the gene. Mutations in *FOXP2* influence both grammar and sequence learning in humans, and motor-skill learning in mice. The indirect pathways that connect genes to mental abilities usually make firm links difficult, but it's plausible that *FOXP2* was favored by cumulative cultural evolution to improve our procedural, motor, or sequential learning abilities, for learning complex culturally emerging protocols for toolmaking. Once in place, these abilities could have been harnessed in our communicative repertoire to permit more-complex grammatical constructions. Or perhaps the sequence was the other way around.[42]

Consistent with this, research applying brain-imaging techniques shows that using both language and tools (manual actions) involve overlapping brain regions. In fact, if you focus only on regions that are specific to either tool-use or language-use, you are left with few differences. The differences that do emerge seem trivial, such as language activating the auditory regions and tool-use tapping motor regions. Brain regions once thought to be specific to grammar now appear to be used for tasks requiring sequential or hierarchical structuring and are tapped for complex manual procedures. Similarly, learning to make both sounds and tools tap the same imitative areas, so vocal imitation is not something totally distinct from other kinds of imitation in the brain.[43]

Even now, culture continues to be the primary driver of human genetic evolution, and the cultural evolution of different linguistic features in different populations may be continuing to alter gene frequencies. One suggestive bit of evidence indicates that the emergence of nontonal languages may have created conditions favoring the spread of two different genetic variants. Many languages overlay pitch or pitch change on top of a set of sounds to distinguish different words. These are tonal languages. In Mandarin Chinese, for example, the sound /ma/ can mean "mom," horse," "hemp," or "scold" depending on the intonations or tonal contours. Nontonal languages, such as English or Spanish (or any European language), only use pitch or intonation at the level of the phrase to convey emphasis, feeling, or mood. The two genes are variants of *ASPM* and *MCPH*, and influence brain growth and development, and these have begun spreading under natural selection in the last 50,000 years or so. Both new variants of these genes are associated with nontonal languages: that is, speakers of nontonal languages tend to have the new variants.

Of course, the history of the spread of human populations means that many elements of genes and languages will be correlated, because they so often move together. Consequently, gene-language correlations are rampant and don't generally point to the action of natural selection. But in this case, it appears that historical relationships, linguistic similarities, and spatial proximity cannot explain the gene-language correlation. And so far, these variants cannot be traced to any other psychological or cognitive differences, such as intelligence, brain size, social skills, or schizophrenia.

Nevertheless, to be clear, *any human child* raised anywhere can learn the local language, but natural selection may be tinkering with the genetics of just how easy or hard it is, which depends on the enduring features of local languages.[44]

Culture, Cooperation, and Why
Actions Speak Louder Than Words

The broad view I've presented in this chapter challenges many claims—old and new—that the emergence of language in our species' evolutionary lineage was THE Rubicon that, once crossed, set us apart from the rest of nature. With the emergence of language, so the story goes, cultural transmission became possible and the problem of cooperation was solved, basically by reputational gossip. While there is no doubt that language plays an immensely important role for our species, there are three main problems with this common view and its overemphasis on language. First, this approach fails to recognize that quite a bit of cultural transmission and cultural evolution is possible without language. Cultural information about tool manufacture, fire making, dangerous animals, edible plants, cooking, and diet (food choice) can all be acquired, at least to some nontrivial degree, without language. Even social norms, such as those related to food sharing, can be transmitted without language. A little way down the culture-gene coevolutionary road, but long before full-blown language, more controlled pointing, facial expressions, and improvised pantomime could have provided communicative tools for cultural transmission, as these still do when people don't share a common tongue. Anything we might call language probably arrived on the scene well after cultural evolution was underway, and had already assembled simple communicative repertoires.

Second, language itself is a culturally evolved product, so it can't *cause* culture. Of course, language can and does greatly facilitate cultural transmission—the flow of cultural information—and opens up whole new pathways, such as the transmission of stories, labeled categories, and songs. But these new pathways were bootstrapped by nonlinguistic forms of cultural evolution, just as the much later spread of writing and literacy was bootstrapped from language to open up new cultural evolutionary avenues.

Finally, language has at its core a rather serious cooperative dilemma: lying, deception, and exaggeration. Lying with language is cheap, at least in the short term, and is a potentially powerful way to exploit and manipulate others. The more complex a communicative system is, the easier it is to lie or shade the truth, and get away with it.[45] If this cooperative dilemma is not addressed, the evolution of language, whether by genetic or cultural evolution, is rather limited. The reason is straightforward. If others are using language to trick or deceive me, I can avoid this by not believing anyone or even by not listening to them at all. If, to avoid being manipulated, everyone stops listening, then there's no reason to try to communicate. Language will go away or remain limited to those situations where deception or manipulation are too difficult.

Thus, for complex communicative repertoires to evolve in the first place, this cooperative dilemma has to have already been at least partially solved. Therefore, language can't be the big solution to human cooperation. Of course, once culturally evolved social norms for honest communication emerge and spread, then language can potentially expand the sphere of cooperation and exchange by helping spread gossip about norm violators and by making the acquisition of social norms faster and more accurate. Many arguments regarding how language solves human cooperation assume—without acknowledging it—that culture already exists because they assume reputations exist and that these provide a policing mechanism for honesty. This presupposes both reputations and social norms about honesty, which are both culturally transmitted and highly variable across societies.

In addition to social norms against lying, the cooperative dilemma of language is further mitigated by our cultural learning abilities in two ways. First, transmitting reputational information does not require language and can effectively be done with nothing more than pointing and facial expressions. For example, I can point at Bill lounging under a shady tree while we are rebuilding his house and make a disgust face or shame display. This will transmit to you my views of Bill's actions. So, the transmission of reputation isn't reliant solely on language. Second, although reputations are subject to manipulation, cultural learners already have cognitive abilities that would have mitigated the impacts of deception by self-interested others. For example, Steve might try to convince you that his rival, John, is a scoundrel so that Steve can gain an advantage on John (suppose John is honorable). However, if you use conformist transmission in acquiring your beliefs about John, you would take input from many people in the community, most of whom would not be John's rival. This would allow you to discard Steve's disinformation and converge on a more accurate version of John's reputation. Thus, our cultural learning mechanisms help prevent lying and deception from destroying the value of reputations.[46]

Nevertheless, the emergence of language would still have created opportunities for individuals, especially those who are successful and prestigious, to manipulate others into doing or believing things that would benefit the manipulator. For example, a manipulator who wants to poison his competitors might try, using language, to spread the idea that blue mushrooms are delicious and nutritious, even though he believes they are mildly toxic. The potential for such manipulative cultural transmission would have created a selection pressure for learners to look for what I call *CRedibility Enhancing Displays* (CREDs). CREDs are actions that a person would be unlikely to perform if he or she believes something different from their verbally stated beliefs or if they prefer something different from their stated preferences. Using CREDs provides learners with a kind of partial immune system, or filter, against manipulators who would exploit the cheap cultural-transmission channel provided by language. A good CRED for actually believing that blue mushrooms are delicious and nutritious is to eat a lot of blue mushrooms and feed them to one's children. If a learner observes a potential model doing this, he should then be willing to weight the model's statements about the nutritional value of blue mushrooms more heavily in forming his own beliefs.

Although research is just beginning, experimental work with children and adults suggests that CREDs are important for the cultural transmission of many beliefs and practices, including food preferences, altruistic behaviors, and supernatural or counterintuitive beliefs.[47] CREDs, for example, help explain why martyrdom can be a powerful force in spreading religious faiths. Anyone willing to die for their supernatural beliefs probably actually holds those beliefs and is not (or was not) engaged in some trickery or manipulation. CREDs also help explain why religious leaders take vows of poverty and celibacy, because these CREDs make them more potent transmitters of the faith. Overall, CREDs provide a partial immunity to the opportunities for exploitation opened up by language, though of course the prevalence of con artists, phony TV evangelists, and other hucksters attests to the persistent problem created by our reliance on cultural learning through language.

Our use of CREDs is why actions speak louder than words and contributes to suppressing self-interested attempts to manipulate or exploit cultural transmission.

Summary

Let me close this chapter with six take-home points:

1. Languages are packages of cultural adaptations for communication. Just as with technologies and social norms, once cultural evolution is unleashed, it will begin generating increasingly complex communicative repertoires adapted to local contexts. We have seen that cultural evolution can even assemble sign and whistle languages or alter the sonority of words to fit the recurrent acoustic environments faced by different speech communities.
2. These repertoires will evolve culturally to be learnable, and this learnability will create shared features among languages as they exploit diverse nonlinguistic aspects of our innate psychology. However, the details of this process will depend on who has to learn the languages—only children or people of all ages.
3. Like toolkits, the size and interconnectedness of populations favors culturally evolving and sustaining larger vocabularies, more phonemes, shorter words, and certain kinds of more complex grammatical tools, like subordinating conjunctions.
4. These patterns combined with anthropological and historical evidence from diverse languages suggests that widely spoken modern languages, like English, probably look quite different from the languages that dominated our species' evolutionary history, just as our technologies and institutions look rather different from those of small-scale societies.
5. Acquiring languages alters aspects of our psychology and endows individuals with new cognitive abilities.
6. As a recurrent product of cultural evolution, languages have produced powerful genetic selection pressures over the course of human evolution, shaping our bodies and brains for speech and cooperative communication. This process continues.
7. Understanding the evolution of language requires seating it within the broader context of the culture-gene coevolution of tools, skills, and social norms.

In chapter 14, we'll see some of the myriad ways in which culture shapes our brains and biology, even when it doesn't mess with our genes. Cultural evolution is a type of biological evolution.

CHAPTER 14

ENCULTURATED BRAINS AND HONORABLE HORMONES

To read these words you are using a culturally constructed network centered in the left ventral occipital temporal region of your brain. This, the brain's "letterbox," is a specialized piece of hardware, or perhaps firmware, for reading.[1] If this brain area suffers targeted damage, literate people become suddenly illiterate but retain their other cognitive and visual abilities. Some with damage to this region, for example, can still understand written numbers and make numerical calculations but can't read. This general patterning remains true even for readers of written languages that use symbols to represent whole syllables or words, like Chinese or Japanese.[2] Brain-imaging experiments show that the letterbox registers the culturally evolved resemblance between "READ" and "read," even though the two sets of markings bear little physical similarly; but, of course, this happens only in readers of English. Thus, the relationship between "R" and "r" is literally etched in the brain circuitry of English readers. Similarly, Hebrew characters activate the letterbox in Hebrew readers, but not in monoliterate English readers.[3]

This means that highly literate people have different brains than those who are illiterate because they've trained their brains to read. Learning to read specializes your brain for visually processing whatever writing system you are working on. The better you can read, the more specialized the wiring of your brain is for reading. In American children, the letterbox begins to appear around age eight but doesn't reach maturity until adolescence, assuming the child continues reading.

Not only does learning to read build a letterbox in the "visual area" between a region specialized to recognize faces and one focused on objects, but it thickens the corpus callosum, the information highway that connects the left and right sides of the brain. Learning to read also modifies the superior temporal sulcus and the left inferior prefrontal cortex, or Broca's area. Overall, this rewiring endows highly literate people with (1) longer verbal memories, (2) broader patterns of activation across the brain to *spoken words*, and (3) a greater awareness of the various sounds that make up words. That is, high levels of literacy improve some cognitive abilities that bear no direct relationship to reading or writing.[4]

However, these enhancements don't come without costs. Skilled readers are probably worse at identifying faces, since jury-rigging the relevant brain areas impinges on the fusiform gyrus, an area that specializes in face recognition. In fact, the well-established neurological asymmetry in face processing, favoring the right side of the brain, may be due to the effects of learning to read, which drives face processing out of the left side and shifts what it can to the right side.[5] I was personally glad to hear this, as I now have an excuse for why I forget faces so often—I've recycled some of my facial recognition neuronal firmware to support my reading addiction.

These are *biological* modifications to our brain, but **not** *genetic* modifications. They are the end result of thousands of years of cultural evolution, which figured out how to effectively modify our brains without messing with our genetics. Reading and writing are cultural products that have evolved by jury-rigging aspects of our genetically evolved neurological systems for object recognition, visual memory, and language. As the neuroscientist Stanislas Dehaene has argued, the brain's letterbox gets wired into a kind of neuronal sweet spot, where it has the tools for fine-grained object recognition and connections to language centers. To see the effects of this, consider three recurrent features of writing systems. First, the specific shapes used to form common characters across alphabetic writing systems, like L, X, O, and T, as well as in many of the characters of nonalphabetic writing scripts, such as Chinese characters like 山 (mountain), are rapidly recognized by our visual systems. This is not surprising since these shapes are also the most frequently occurring in the natural world. Second, the meanings or sounds represented by characters in writing do not depend on the absolute size of the character; after all, the apparent size of objects in the real world depends on their distance from the viewer. Children are never taught that letter size is irrelevant; they automatically know or infer this rule. They are, however, occasionally taught that relative letter size is sometimes important, as in Japanese Hiragana. Finally, letters or characters in writing are spaced optimally for speed. If the letters a r e t o o f a r a p a r t , your reading speed plummets.[6]

Like the other cultural adaptations we've discussed, the symbols used in writing probably didn't begin with basic characters optimized for our object recognition systems, and no one figured out that, by default, children automatically infer size invariance for letters. In fact, many writing systems began with iconic pictograms, probably because, for example, wavy lines that look a bit like water or wheat are easier to remember than arbitrary symbols. Egyptian hieroglyphs used representations of animals, objects, and tools, some of which eventually came to represent phonemes. Such recurrent patterns emerged as cultural evolution selected elements of writing systems that better fit the architecture of our brains, but of course, many other forces influenced cultural evolution. And there may in fact have been entirely different cultural evolutionary avenues to creating highly readable writing systems—systems that fit our brains in quite different ways. Let's consider the Americas, where writing emerged independently from the Eurasian systems.

In the Americas, the use of stylized faces as symbols for concepts, dates, proper names, and syllables emerged early and appears to have increased over time. Among the Maya, for example, stylized faces endured for six centuries as a central part of the writing system (see figure 14.1). This unusual element in writing harnesses our innate proclivities, including dedicated brain regions, for recognizing and remembering faces. It also seems likely that skilled Mayan readers probably got better at recognizing and remembering faces in nonreading contexts—unlike me and other obsessive English readers. This cognitive bias may be reasserting itself in the cultural evolution of modern writing systems, as I suspect that stylized faces are making a comeback ☺.[7]

Reading, then, is a cultural evolutionary product that actually rewires our brains to create a cognitive specialization—an almost magical ability to rapidly turn patterns of shapes into language. Most human societies have not had a writing system, and until the last few hundred years, most people did not know how to read or write. This means that most people in modern societies (those with high reading proficiency), because they have learned to read, have different brains with somewhat different cognitive abilities than most people in most societies across human history.

This example underlines a crucial point, which we'll explore below. *Cultural differences are biological differences but not genetic differences.* Human biology, including our brains, involves much more than genes. Oddly, many people, including scientists and science journalists who should know better, often treat cultural differences as if they were nonbiological and nonmaterial, almost otherworldly. This confusion emerges when people think that showing something is "in the brain" or driven by hormones means it's genetic. This is not the case, and the goal of this chapter is to relieve you of these misconceptions (if you have them). Recent evidence clearly shows how culture can shape biology by altering our brain architecture, molding our bodies, and shifting our hormones. Cultural evolution is a type of biological evolution; it's just not a type of genetic evolution.

Before proceeding, let me emphasize an important caution. Throughout this book so far, I've been laying out how cultural evolution has shaped genetic evolution over tens or hundreds of thousands of years. Now, I'm making a different point: growing up in a culturally constructed environment shapes our bodies and brains over development nongenetically. Clearly, these two processes are related and interact, and this is what so much work in the evolutionary and social sciences has not sufficiently taken into account.

Cultural evolution has figured out many ways to shape our biology. At the

Figure 14.1. Mayan writing from the temple of the foliated cross, Palenque (Mexico).

most basic level, cultural learning shapes the reward circuitry in our brains so that we come to like and want different things. These differences are neither superficial nor unimportant, as we saw with chili peppers in chapter 7. Cultural learning influences what we want to eat, the characteristics we prefer in sexual partners, and how much something hurts. The evidence below shows how cultural learning makes food and drinks like wine taste better and makes electricity hurt less. Further, by shaping both the incentives in our social world and our motivations, culture causes us to train our brains in particular ways. In some societies, this involves learning to read, tracking gender or status differences, or using fractions. In other societies, it involves spotting and identifying animal tracks, seeing clearly underwater, or recognizing individual cows within large herds. As part of this process, cultural evolution sometimes constructs what amount to mental prostheses that augment our brains and cognitive abilities, like the physical and mental abaci that permit children to make mental calculations faster than calculators (see chapter 12).

Wine, Men, and Song

Researchers applying evolutionary principles to understand sexual attraction have shown that humans tend to prefer certain attributes in their romantic partners and mates. Nevertheless, selecting a romantic partner or mate, especially a long-term one, is precisely the kind of problem that cultural learning is well suited to help solve, because judging mate quality is difficult, uncertain, and time intensive. By carefully observing who others are attracted to or are inclined to start a relationship with, learners can save their investments of time and energy and can thus better focus their efforts. To test this possibility, the setup of these experiments is generally pretty similar. You get people to rate single images or videos of a "target" (say, "Ted"), and then expose them to another image or video of the target receiving some kind of attention from a person of the opposite sex (the "model," say Stefania). Depending on the cues, participants will often then rate Ted as more attractive and reveal more interest in a long-term relationship with Ted. As expected from chapter 4, the attraction-enhancing effects emerge or become more pronounced when the model herself, Stefania, is (1) more attractive, (2) older or more experienced than the participant, and (3) smiling at Ted as opposed to having a neutral expression. The effect on people's desire for specific individuals seems clear, but some work also suggests that learners generalize from preferences for specific individuals to traits. So, for example, if Ted dresses in all black and is preferred by Stefania, then a learner may later rate all those who dress in black as more attractive (not just Ted). Empirically, these results have been shown for traits related to dress and hair, and even to a person's eye spacing.[8]

By adding brain-scanning technology to these kinds of experimental setups, neuroscientists have examined the process by which people change their ratings of facial attractiveness in response to cultural learning. In one experiment, male participants rated 180 female faces on a seven-point scale from 1, unattractive, to 7, attractive. After each rating, they were then shown what they believed to be the average rating for that face by other men. In reality, however, on 60 random faces this rating was generated by the computer to be 2 to 3 points higher or lower than the participants' rating. The rest of the time, the "average rating" was calculated to be close to the participant's own rating. Then, a half an hour later, participants underwent brain scanning while they rated all 180 faces again, though no average scores were provided this time. The questions are, how did seeing others' attractiveness ratings influence their subsequent ratings of the *same* faces, and what was going on in their brains?[9]

As usual, participants raised their attractiveness ratings when they saw higher averages from others and reduced them when they saw lower ratings. The brain scans reveal that seeing the different ratings of others altered their subjective evaluations of those faces. Combined with data from other similar studies, it appears that shifting to agree with others is internally (neurologically) rewarding and results in enduring neural modifications that change preferences or valuations. In short, cultural learning changes how people perceive or experience faces based on other people's preferences. These are biological and neurological—but not genetic—changes to what people find attractive.

Existing research suggests similar processes probably occur with wine, music, and other tastes. Wine provides a great case example. The price of a wine, or anything else, represents an aggregate of many other people's judgments. Consequently, learners should attend to price in figuring out what to like, as well as to the preferences of prestigious experts. While in a brain scanner, participants tasted five wines that were labeled and referenced using prices per bottle, specifically, $5, $10, $35, $45, and $90. However, unbeknown to the participants, they actually only tasted three different wines. Two of these three were labeled as "$5" or "$45" and as "$10" or "$90." As usual, people rated the more expensive wines as better (more pleasurable), even though they were actually exactly the same wine.[10]

Here, however, we can have a look at people's brains. By comparing the scans from the same wines at different prices, the results show that people drinking the more expensive wines experienced more activation in their medial orbitofrontal cortex, a region associated with the experience of pleasure or pleasantness for odors, tastes (food and drink), and music. Thus, this work suggests that while price does not affect primary sensory regions of the brain, it changes the valuation of the input from those regions. Sensory input is what it is, but cultural learning affects whether we perceive the same sensory input as better or worse.

This is particularly interesting for wine because in thousands of *double-blind* taste tests with wines ranging from $1.65 to $150, Americans *without* wine training prefer wines that are actually *less* expensive. It takes some training to get people's wine preferences to correlate positively with price.[11] In terms of rational gift-giving strategies, this suggests that when giving wine as gifts to Americans without wine training, you should buy cheap wine, remove any price indications, and tell them it's really expensive wine. This will maximize their pleasure and yours (saving money), assuming you are okay with lying to make others happier.

Overall, this research makes it clear that people's preferences, or tastes, are strongly influenced by observing and inferring the tastes and preferences of others, and that price is one of the cues people use to set their preferences. These effects represent neurological changes, which alter what people find internally rewarding. It would thus be a big mistake to assume people's preferences are fixed or stable. We evolved genetically to have (somewhat) programmable preferences, and modifying our preferences via cultural learning is part of how we adapt to different environments.

Modifying the Hippocampus by Driving in London

In 2009, I was heading to a conference in London to speak at one of the Royal Society's celebrations of Charles Darwin's *Origin of the Species* (1859). After taking the tube (subway) from Heathrow Airport, I jumped into a taxi. We drove for a mile or so and found ourselves stuck in heavy traffic. The driver turned to me and explained that it would take forever to get where I was going because, while my hotel was not far, there was no easy way to get there from our current location. Instead, he recommended that I hop out of the cab, cross the street, enter a pedestrian alleyway, duck under the overhang, make the next left, turn right into another alley ... [a bunch more directions I have now forgotten] ... and my hotel would be on the left. Sure enough, his directions were right on, and I was impressed with his mental map of the city.

Little did I realize at the time, but my driver's brain, and specially his hippocampus, had probably been modified and specialized for London. Becoming a London taxi driver requires passing a stringent set of examinations on how to get around within a 6-mile radius of Charing Cross train station. Passing this set of examinations usually takes between three and four years, as London's downtown is a maze of 25,000+ complex and irregular streets.[12]

Several studies have now focused on London taxi drivers, and all tell the same story. Successfully training to become a taxi cab driver increases gray matter in the posterior hippocampus. The hippocampus is the brain structure where humans and other species store spatial information. More years of experience driving a taxi means more gray matter in this part of the hippocampus. This renovation in brain architecture allows cabbies to demonstrate better cognitive skills at remembering London landmarks and judging the distances between landmarks. But although most other cognitive abilities are unaffected, these new abilities are not without costs, as cabbies get at bit worse at remembering complex geometrical figures.

Being born into a world governed by social norms and reputational consequences means that individuals have immediate incentives to stay in line, follow the rules, and figure out how to excel in locally valued domains, which might involve golf, accounting, reading, abacus use, blow-gun hunting, or the performance of shamanistic rituals. Essentially, social norms create a training regime that every child goes through. Children internalize local standards and engage their minds and bodies according to local norms, equipping themselves with locally valued physical skills and mental tools.

Taxi driving is the tip of an iceberg. These kinds of studies, which are still in their infancy, have already shown that learning to juggle, speak German, and play the piano all have specific effects on gray and/or white matter in various parts of the brain. Such results suggest that our brains are specialized for the skills and demands of the worlds we inhabit. However, unlike other animals, cultural learning and cultural evolution supplies us with the incentives (e.g., reputational improvements), tools (e.g., color terms, books, abaci, maps, and numbers), and motivations for modifying our brains. Recognizing this, the emerging subdiscipline of cultural neuroscience has now begun to show the impact of day-to-day culturally transmitted routines, practices, norms, and goals on our brains. These cultural niches result in different cognitive abilities, perceptual biases, attention allocations, and motivations. Let's consider some work by Trey Hedden and his colleagues.[13]

People from different societies vary in their ability to accurately perceive objects and individuals both in and out of context. Unlike most other populations, educated Westerners have an inclination for, and are good at, focusing on and isolating objects or individuals and abstracting properties for these while ignoring background activity or context. Alternatively, expressing this in reverse: Westerners tend not to see objects or individuals in context, attend to relationships and their effects, or automatically consider context. Most other peoples are good at this.[14]

One consequence of this characteristic is that Westerners are relatively better at judging the *absolute* length of a line, independent of the size of the frame surrounding it. Meanwhile, people who grow up in many East Asian societies (hereafter shortened to "East Asians") are better at judging the *relative* length of a line segment within a frame. That is, they are better at scaling the line to the frame size than Westerners. Because of such results, psychologists say Westerners are less "frame dependent" than most other populations, including East Asians. Figure 14.2 illustrates the difference between absolute and relative judgments. I like this task because it's perceptual and has an objectively correct answer.

Hedden and colleagues put both European-descent Americans and East Asians living in the United States into a brain scanner (fMRI) while they made judgments about whether a newly presented line segment matched the preceding line segment based on either absolute or relative criterion. Crucially, the tasks were designed to be relatively easy, so that with some mental effort anyone could do well. If you make the task harder, then the Americans would outperform most others on the absolute task while getting beaten on the relative task. The ease of the task equalizes performance, which means that any differences in brain activation are not due to different final behaviors or judgments, but only to the mental effort involved and the neurological resources tapped to solve the problem.

Comparisons of the brain scans for each group show that European-descent Americans had greater activation in brain regions associated with effortful control and attention when doing the *relative* task, the task Westerners are typically bad at. The activated brain regions were

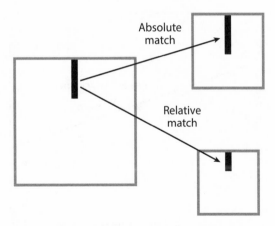

Figure 14.2. Line judgment task. The thick line inside the frame can either be matched with a line of the same absolute length, or the same length relative to the frame.

located mostly in the frontal and parietal lobes. Meanwhile, East Asians showed more activation in these same brain areas when doing the *absolute* task. The upshot is that Americans and East Asians revealed different patterns of brain activation for the identical task—they are neurologically different.

Now, one of the main points of this book is the importance of culture-gene coevolution during our species evolutionary history, which means that seeing differences like this does not immediately imply *cultural* differences. Very different social structures sustained for hundreds or thousands of years could—in principle—have led to genetic differences among populations—remember our discussion of the potential impact of tonal languages on specific genes.

In this case, however, genes play at most a tiny role, and I suspect no role. Here, Hedden's team used a questionnaire to assess how acculturated East Asian participants were to life in the United States. The result, shown in figure 14.3, is a striking negative correlation between acculturation and brain activation patterns. The brains of Asians who felt more acculturated to the United States appeared to find the absolute task—the one Westerners are typically better at—less effortful and engaged in no more effort than many European-descent Americans. Of course, this is not decisive, since genes may be influencing only rates or degrees of acculturation. But when combined with studies of immigrants to the United States and Canada showing that most differ-

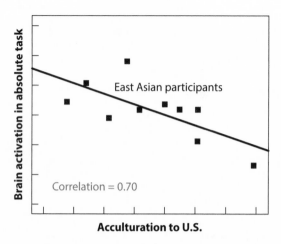

Figure 14.3. Relationship between brain activation in key areas with acculturation to the United States among East Asian participants.

ences are wiped out in a few generations at most, these psychological differences are most likely, or at least primarily, the product of cultural evolution.[15]

Honorable Hormones

You are walking down a narrow hallway. Ahead, blocking your path, is a large man with a drawer pulled out from a filing cabinet. As you approach, he's forced to slide the drawer back in and step against the filing cabinet to permit you to pass. As you pass behind him, he bumps you with his hip and calls you an "asshole" as you walk away. What's your reaction? How much would this bother you?

Well, the answer probably depends on where you grew up. If you are from what have been called "honor cultures," you would likely experience this as a challenge to your manhood (assuming you are male). Both your testosterone and cortisol would spike. Testosterone is a hormone associated with readiness for aggression, and cortisol is associated with stress. Your body would be preparing for a fight. The next person you meet will experience an "I'm not taking any shit" attitude and a very firm handshake to match.

If you are not from an honor culture, you would probably not be much affected by the odd encounter. Your hormones wouldn't spike, and the next person you met won't be put on notice that a new sheriff is in town.

The world is full of cultures of honor, societies with complex sets of social norms that oblige and motivate men and their kin to defend their properties, wives, and families with violence. The social norms here specify that personal or family insults, property damage or theft, and any endangerment to kin jeopardizes a man's reputation unless he replies with immediate and substantial violence, violence that would seem rather disproportionate to those of us from non-honor societies. Flirting with an honorable man's wife or girlfriend will get you punched in the face; anything more might get you killed. The cultural evolution of such norms makes good adaptive sense in a world without formal or effective policing institutions, where property such as cows, horses, sheep, and goats can easily be stolen.

I didn't make up the "asshole guy" at the filing cabinet. This famous experiment was conducted at the University of Michigan by the psychologists Dick Nisbett and Dov Cohen. They didn't have to look far for an exotic honor society, either. Many regions of the Deep South in the United States were colonized by Scotch-Irish immigrants (from Ulster county Ireland) as well as Highland Scots, who brought an honor culture with them. Of course, many such immigrants also settled in the northern parts of the United States, but there they were assimilated in a few generations. In the South, the early dominance of Scotch-Irish in the founding populations has permitted a culture of honor to persist, especially in the rugged hills, swamps, and other rural regions. In their classic book, *Culture of Honor*, Dov and Dick mobilize an immense range of evidence showing how these honor norms and their psychological consequences influence everything from moral judgments and child socialization to gun laws and the spin found in newspaper articles.[16]

More recently, the economist Pauline Grosjean has highlighted the importance of this phenomenon and used it to explain why the Deep South still has twice the murder rate of the rest of the United States. The two counties with the highest murder rates are in Texas and Georgia. At the state level, the Carolinas have the highest murder rates. Using the first U.S. census, from 1790, Pauline obtained the number of settlers who were immigrants from Ireland and Scotland for 150 counties. There had been a massive migration of Scotch-Irish and Highland Scots in the eighteenth century prior to the census. For the Deep South, she then shows that counties that had more Scotch-Irish and Highland

Scotch settlers in 1790 now have much higher homicide rates in the twenty-first century, even after statistically removing the current influences of poverty, race, inequality, etc. This is especially true in more remote counties with rugged terrain, where these immigrants and their descendants could maintain their herding lifestyles and culture of honor far from the formal institutions of the newly formed states.[17]

Wait. Maybe this is genetic, and maybe those Highland Scots brought some "aggression" genes with them, which have persisted in rural parts of the Deep South. Though it's good to put all hypotheses on the table, this one seems rather unlikely since neither high levels of violence nor a culture of honor seems to have persisted back in Scotland (which is loaded with Scots). Scotland has a murder rate more like Massachusetts, about one-third that of America's Deep South. Moreover, in New England and especially the Mid-Atlantic states, where many Highland Scots and Scotch-Irish also settled, there's no relationship between settlers in 1790 and twenty-first century homicide rates. In these places, settlers intermixed with English, German, French, and Dutch settlers and lost their culture of honor.

Here's my point: a package of social norms, which culturally evolved in a world with weak formal institutions, persisted in certain parts of the United States. These social norms evolved to harness men's hormones in ways that shape their behavior, to foster violence in particular "honor" contexts, such as those involving threats to family and property. This culturally constructed biological reaction leads to higher rates of certain specific types of violent crimes in the Deep South. It's a biological difference, but not a genetic difference.

Chemically Inert but Biologically Active

Placebos open a window on culture and biology. Most people have heard about placebos in the context of testing new medicines. In a randomized control trial, one randomly assigned group gets the drug under testing and the other group gets a sugar pill or other inert substance. Both groups are told they will get either the drug or the placebo (sugar pill) based on a coin flip, but they don't know which. The common assumption is that this addresses any response bias, which might cause people to inaccurately report getting better or worse based on their opinions about the drug in question. Since the placebo is chemically inert, it can't actually "do" anything, right?

Wrong. Decades of research now make it clear that this is not the case. Depending on a person's beliefs, desires, and prior experiences, taking a placebo or experiencing any "sham" medical procedure including fake surgery can activate biological pathways in the body. Often these pathways are the very same pathways triggered by the active chemicals in popular drugs. Placebos can reduce pain, activate the immune system, mitigate irritable bowel symptoms, improve motor coordination in Parkinson's patients, and ameliorate asthma. However, the action and effectiveness of a placebo often depends entirely on how much faith a patient puts in a particular placebo or medical treatment. The more you believe it will work, the more it may actually work. Not only that, there appears to be a synergistic interaction between the size of the placebo effects and the size of the chemical effects; that is, the more one believes a drug like morphine will reduce pain (measured by using placebo-morphine), the more effective real morphine actually is. Some drugs don't work at all if administered *without* the patient's conscious knowledge—that is, the drug requires some placebo effect to catalyze the chemical effects.[18]

Culture enters powerfully here because our beliefs and expectations are either set by our own direct experience (what placebo researchers call "conditioning") or by cultural learning. Cultural learning could set the beliefs we bring into the doctor's office, or they could be set by the doctor herself after we enter. Moreover, culture can set up a feedback loop. Suppose we first acquire expectations from our social milieu that a medical treatment is very effective. Then we are treated and get better, in part due to the placebo effect created by those initial expectations. The next time, we bring in both the conditioning from the last treatment (our direct experience of getting better) and again our culturally acquired expectations. This kind of feedback could create either virtuous or vicious cycles, making people more or less likely to get better after a treatment.

Because of this phenomenon, different medical treatments have different levels of effectiveness in different countries. For example, placebo treatments for ulcers in Germany are twice as effective as the same treatment is in neighboring Denmark and the Netherlands. Germans got better 59% of the time on this placebo, whereas the others got better only 22% of the time. Meanwhile, although placebos can lower diastolic blood pressure by 3.5 mm Hg (compared to about 11 mm Hg for chemically active drugs) in many places, this does not work in Germany. Germans worry a lot about having low blood pressure, which no one else seems very concerned about, so the placebo effect may be inhibited by this culturally transmitted concern.[19]

The biology of placebo processes are best mapped out for pain. Some chemically active pain relievers work by firing up the opioid neurotransmission system in brain regions like the dorsolateral prefrontal cortex, anterior cingulate cortex, and nucleus accumbens. Placebo pain relievers fire up these same systems and activate the same brain regions. This effect suggests that placebos are probably doing the same thing biologically as the chemicals; thus, they are *chemically inert but biologically active*. To say placebo effects are "in your head" is merely to group them with so many other chemically active treatments.

Placebos also help patients with Parkinson's disease. Parkinson's patients suffer from three motor symptoms: tremors, muscle rigidity, and slowed movements. When patients believe they are receiving an established and powerful Parkinson's drug, and told that they should experience motor improvements, brain scans show that the placebo treatment causes a massive release of dopamine in a part of the brain called the striatum. Patients who experience a large dopamine release report more motor improvements.[20]

More broadly, the biological pathways used by a range of placebo treatments include suppressing immune responses, releasing hormones like serotonin and dopamine, altering brain activity in particular regions, conditioning opioid receptors in respiratory centers, and reducing β-adrenergic activity in the cardiovascular systems. This last effect of placebos parallels that created chemically by the class of drugs known as beta blockers, which are used to treat many heart problems.

Here's some food for thought. Survival in the years after a heart attack is increased if a patient follows his doctor's orders and takes his medication. In one study, patients who failed to take at least 80% of their medication were dead five years later one-quarter of the time. Those who managed to take 80% or more of their medication were dead only 15% of the time. That makes sense.

Now consider that those who failed to take 80% of the *placebos* were dead 28% of the time, whereas those who strictly adhered to taking their placebos were dead only 15% of the time. So the chemicals in the drug didn't do anything, but it might appear that they did since staying on the medication improves survival. Both believing a medicine will work and being able to stick to a treatment regime doubles one's chances of survival after a heart attack, even if the treatment is chemically inert.[21]

Of course, other studies show that things like beta blockers actually do work better than placebo beta blockers, but strictly sticking to a placebo still *improves* people's chances of survival as much as a beta blocker. Self-regulation, which can improved through religious rituals, can extend your life.

It Hurts So Good

Cultural learning is so powerful that it can both reduce our subjective experience of pain and alter our physiological reactions to pain, and the threat of pain. It can even make pain desirable. For example, runners like me enjoy running, but normal people think running is painful and something to be avoided. Similarly, weight lifters love that muscle soreness they get after a good workout. It feels good to be sore. Parents, like me, have probably had the experience of watching their child fall and then look up at them for a reaction. If the parent has a smile or looks unconcerned, the child may just stand up and press on. If the parent flashes a grimace, empathetically feeling the fall, the kid is more likely to burst into tears and need a hug. Some pain is good for us, so we humans have to learn to distinguish the good pain (working out) from the bad pain (stab wounds). Experimental work shows that *believing* a pain-inducing treatment "helps" one's muscles activates our opioid and/or our cannabinoid systems, which suppress the pain and increase our pain tolerance. In contrast, believing the identical treatment is damaging our tissues results in a different biological response, which lowers our pain tolerance.[22]

My UBC colleague, the psychologist Ken Craig, has directly tested the relationship between cultural learning and pain. Ken's team first exposed research participants to a series of electric shocks that gradually increased in intensity and thus painfulness. Some participants observed another person—a "tough-model"—experience the same shocks right after them, and some did not. Both the participant and model had to rate how painful the shock was each time. The tough model, however, was secretly working for the experimenter and always rated the pain about 25% less painful than the participant did. Then, after this, the model left and the participants received a series of random electric shocks. For this new series of shocks, the participants who had seen the tough model rated them half as painful as those who didn't see the tough model. This is interesting, but I'd worry that the participants were just saying the shocks were less painful so they didn't look wimpy to the experimenter compared to the tough model.

That's not it. Ken showed that this was not some subjective effect on reporting. Those who saw the tough model showed (1) declining measurements of electrodermal skin potential, meaning that their bodies stopped reacting to the threat, (2) lower and more stable heart rates, and (3) lower stress ratings. Cultural learning from the tough model *changed their physiological reactions* to electric shocks. The effect of observing a tough model and inferring their underlying experience is a more potent inducer of placebo effects than mere verbal suggestions. In fact, it's about as effective as direct conditioning.[23]

The Biological Power of Witchcraft and Astrology

A nocebo is the opposite of a placebo. It's a "treatment" in which the "patient" or "victim" has an expectation of getting worse in some way. There is much less research on nocebos than placebos, probably due to the ethical issues surrounding making people sicker and the lack of (obvious) clinical applications. However, the writing is on the wall. Giving people a chemically inert "treatment" that the receiver believes will make them sick often causes a biological response that actually makes them experience negative biological effects. Pain-inducing nocebos activate cholecystokinin and deactivate dopamine in the brains of victims while bringing on anxiety via what's called the hypothalamic-pituitary adrenal axis (HPA). This effect can not only make induced pain worse, it can make normal tactile stimulation painful while increasing anxiety.[24]

Throughout the world today, from Africa to New Guinea, and across much of human history, people have believed that the strong emotions of others, as well as the active casting of magical spells, could cause sickness, injury, and death. Such belief systems go under the labels of witchcraft or sorcery. Of particular concern in many places is being envied by others, which is associated with the "evil eye" from Chile to the Middle East. The reason is often simple: people believe that being envied will negatively affect their health or luck. Because of concerns about envy, people hide their successes and avoid excelling too much or standing out. Of course, given what we now know about nocebos, they were quite sensible to behave this way, given their witchcraft beliefs. If they had excelled, other people would have experienced envy toward them. Accurately perceiving this envy from others while holding witchcraft beliefs could very well provoke biological responses in their bodies, potentially leading to illness and sometimes death in pathogen-rich environments. Witchcraft, then, can actually cause material responses in our bodies if we believe it can.

To see the power of cultural beliefs, consider California in the late twentieth century. In traditional Chinese medicine and astrology, a person's fate is linked to their year of birth, and each birth year is associated with one of the five phases: fire, earth, metal, water, and wood. A person born in a particular year is believed to be more affected by the phase associated with their birth year than other people. And each phase is associated with particular bodily organs or with specific symptoms. Fire, for example, is associated with the heart. Earth is associated with lumps, tumors, and nodules. Thus, a person born in 1908, an earth year, is (believed to be) more susceptible to tumors.

If people believe these associations, then they may act as nocebos. David Phillips and his colleagues tested this idea by comparing the age at death for Chinese and Euro-Americans in California between 1969 and 1990. The prediction is that the combination of getting a particular disease and being born in a year that is believed to be associated with the relevant symptoms or organs of that condition will result in an earlier death. Chinese Americans with an unfavorable disease–birth-year combination can be compared against both Chinese Americans with different disease–birth-year combinations and with white Americans with the same diseases but who presumably don't put much weight on traditional Chinese astrological beliefs.[25]

Figure 14.4 shows the results for bronchitis, emphysema, and asthma, which are associated with the lungs and thus with birth years ending in 0 or 1. The height of the bars shows how many months a person with the unfavorable disease–birth-year combination *loses* from their life. As expected, Americans of European descent in California are not affected, which is crucial since it eliminates the concern that Chinese astrology might be generally true. Looking at all Chinese Americans, females lose four years of life and males lose five. Since many Chinese Americans may no longer put much stock in traditional Chinese astrology, it's worth looking at those living in the larger Chinese communities in Los Angeles and San Francisco and those born in China, because these populations are more likely to have retained traditional beliefs. For men, the picture doesn't change much, but the years of life lost by women go as high as nine years. Similar patterns emerge for cancer and heart attacks.[26]

Growing up in the social and technological worlds built by cultural evolution changes our biology in profound ways, even when it has not yet had time to alter our gene frequencies. Unconsciously and automati-

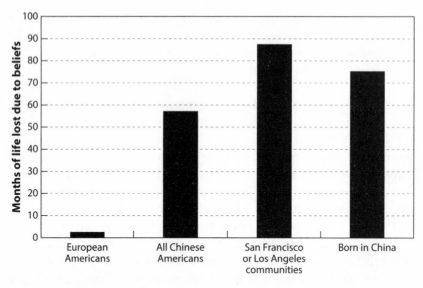

Figure 14.4. Months of life lost due an unfavorable disease–birth-year combination for different populations.

cally, we culturally learn motivations, preferences, and values that change our brains in ways that alter what is neurologically rewarding (what we "like") and endow us with our rapid intuitive responses, as we saw in chapter 11. This does not mean we can culturally learn to like anything, but it's worth noting we can learn to enjoy certain types of pain, including muscle soreness and the burn of a chili pepper. We also culturally acquire mental models or beliefs about how the world works, which direct our attention and generate our expectations. These explain why numerous medical treatments (which are purely placebo effects) and witchcraft often actually work and *cause* important biological effects. Cumulative cultural evolution also produces packages of technologies, practices, and social norms that provide us with the mental and physical tools, as well as the reputational incentives, to change our brains in ways that create whole new cognitive abilities or hone existing capacities. I've illustrated these effects in discussing abilities related to counting, colors, the mental abacus, frame dependence, reading and London cab training, and mentioned many others along the way.

Recognizing the cultural nature of our species means both realizing that human populations are likely to be psychologically rather different along numerous dimensions related to institutions, technologies, and practices and that these psychological differences are ultimately (nongenetic) biological differences. Equating *genetic* explanations with *biological* explanations and distinguishing these from *cultural* explanations makes no sense.

Over the last fourteen chapters, we have explored how natural selection has shaped our psychological capacities for culture; how learning from others across generations produces sophisticated technologies, intricate institutions, complex languages, and vast bodies of knowledge; and how these products of cultural evolution have shaped our bodies, brains, and psychologies, both through development in the short run and by driving our genetic evolution in the long run. Now, in chapters 15 and 16, I will turn to the questions of when this trajectory of culture-driven genetic evolution first launched and why it was our lineage that took off. In chapter 17, I'll bring all these lines together to tackle a series of big questions about the nature of our species, human cooperation, innovation, and the future.

CHAPTER 15

WHEN WE CROSSED THE RUBICON

The crucial event in the history of our species involved crossing the threshold into a regime of cumulative cultural evolution, which has driven human genetic evolution ever since. This is the process that made us uniquely human, but I have not said much about when our lineage made the crossing. When did our ancestors first begin to possess bodies of know-how or tools of sufficient complexity that no single individual could have invented it from scratch on his or her own in one lifetime?

Well, first, there's no reason to suspect it was a singular moment or event. Instead, our lineage probably danced around the starting line for a long time. One group would cross the threshold and begin to very slowly accumulate cultural know-how, only to have those cultural products lost when the group was broken up or decimated by environmental shifts, diseases, or conflicts with nearby groups. Setbacks would also result from migrations to new environments or environmental changes, because any emerging cultural adaptations might not fit the new circumstances or the relevant materials or resources might not be available. For example, groups might learn to control—but not start—fires for cooking. Once they move away from their natural fire sources and their fires go out in a rainstorm, that's it, no more fire. Or a little cumulative cultural evolution might create a nice stone chopper made from a certain kind of volcanic rock, but once the group migrated to a new region, perhaps driven there by climatic shifts, they'd lose access to that type of rock and eventually forget how to make those choppers. With low-quality imitative abilities, error-prone copying, social intolerance, and little or no teaching, accumulating and sustaining cultural adaptations would have been very delicate in the early days. So, when were these early days?

To triangulate in on them, we have three main lines of evidence. Studies of living apes, when used with care, can help us make inferences about what our last common ancestor with chimpanzees was like. That's our starting point. From there, we rely on detailed analyses of the stones and bones unearthed by paleoanthropologists, which permit us to infer aspects of the lives and abilities of our ancestors and how these have changed through time. Finally, comparative studies of genes from humans, apes, and even ancient DNA taken from bones helps us figure out which genes have changed and how they've changed. From analyses of these stones, bones, and genes, we'll look for evidence of cumulative cultural evolution and map out a crude—and *admittedly quite speculative*—time line.

Let's start at the beginning: how cultural was our common ancestor with chimpanzees? Genetic evidence indicates that our lineage split from the line leading to chimpanzees between 5 and 10 million years ago. Since neither chimpanzees nor any other primate have any significant cumulative culture, it's a good bet that our last common ancestor also lacked cumulative culture and wasn't any more cultural than modern chimpanzees. The upshot of this is that we can immediately put down a bookend: our last common ancestor with chimpanzees who lived 5 to 10 million years ago had not crossed the threshold to cumulative cultural evolution.[1]

After about 4 million years ago, the bones tell us that an ape that walked on two legs with a brain somewhat larger than a chimpanzee appeared in Africa. There are a number of species of these apes, but for simplicity, I'll refer to them collectively as *Australopiths*. We don't have tools that can be directly and uncontroversially connected to these guys, but we do have three kinds of indirect evidence: (1) their brain size combined with data on what bigger brains usually imply for primates, (2) cut marks made by stone tools on animal bones, and (3) anatomical changes in their hands, which fit into the account I developed in chapters 5 and 13 on the coevolution of tools and dexterity.[2]

Let's start with brain size. Both field and laboratory studies suggest that primates with larger brains are better at individual learning, social learning, and several other cognitive skills. Among large-brained apes like chimpanzees and orangutans, this research suggests that greater social learning leads to more frequent tool use. Young chimpanzees are more likely to learn to use probes to extract termites, ants, and honey from otherwise difficult-to-access locations and to use stone or wooden hammers to crack nuts open, if they can be with others who are doing these activities. At least one group of chimpanzees even adds a "brush head" to stick-probes that helps them to more efficiently extract termites. Since Australopiths had somewhat bigger brains than chimpanzees (440 cm^3 vs 390), we can infer that Australopiths were better at social learning and were thus probably using more tools, more often, and in more kinds of tasks than any living ape. Tool use by Australopiths was also likely enhanced by the fact that they were primarily ground dwellers (discussed in chapter 16) who walked comfortably on two legs, instead of knuckle-walking like living apes, so they would have had an easier time carrying tools around.[3]

About 3.4 million years ago in Ethiopia, somebody was using stone tools to cut and scrape the meat off a cow-sized ungulate (like a horse or zebra) and a goat-sized bovid (think baby antelope). Stones were also used to smash open bones to get at the rich and nutritious marrow inside. Based on the current evidence about who inhabited this region of Ethiopia at that time, it was probably Australopiths doing this food processing.[4] Now, we don't know if these were stones used as tools or if they were stone tools—that is, stones that had been crafted to have sharp edges or other useful features. The Australopiths might have simply searched around for sharp-edged stones and used those. As a point of reference, chimpanzees do use stones to smash open nuts, but they don't use stones to butcher the animals they kill (and they do hunt), nor do they use stones to crack open bones to get marrow. They also don't get access to anything the size of a cow, so they hunt only small game. I will be summarizing these pieces of evidence in table 15.1 as I go along.

Roughly 200,000 years later, fossils from an ancient hand suggest that the use of such stones as tools may already have been at work shaping the evolution of the anatomy of some Australopith species, tinkering away via natural selection to improve their "precision grip." This is the delicate grip—between our thumbs and fingers—that we use to carefully manipulate tools. It's the configuration that allows us, but not other apes, to thread a needle and to effectively craft stone tools, which were soon to appear in the paleoarcheological record. Unlike chimpanzees, the thumbs of (at least) some Australopiths had lengthened and the fingertips had widened, which substantially improves the precision grip. The bones also show an attachment area for a muscle called the flexor pollcis longus. This muscle, which strengthens the thumb when flexing, is either missing or underdeveloped in living apes. Alongside these changes are others that stabilized, strengthened, and honed the precision grips of Australopiths, including adjustments that gave their hands the distinctive human ability to "cup" and to perform other kinds of grasping movements that apes cannot.[5] These anatomical adjustments in Australopiths appear to be a coevolutionary response to a world in which making and using tools is already more important than anything we see in living apes.[6]

Of course, we need to remain skeptical. This need not be culture-driven genetic evolution. The tools that drove this evolutionary change in the hands of our ancestors may be purely a product of individual smarts—maybe each Australopith is figuring out how to make tools by himself or herself, and social learning played no role. This is possible, but it would mean these apes were unlike both other living apes and humans, whose use of tools, even simple ones, is increased by social learning.

By 2.6 million years ago, the first stone tools appear in the paleoarcheological record. Known as Oldowan tools (after Olduvai Gorge in Tanzania), these are not just stones used as tools, but stones crafted and shaped to be choppers, scrapers, hammerstones, and awls. Detailed analyses and experiments that involve making and using new versions of these tools suggest that different tools were variously specialized for cutting the meat off of large mammals, breaking bones open to access the marrow, scraping hides, cracking nuts, and slicing through tough grasses or reeds. The remains of animal bones left at these sites make it clear that these guys weren't just messing around. They were butchering giraffes, springboks, buffalos, and even the occasional elephant. These tools were made from a variety of stone materials, including obsidian, limestone, and quartz, which were often transported 10 to 20 kilometers from their source.[7]

Keep in mind as I go along that I'm giving the current earliest dates for these tools and anatomical changes. Since the chances that researchers happened to find the actual earliest case of anything is for all practical purposes, zero, we can be fairly confident that whatever we are talking about probably emerged before the date given.

The earliest Oldowan tools don't look impressive, but they are deceptively difficult to learn to make well. The percussive blows used to make them require accurate and powerful targeting at the correct angle. Skilled Oldowan toolmakers, even at this primitive level, needed to have mastered multiple techniques with different materials, including the

Table 15.1. Evidence Relevant to the Beginnings of Cumulative Cultural Evolution

Millions of years ago	Inference to abilities	Evidence	Implication	Most likely species
3.4	Use of stones as tools to process game	Cut marks from stone tools on animal bones	Tools used for butchery in a more sophisticated way than by any living ape	Australopiths
3.2	Emergence of precision grips for tool use	Anatomy of hand bones	Sufficient reliance on tool use to drive anatomical changes	Australopiths
2.6	Fracturing techniques for making stone tools ("Oldowan tools")	Intentionally shaped stones and bones with cut marks	Substantial food processing, with tools used for butchery as well as cutting grasses and sawing wood	Early Homo or Australopiths
2.3	Larger brains with improved sequential and hierarchical processing for tools, language, etc.	Larger crania that show development in Broca's cap	Increasingly complex cultural practices demanding more sophisticated abilities	Early Homo
2.0	Brain lateralization—hemispheric division of labor beginning to emerge	Stone tools made by right-handers	Demands being placed on brains for learning tools and communicative repertoires	Early Homo
1.8	Sharpened bones as tools to access termites, etc.	Detailed use-wear analyses of bones	Access to high-quality food sources and possession of non-stone technologies	Australopiths or Early Homo
1.8	Bigger brains, better at acquiring, organizing and transmitting culture	Larger crania (800 cm3)	Increasing dependence on culturally acquired know-how	*Homo erectus*
1.8	Ability to move into and adapt to diverse, novel environments	Skulls found in Asia and the Caucuses	Reliance on cultural evolution to adapt to novel environments	*Homo erectus*

Table 15.1. (*cont.*)

Millions of years ago	Inference to abilities	Evidence	Implication	Most likely species
1.8	Dependence on food processing, possibly some cooking	Smaller teeth, jaws, and guts; facial shortening	Cultural know-how to find, extract, and process high-quality foods driving anatomical changes	*Homo erectus*
1.8	Accurate and fast throwing abilities, with practice	Anatomical changes in shoulders and wrists	Cultural transmission facilitating the acquisition of throwing skills, which in turn shape anatomy	*Homo erectus*
1.8	Long-distance endurance running or scavenging	Anatomical changes	Requirement for water containers and tracking skills	*Homo erectus*
1.75	Fancier tools: techniques for creating large stone slabs	Large hand axes, cleavers, and picks	More controlled force and the production of large "blanks"	*Homo erectus*
1.7	Greater reliance on acquiring and processing high-quality foods	Narrowing of the pelvis and reduction of cheek teeth	Techniques for the acquisition and processing of meat, marrow, and underground roots/tubers	*Homo erectus*
1.6–1.2	New techniques, materials (stone types and bone) and standardization	Variety of large hand axes, cleavers, and picks	More refined tools worked on both sides, with greater symmetry and edge thinning	*Homo erectus*
1.4	Better precision grip	Changes in finger bones	Sufficient tool use to drive emergence of modern hands	*Homo erectus*
0.85	More sophisticated techniques for making stone tools	3D symmetry and tool thinning	Soft-hammer techniques and more-complex procedures	*Homo erectus*
0.75	Knowledge of cooking, quarrying, plants, fish, and turtles	Hearths, animal and plant remains, tools, big slabs	Dependence on substantial cultural know-how	*Homo erectus*

ability to repair stone cores, which are routinely damaged during the manufacturing process. Even when tutored by human experts and motivated by the need for a cutting edge to access some food, apes cannot learn to make stone tools in the manner, or with the skill, of the Oldowan toolmakers. The famous language-trained bonobo (a type of chimpanzee), Kanzi, was given stones and shown the Oldowan technique by an expert human stone knapper. Kanzi, however, lacked the dexterity to use the technique, so he invented his own approach, which involved using his full force to smash the stone into the floor or another stone. He would then search the refuse for any stone pieces with a sharp cutting edge. These edges got the particular job done that he wanted, but analyses of the stone assemblages he created show they lack many of the key features of Oldowan stone tools.[8]

By around 2.4 million years ago, a bigger-brained (about 630 cm^3) bipedal ape appeared in Africa. These apes, and there may have been more than one species, are typically considered the first members of our genus, *Homo*, so I'll refer to them collectively as Early Homo. Early Homo had smaller jaws and cheek teeth, which had thinner enamel. These characteristics suggests that their anatomy was responding to the spread of know-how about food processing, of which those Oldowan tools likely played a role. Stone tools probably accompanied wooden tools, which were part of a body of cultural expertise. Such know-how would have facilitated locating and digging up underground tubers and roots, scavenging and processing (cutting, chopping, and pulverizing) meat and bone marrow, and finding and extracting termites, ants, eggs, rodents, and honey. Knowledge of seasonal spawning, along with clubs and simple spears, may have permitted Early Homo to obtain large fish, because fish bones appear in many early sites. Similarly, turtle shells suggest that Early Homo may have figured out how to "flip" turtles and then penetrate their shells. The access to, and processing of, these richer resources would have opened the door for brain expansion, as new energy sources became available and less energy was needed for the grinding, cutting, and chopping done by big teeth and jaws, or for the slow breakdown and digestion that occurs in the mouth, stomach, and small intestines.[9]

At some point during this early phase, the gene *MYT16* probably got switched off in our lineage by a mutation on chromosome 7. In living primates, this gene helps build the massive muscles that wrap around primate skulls and allow them to chew tough foods and to bite really hard. In humans this gene is deactivated, because we no longer need the high-powered muscle fibers that it helps build. This genetic deactivation would have been selected for in an ape only if that ape had the tools and techniques for acquiring high-quality foods or for processing them (using stone choppers), as well as for offense and defense (spears, clubs), thus making those big jaw muscles merely extra baggage. Because such muscles are costly to maintain, natural selection will drop them if it can—if it has another solution.[10] Australopiths are interesting in this domain because some species, which coexisted with Early Homo, developed massive jaws and teeth. This change suggests that natural selection solved a similar food-processing problem in later Australopiths—via powerful jaws and teeth—that cultural evolution was solving in Homo.

There is reason to suspect that the overall body of technical skills and know-how was actually much larger than merely making and using stone tools. By 1.8 million years ago, it appears that someone was using bones and horns as tools, perhaps even sharpening these on stones. Careful analyses of the wear on these bones suggests that they were likely used for digging, which probably included opening up termite mounds and excavating subterranean roots and tubers. Some appear polished, possibly having been rubbed against an animal hide. To be clear, these are not the finely crafted bone implements that later signal a rapidly expanding body of Paleolithic cultural knowledge.[11] Nevertheless, they do suggest that the accumulation of "bone know-how" began early. These bits of evidence also point to what I think is an obvious inference: both Early Homo and at least some Australopiths must have had a rich repertoire of tools made from wood, reeds, and hides, not to mention other kinds of socially learned expertise.

Amazingly, we can also make some inferences about the ancient brains of our ancestors around 2 million or more years ago. Analyses of Oldowan tools suggest that 90% of these toolmakers were right-handed. This is odd because apes don't have a preferred hand, and our handedness comes from a division of labor between our brain hemispheres, with the left half focusing on language and tool use in most people. Consistent with this, Early Homo skulls tentatively suggest an expansion in regions known to be important for speech, gesture, and tool use and show an emerging physical separation between the hemispheres. It seems the neurological division of labor that characterizes modern human brains may have already begun to emerge in a pattern consistent with a culture-driven response to the presence of tools and the synergy between tools for things like processing food and those for communicating.[12]

Let's return to our interspecies game of Survivor from chapter 1. Would you have known how to knap stone flakes to aid you in cutting the tough hide and thick tendons on an animal carcass? Would you have known which type of rock to use and where to find it? You can't just pick up any old rock, and chucks of flint, quartzite, and chert are not usually just lying around.

Experiments show that humans are massively aided by opportunities for cultural transmission in learning to make stone tools. However, I suspect that given enough free time and motivation, you could eventually figure all this out. This means that this skill alone is probably not cumulative cultural evolution, at least for us modern humans, since the products of cumulative cultural evolution are sufficiently complex that single individuals cannot figure them out in their lifetime. Of course, it would still be much less costly for you if you could learn from an expert. Nevertheless, for our small-brained ancestors, such tools and techniques would probably have been much harder to figure out by themselves, so these might have constituted cumulative cultural products.[13]

In this light, Oldowan tools might have persisted and at times been widespread because they were usually socially learned *and* possible for a single individual to occasionally invent on his own. What unifies Oldowan tools as a type is the use of an artificial (conchoidal) fracturing technique on certain stone types. Under the right conditions, with the right types of stone around, it's plausible that smart, terrestrial apes would have figured out this technique. Once produced, other individuals could then have acquired it by social learning, making it more common. The tools persisted through time because, even though populations may have collectively lost the technique and know-how from time to time, someone would soon have reinvented them, and then they could spread locally again. This process could have happened repeatedly.

My sense is that, at this point, we are seeing populations that were dancing on the threshold of cumulative cultural evolution. Forms of social learning were probably important to them for acquiring skills and know-how, and the consequence is that groups had much more know-how than they would have had without such learning. At this stage, social learning may have permitted individuals to acquire different kinds of adaptive know-how, about things such as making stone tools, sharpening bones, and locating key resources (like spawning fish or nesting turtles), than they could have otherwise acquired on their own. But such learning didn't yet permit the sustained accumulations that generate sophisticated tools or complex cultural packages. Any particular cultural adaptation could be figured out in one lifetime, but all of them probably couldn't.

Accumulations come—two steps forward—but then they go—two steps backward—because groups split up, environmental shocks hit, or bad luck strikes. If this is the case, we should expect the archeological record to show variation among sites, with some sites revealing more-complex technologies than others; however, we shouldn't expect a continuous trend toward greater complexity over time. The former is what it would look like when groups are dancing on the threshold of cumulative cultural evolution.

Based on current readings of the Oldowan data by some researchers, this may be the case. Different groups at different times used different knapping strategies and preferred different stone types. Some groups even made superior choppers, which were sharpened to an edge by flaking both sides. But any trend of improvement over time in stone tools is somewhere between fuzzy and nonexistent from 2.6 million years ago to roughly 1.9 million years ago.[14]

By 1.8 million years ago, selection pressures generated by culture-gene coevolution had intensified, and the pace of change picked up. In Africa and then rapidly across Eurasia, a new species of the genus *Homo* was on the move with bigger brains (800 cm^3); a much more modern physique, including a narrower pelvis and longer legs; and often fancier stone tools. For simplicity, I will refer to all varieties of this guys— whether in Asia, Africa, or Europe—as *Homo erectus*.

The anatomy of *Homo erectus* indicates that he'd become dependent on food processing, as discussed in chapter 5. Compared to Early Homo, his cheek teeth, face, and jaws are smaller, all of which indicates that he had less tough chewing to do. Stomachs and intestines don't fossilize, but rib cages and the surrounding bones sometimes do. *Erectus* had a barrel-shaped rib cage like us, which is unlike the funnel-shaped rib cages of chimpanzees and gorillas. This anatomy implies that *erectus* had lost the large guts that apes have and use to digest the raw foods they eat. Thus, *erectus* was becoming increasingly addicted to some combination of the high-quality, hard-to-get, processed foods our species depends on. The cultural products that facilitate this evolutionary change likely included (1) the weapons, strategies, and skills for hunting and scavenging large game; (2) the manufacturing skills, tools, and know-how for butchering big animals; and (3) the knowledge and techniques to locate and unearth underground tubers and roots or obtain honey. There's also a good case to be made that *erectus* had some control of fire and perhaps cooked. How often and how much *erectus* depended on fire is a running debate, but there's some evidence of fire by around 1. 5 million years ago, better evidence at 1 million years ago, and good evidence of both cooking and fire at around 800,000 years ago.[15]

In Africa, *Homo erectus* co-emerges in the paleoarcheological record with new large stone tools, which are distinct from Oldowan tools in that the former required that large stone slabs of certain materials be quarried. Once these "blanks," as they are termed, are created, they can be worked on one or both sides to produce hand axes, cleavers, and picks. Making these tools was getting pretty complex, especially if you include locating the stone (which is often far away), quarrying and transporting the large blanks, and then actually crafting the axe, cleaver, or pick. Just the final part, knapping the blanks, seems to require hundreds of hours for modern humans to learn, even with help. It's far from clear how often an individual could figure all this out on their own in a lifetime. Despite their larger brains, lost European explorers in need of cutting edges didn't just quickly whip up some stone hand axes. In small-scale societies, populations that still make stone tools, which have been studied by anthropologists, rely on years of both cultural transmission and individual practice to make quality stone axes or adzes. And precisely how one quarries those large stone blanks is not obvious, to say the least.[16]

During this period, anatomical changes in the shoulders and wrists of *Homo erectus* suggest the emergence of powerful throwing abilities, a skill useful in hunting, scavenging, and attacking other people. Only humans can throw fast and accurately. By contrast, although chimpanzees do occasionally throw stuff, their tosses are slow and usually inaccurate compared to their physically weaker cousins (us).[17] However, since early throwers probably threw rocks and simple spears, it's not immediately obvious that the evolution of the anatomical adjustments for throwing would have been powerfully spurred by cultural evolution. However, here's the thing: humans don't automatically develop powerful and accurate throwing. People who don't practice throwing while growing up, like most girls in my elementary school, turn out to be rather bad throwers as adults. Now imagine that you are a child growing up in a world where no one throws things, or when they occasionally do, their throws are slow and off target. You might experiment with throwing rocks or sticks at small game or birds, but you'd very likely miss, get frustrated, and give up to pursue some more fruitful avenue. But if you could watch and emulate older models who success-fully brought down birds with fast and accurate pitches, you'd be more likely to persevere through a long period of practice. You also might notice some of the older kids engaging in target practice, which you could copy. Of course, these observations will influence you only if you are a keen cultural learner. To get a throwing tradition off the ground, one or a few unusual or lucky individuals would have to persevere through the practice period without relying on cultural learning, but that's a much less demanding requirement than needing everyone to develop throwing on their own. So cultural evolution could have cre-ated conditions favorable to throwing as a learned practice without hav-ing to first create fancy projectiles.

Similarly, as described in chapter 5, the presence of a combination of know-how related to animal behavior and tracking, as well as the mak-ing and use of water containers, may have opened the door for *erectus* to engage in long-distance persistence hunting or scavenging. Although some of the anatomical adaptations for endurance running appear in Early Homo, *erectus* is loaded with them, including most notably long legs, narrow hips, a large butt, and better heat loss from the head.[18]

Things accelerated from 1.6 to 1 million years ago, as new techniques and materials increasingly appeared among the remains of *erectus* soci-eties. Hand axes, including those from the same site, go from mostly worked on one side (unifacial) to being worked on both sides (bifa-cial). These axes got more refined over time, showing greater symmetry and sharper edges, and are sometimes made of large bones (see figure 15.1).[19] Experimental efforts by modern stone knappers reveal that these ancient tools imply the spread of new techniques that were being added to an expanding knapping repertoire. The new techniques require more-complex recipes, including procedures that prepare the stone for knapping.[20]

As one would expect if cumulative culture was just emerging, there were both variations in techniques and practices across sites, but also uneven accumulations, with technological disappearances being com-mon. Sometimes, even frequently, the large and distinctive hand axes vanished from the record, leaving only simpler stone tools (Old-owan).[21] Populations could lose techniques and their associated tools if they lost contact with the relevant raw materials for a generation or two. And, as with other cultural adaptations, we also should expect fire control and cooking to drop out of the repertoire of specific groups for periods of time. As noted earlier (chapter 5), the knowledge of how to

Figure 15.1. Hand axe made from mammalian long bone from 1.4 million years ago (Ethiopia).

make fire even occasionally dropped out of the repertoire of modern hunter-gatherers.

By 1.4 million years ago, anatomical evidence from fossil finger bones indicates another response to the increasing importance of tools and new tool-related techniques on our lineage's fine dexterity and precision grip. The base of *erectus*'s middle finger had been reshaped to be indistinguishable from ours and different from that of both living apes and the Australopiths. The new design stabilizes the joint during a precision grip, and reinforces the whole hand and wrist.[22]

Although researchers have long argued for an immense period of stasis in the tools and know-how of *erectus*, the emerging evidence is beginning to indicate that this may not be the case.[23] By 850,000 years ago, *erectus*, now with a bit larger brain, is sometimes thinning his large cutting tools and creating greater symmetry. Efforts to reconstruct the techniques used in preparing these beautiful tools suggest the use of a soft-hammer technique, which would have required first making a soft hammer out of bone or antler. Such a change would have required the use of longer procedures, with more steps, and greater technical skill. A bit later, it also appears that different sites used different manufacturing methods and quite different materials to achieve the same ends.

After about 900,000 years ago, the pace of environmental fluctuations picked up. Based on mathematical modeling, this change should have increased the selection pressures for social learning, especially in a species already somewhat dependent on culture. However, shifting environments would have also impeded, stifled, and sometimes set back cumulative cultural evolution. As cultural evolution built adaptations attuned to local circumstances and resources, more frequent environmental shifts would have rendered these cultural adaptations useless by changing the available plants and animals, and altering local weather patterns. Populations would have inhabited environments for several hundred years, building up a suite of cultural adaptations, only to have to move or readapt after an environmental shift. Under such conditions, we should expect more-complex cultural products to pop up from time to time, but it should be patchy, with lots of fits and starts. Since fancy tools, for example, are not the direct product of *erectus*'s big brains or intelligence, any more than our fancy tools are a product of our own smarts, we shouldn't expect to see equally fancy tools everywhere he went (see chapter 12). We shouldn't expect this any more than we would expect the early European explorers who sailed across the globe after 1500 CE to have encountered societies with equal levels of technological complexity or sophistication. Generating and sustaining technological complexity requires collective brains linked by social networks and galvanized by social norms.

Nevertheless, the signature of cumulative cultural evolution seems to emerge at certain times and places. For example, between the Golan Heights and the Galilee mountains (Israel) on the shores of an ancient lake, the site of Gesher Benot Ya'aqov has yielded rich insights into the life of one society around 750,000 years ago. The extensive remains indicate the existence and persistence of hearths and areas for both stone-tool manufacturing and food processing. The inhabitants controlled fire and made a variety of stone tools, including hand axes, cleavers, blades, knives, awls, scrapers, and choppers. Made from flint, basalt, and limestone, tool manufacture was done on-site, often from giant slabs carried in from a distant quarry by a team. Some of the basalt slabs have notches, indicating the use of levers as part of the quarrying process. The basalt is of the highest quality and well quarried, suggesting that someone had a storehouse of know-how on the topic.[24]

The group's diet was diverse and also would have required extensive local knowledge. The stone tools were used to process the carcasses of elephants, deer, gazelles, and rhinos as well as boars and rodents. The cut marks left on deer bones are not much different from the same marks made hundreds of thousands of years later by late Paleolithic hunters. The inhabitants of Gesher Benot Ya'aqov also, somehow, obtained freshwater crabs, turtles, reptiles, and at least nine types of fish, including carp, sardines, and catfish. Some of these fish were big, longer than a meter. On top of this, there were seeds, acorns, olives, grapes, nuts, water chestnuts, and various other fruits. This bounty included the submerged prickly water lily, which grows well away from shore. It also appears that they were cracking nuts open and roasting acorns to remove their shells and perhaps reduce the bitter tannins. They may even have made "popcorn" by roasting the seeds from the prickly water lily, as has been done for thousands of years in India and China.

Clearly, cumulative cultural evolution is up and running at this point, generating more know-how than you, me, or our lost European Explorers could have ginned up in a lifetime. If you aren't sure, go quarry some high-quality basalt slabs (you'll need a lever, I think), make a beautiful hand axe with full symmetry (remember to make that antler or bone hammer first), bring down an elephant (trust your Paleolithic instincts?), butcher the carcass (use the hand axe), make a fire or find naturally occurring flame, and then whip up a raft to paddle out and pick some prickly water lily (you can spot this plant, right?). Then, enjoy the elephant steaks and "popcorn."

Of course, this is not to imply that these ancient humans were like us, but merely that they had crossed the Rubicon and embarked on a genetic evolutionary trajectory that was primarily driven by culture and its products.

If you have any remaining skepticism that the cumulative cultural evolutionary threshold has been crossed by this point, in the next 300,000 years after the activities at Gesher Benot Ya'aqov, *Homo erectus* changed sufficiently, including a brain expansion to 1200 cm^3, to justify a new species name, *Homo heidelbergensis.* This period revealed the first evidence of projectile weapons, including wooden throwing spears with stone points, and the emergence of a variety of techniques for producing stone blades. These techniques are consistent within sites or populations but did vary between populations. Distinct tool traditions and composite tools that exploited natural glues weren't far behind. *Heidelbergensis* also appears to have had ears calibrated like humans to speech sounds, unlike other apes, which suggests that a culturally evolved spoken communicative repertoire had been hammering away at our genes (see chapter 13).[25] Now, many of these bits of know-how and technologies seem to drop in and out of the record and don't make a sustained an enduring appearance until the last 100,000 years. But this is just what you'd expect in a species dependent on a collective brain that was subject to shifting environments, intergroup competition, and challenging ecologies that constrained group size and fractured the social ties between bands.

The upshot of all this is that, based on current evidence, Australopiths probably began to aggregate cultural information more intensively than any other living ape (except us), but that no particular tool or element of know-how was more sophisticated than an individual could invent in his or her lifetime. However, taken together, these aggregations may have at certain times and places been more than any one individual could have figured out. These populations, however, likely stood on the precipice of true cumulative cultural evolution, with slow advances and common retreats. This aggregation of cultural know-how drove the evolution of Early Homo, expanding his brain and reducing his teeth and jaws. By 1.8 million years ago, however, the threshold had probably been crossed, and cumulative cultural evolutionary products were driving the genetic evolution of our genus, shaping our feet, legs, guts, teeth, and brains. Albeit slow by later standards, toolkits improved and techniques were added gradually, though of course, cultural losses and technological setbacks continued, and would continue into the modern world. By 750,000 years ago at Gesher Benot Ya'aqov, there's little doubt that we are dealing with a cultural species who hunts large game, catches big fish, maintains hearths, cooks, manufactures complex tools, cooperates in moving giant slabs, and gathers and processes diverse plants.

The bottom line: cumulative cultural evolution is old in our species' lineage, dating back at least hundreds of thousands of years, but probably millions of years. Now, let's explore why it was our lineage that crossed the Rubicon.

CHAPTER 16

WHY US?

Why was our lineage the one that crossed the threshold of cumulative cultural evolution to enter into a long period of culture-driven genetic evolution? Why did this process only begin in the last few million years? Why not 10 or 200 million years ago? And why haven't other species experienced similar culture-gene coevolutionary trajectories?

The first thing to realize is that our species is probably not the only one whose brains and bodies have been shaped by the importance of social learning. Biologists and primatologists like Kevin Laland, Andy Whiten, Carel van Schaik, and their collaborators have made the case that the evolutionary expansion of brains in many species has been influenced by the tendency of social learning to spread useful behaviors, like tool use and food choices, through populations. Through some combination of individual experience, behavioral flexibility, and social learning, animals with larger brains survive and reproduce more effectively in new environments and can solve more novel problems. Species with bigger brains have higher rates of behavioral innovation and do more social learning. Because bigger brains are costly to develop and program, this approach also explains why bigger-brained species tend to have extended childhoods (juvenile periods) and longer lives.[1] Bigger brains require longer childhoods in order to permit offspring enough time for learning. They also require longer lives for mothers, so they can both care for their offspring during these extended childhoods and still have enough time to have more offspring.

So in one sense our species is just part of a much larger pattern created by an evolutionary process that helps explain why some species have bigger brains than others do. But don't worry, in another sense we are still quite special. In no other living species has this process sparked substantial cumulative cultural evolution and the autocatalytic culture-gene interaction that has powerfully driven our genetic evolution. The reason why other species haven't experienced this may lie in a kind of *start-up problem*.[2] Once cumulative cultural evolution is up and running, it can create a rich cultural world, full of adaptive tools, techniques, and know-how that can more than pay the costs of building and programming up a larger brain designed and equipped for cultural learning. However, in the beginning, there won't be much out there to culturally acquire, and what is there will be simple enough that it will still be learnable by one's own individual learning efforts (without social learning), by trial and error for example. Thus, natural selection may not favor larger brain size or complexity, because brains are costly to develop and program. And even if the costs of brain expansion are manageable, natural selection will then tend to invest in improving animals' *individual* (asocial) learning abilities, since individual learning allows innovation and behavioral flexibility even when no one else around is doing anything worth learning (when social learning isn't useful).[3] Moreover, by improving individuals' abilities to learn on their own, from their environment, natural selection also improves those particular forms of social learning that are parasitic on individual learning, in which individuals are exposed to tools and resources (e.g., termites) by hanging around others but then figure out how to use the tools or access the resources on their own. When natural selection acts on individual learning, it increases some simple forms of social learning to a certain degree, as a by-product.

Thus, the *start-up problem* means that the conditions under which there's enough cumulative culture to begin driving genetic evolution are rare, because, early on, there wouldn't have been enough of an accumulation to pay the costs of bigger brains. And if there had been, the most adaptive investment would be in improved individual learning, not better social learning or eventually cultural learning. So in order for natural selection to favor improved social learning, there has to be a lot of cultural stuff to learn, but if you don't have much social learning, there's unlikely to be much of an accumulation out there to tap.

To circumvent the start-up problem, you need to somehow get so much accumulated know-how into the minds and behaviors of others that individual learning, even with its indirect effects on simple social learning, can't do the job of keeping up with *much better* social learners. The start-up problem explains why we don't see other species with substantial cumulative cultural evolution. It's also why there is a Rubicon, which needs crossing, rather than a convenient evolutionary entrance ramp onto the culture-gene coevolutionary highway. Now, the question is, specifically, how did we first jump-start the cumulative cultural evolutionary engine? Below, I bulldoze ahead and layout the best account I have come up with, which fits the current balance of the evidence; however, this view of human evolution is new enough, that my account must remain rather speculative. I'm confident that my many fine colleagues will correct my inevitable mistakes in the future.

Two Intertwined Pathways Create a Bridge

There are two evolutionary pathways around the start-up problem. First, as I mentioned, if we can somehow expand the size and complexity of a species' cultural repertoire *without* altering its brain size, then there will be more good adaptive stuff in the world to learn from others. If that happens, then having genes that improve social learning will pay for themselves because there will already be plenty of stuff out there to socially acquire. Or we can somehow lower the costs of bigger brains. These costs are in part actually incurred by the mother, since she has to supply longer periods of care while her offspring are booting up that bigger brain for life as an adult. I think both of these avenues were important and were mutually reinforcing. Figure 16.1a sketches the major causal connections in the *know-how pathway*, the first of these pathways. Figure 16.1b overlays the *sociality-care pathway*. Note that these figures only sketch the major causal connections, so some of the connections described below are not depicted. I'll begin by discussing why starting with a large ground-dwelling primate is important, then add predation and intergroup competition to this, and finally bring in fluctuating environments. These three exogenous factors, highlighted in bold in the figure, sparked the evolutionary processes that run through our two intertwined pathways. Now, let's examine the two pathways.

Large, Ground-Dwelling Primates Produce Bigger Cultural Accumulations

Primates evolved hands for eating and hanging around in trees as well as for traveling. But when they descend from the trees to the ground, the possibilities created by having hands open up. For some primates, spending time on the ground—terrestriality—fosters the development of more tool types and more-complex tools, and a greater spread of those skills by social learning. The advantages of terrestriality favoring this are straightforward. On the ground, individuals can have both hands free and often have greater access to more types of resources, such as insects (e.g., termites and ants), nuts, stones, sticks, reeds, grasses, various leaves, and water. Without both hands and various items at the ready, tools with different parts or procedures involving multiple tools or steps are harder to figure out. This is clear for chimpanzees, who are semiterrestrial. The tools and procedures they use on the ground are typically more complex than those they use in the trees. And their few complex tree tools are merely reapplications of tools first developed and practiced on the ground. In addition, on the ground individual learners have access to more individuals, more easily, and with less obstructed views. As a result, by hanging around others on the ground, they are more likely to figure things out individually. They can also benefit from tools left lying around by others, which primates readily pick up and monkey around with. In the trees, by contrast, tools typically fall to the ground and thus can't spark learning in other tree dwellers.[4]

The advantages of terrestriality can be seen by comparing the degree of tool use among living primates in situations that encourage or compel them to spend more time on the ground. These include both natural experiments in the wild and comparisons with primates in captivity. Monkeys, for example, generally don't use tools; however, when forced into habitats with few trees, where they must spend more time on the ground, monkeys are more likely to start using tools. Among apes, captive orangutans spend more time on the ground than they do in the wild, and they use tools more frequently and make more-complex tools. In the wild, this may also explain why the tools of largely tree-dwelling orangutans are typically less complex than those of chimpanzees, who spend much more time on the ground.[5]

Here, the effects of spending time on the ground augment the advantages large apes already have over other primates in solving problems on their own, for example, by figuring out how to make tools. So, all other things being equal, bigger-brained primates figure out more stuff that a good social learner might benefit from, and bigger-brained primates who live on the ground generate even more such useful stuff. Remember, we are looking at the conditions under which there might be enough valuable information contained in the behaviors of others

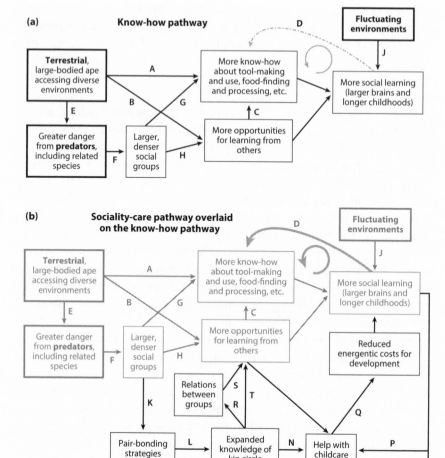

Figure 16.1. The top diagram shows the know-how pathway. The bottom diagram overlays the sociality-care pathway over top of the know-how pathway.

in one's group that natural selection will start favoring larger cultural brains.

Fossil evidence from the likely ancestors of Australopiths indicates that a move out of the trees and onto the ground was well underway by 5 million years ago. By 4.4 million years ago in Africa, there were apes with chimpanzee-sized, or a bit smaller, brains who were adept at both walking on the ground and clambering in the trees.[6] Based on brain size and terrestriality, we'd expect these apes to be good individual learners, and because they have a lot of ground time, they should have a decent repertoire of learned behaviors and skills, perhaps not much different from that of chimpanzees. If some other factor were to increase this terrestrial apes' brain size to match that of chimpanzees, then we'd expect a richer cultural world than observed in chimpanzees due to this ape's greater terrestriality—more ground time, more tools.

The effect of terrestriality on the size of cultural repertoires is depicted in Figure 16.1a by the arrows marked A and B. Arrow C shows how more access to others on the ground will influence the presence of know-how, due to the increased opportunities for social learning. Large-brained apes will have at least some limited social learning abilities because they come free, as a by-product of selection for individual learning. More know-how and more opportunities for social learning will increase the strength of natural selection for genes that make individuals more capable of using social learning to take advantage of this situation. Improved social learning eventually demands larger brains and longer childhoods. The dashed line, D, captures the major feedback we are studying, and the circled arrow represents the possibility of an autocatalytic reaction.

Predation Favors Larger Groups, Which Favor Greater Cultural Accumulations

Increased terrestriality would have exposed these apes to greater danger from predators, as trees are where primates go to avoid many dangers. In Africa, our early ape ancestors would have faced substantially bigger dangers from predators than any modern, jungle-dwelling ape (aside from the dangers posed by humans). The number and variety of large carnivores at this time was about twice that of current levels. On top of the lions, leopards, cheetahs, and hyenas of today, there were several big saber-toothed cats, including two the size of lions and one that probably specialized in ambush hunting from dense vegetation. There were also wolf-like wild dogs and other still-living species of hyenas hunting in packs, as well as an increasing number of venomous snakes. In addition, as we saw for chimpanzees in chapter 10, other ape groups may have created a further predatory threat, as different groups competed for access to fruit trees, recent kills of large game, defensible caves, valuable stone tools, or quarrying sites. And since we eventually see multiple species of Australopiths, there might have been predation from other ape species as well.

Faced with increased predation, mammals often respond behaviorally by forming larger groups, since there's safety in numbers. This behavior suggests that our ancient ancestors might have formed larger groups as a defensive tactic, since they spent more time on the ground.[7] This is shown by pathway E–F in figure 16.1a.

As a by-product of this defensive strategy, larger groups might have caused an increase in the size and complexity of toolkits, skills, and learned bodies of know-how, as larger groups generated, spread, and preserved more innovations and ideas—shown as arrow G in figure 16.1a.

Further fueling the same effect, the pressure from predators could have also suppressed the broad dispersal of individuals for foraging that we see in living apes. By inhibiting dispersals, young apes in the group would have gained more intensive access to more individuals (arrow H). Overall, the increase in the number and complexity of the tools and types of local know-how—including how to find, extract, and process high quality foods and to access water—created by these effects would have increased the selection pressures for building brains better at acquiring, storing, organizing, and retransmitting this cultural information. These abilities would have favored larger brains and longer developmental periods—childhood—for learning all the information available.

Shifting Environmental Conditions Favor More Social Learning

As the above evolutionary processes were unfolding, climatic shifts were causing environmental conditions to fluctuate more rapidly. Mathematical models of evolutionary processes show that a greater reliance on social learning over individual learning is favored when environments destabilize and begin changing over centuries or millennia.[8] After about 3 million years ago, the climate became increasingly variable and showed systematic patterns of cycling until about 10,000 years ago. This pattern suggests that climatic fluctuations intensified the selection pressures for social learning, as forests, lakes, savannahs, and woodlands repeatedly expanded and contracted as the world became generally cooler and drier. Shifting environments would have given our reliance on social learning a first kick between 2 and 3 million years ago, which is when our genus emerged (arrow J). Of course, this kick would have impacted many species, so alone it can't explain why our lineage crossed the Rubicon. But with this kick in mind, let's return to the unfolding evolutionary process.

The Sociality-Care Pathway

While the combination of fluctuating environments and the enrich- ment of the cultural repertoires created by larger, terrestrial groups might be enough to cross the critical threshold (i.e., create feedback loop D), there remains an important obstacle. To create bigger-brained primates, moms need to invest more time and energy in gestation, lacta- tion, and care for their offspring while they are growing up and doing all that social learning and practicing to tap into the cultural pool of know-how. This means that mom will have to somehow obtain more calories and increase the spacing between her births so she can care for each offspring longer, while they are learning. To accomplish this, she will have to live longer, otherwise she will have fewer total offspring. We can see natural selection operating in this way in living apes. Chimpan- zee moms have to nurse for 5 years, and so have 5 to 6 years between births. The problem is that species who push this too far are more likely to go extinct when droughts, diseases, floods, or famines—shocks—hit that increase adult mortality rates, because the population can't recover fast enough from the shock. For example, if females need to live to an average age of 30 years for a population to reproduce itself and a drought suddenly pushes the average age of death down to 20 years, the popula- tion will start shrinking and may disappear entirely. So, as natural selec- tion pushes brain sizes up to increase individual fitness, it makes the species more vulnerable to extinction from shocks. This sets a limit to how far natural selection can increase brain sizes by reallocating a moth- er's efforts and extending her life. Modern apes are probably up against this limit.[9]

There is, however, a work-around, and it's conceptually simple: other individuals have to help mom raise her kids. If a mother can get some help, she need not space her births so far apart or worry about getting enough calories for herself. For our lineage, an evolutionary pathway to this solution opened up once groups got larger and began to rely a bit more heavily on social learning.

CHAPTER 16

Pair-Bonding, Social Learning, and Families

Increasing group size and social learning about local resources may favor pair-bonding strategies by both males and females. Let's first focus on the effects of group size, as shown in figure 16.1b. In many species, males use force or coercion against other males to secure as many mating opportunities with as many females as they can. A male's dominance ranking is the primary predictor of his mating success and long-term fitness. However, as the density of males increases in a group, the payoffs from using dominance go down, since males have to fight off more competitors and keep track of more females. When this happens, one strategy males can pursue in light of the increased competition in larger groups is pair-bonding. As I explained in chapter 9, in using pair-bonding strategies, males seek to develop ongoing dyadic relationships with females by offering things like meat, protection for her and her offspring, and potentially care for her offspring. In exchange, he gets preferred sexual access to her and more reliable knowledge about her cycles so he knows when she can get pregnant, although, of course, this means he has to hang around her, at least sometimes.

Intriguing evidence for the effect of group size on the emergence of pair-bonding has been gathered for the Ngogo chimpanzee group discussed in chapter 10. As noted, the Ngogo group is unusually large compared to typical wild chimpanzee troops. Due perhaps to the intensified competition of the larger group, detailed analyses of male and female movements across their large territory indicate that some males and females are moving around in pairs. This is odd since chimpanzees are notoriously not pair-bonders. Genetic studies show that these paired males got to preferentially father the offspring of the female he was with. These pairs tended to be relatively physically close to one another and groomed each other. The effects of these relationships were often enduring, with one male fathering three of a female's four offspring over sixteen years. While only about one-quarter of males engaged in this pairing behavior, about half of females did (the community has more males than females). Not surprisingly, it wasn't the dominant males who used this relationship-building strategy, and dominance rank remained an important determinant of male fitness.[10] Nevertheless, these observations suggest that the seeds of pair-bonding may begin to sprout when groups get large.

Now, let's consider how the social learning abilities of large apes influence matters. Social learning means that males may come with a kind of cultural inheritance that may be of value to females. In chimpanzees, males remain in their home group (where they grew up) whereas females are generally immigrants. Analyses show that male chimpanzees tend to follow the foraging patterns of their mothers and may well have acquired tips about when and where to find food in their territory and how to extract it. Thus, by hanging out with a locally experienced male, females get access to his local knowledge (actually, his mom's). In general, in a species with social learning skills in which females arrive from other groups, males have something to offer—local knowledge. This means that females should choose males who were particularly good social learners while growing up and then hang around them for extended periods (to learn).[11] Thus, in a large ape group in which females are immigrants, females will benefit from pair-bonding by getting access to local knowledge (along with the usual protection, food, etc.), and males will benefit from pair-bonding by mitigating fierce male-male competition. And males who are particularly good social learners may disproportionately benefit, since they will be preferred by females shopping for quality local knowledge.

So here's where we're are now: predation favors the evolution of larger, less diffuse groups (arrow F in figure 16.1b), which in turn favors the spread of a pair-bonding strategy (arrow K) as a response to the increased density of individuals and the social inheritance of local knowledge. Next, let's examine how pair-bonding extends the circle of kinship in apes.

Pair-bonding in large primate groups will increase the recognition of blood relatives—particularly siblings, half-siblings, fathers, and perhaps fathers' brothers (uncles). The primatologist Bernard Chapais argues that this happens because primates primarily (though not entirely) figure out who their relatives are by noting who else is hanging around mom. Mom, of course, is unmistakable: she's the one you nurse from. In living great apes, this approach to identifying relatives doesn't produce very much usable kinship. To see why, consider a male chimpanzee. Growing up, he will meet only those brothers or sisters who are close to him in age, though many of these will actually be his half-siblings from different fathers. Any sisters or half-sisters he meets will soon leave to permanently emigrate to another group, and he likely won't see them again. His brothers, as soon as they are old enough, will head out to be with other males, presumably to start learning how to be a successful male and begin building important alliances with other males. So, our young male will see his brothers and half-brothers later in life, but he may not recognize them unless they were close to him in age. He can't spot his grandmothers, since his mother's mother lives in another group, and he has little hope of identifying his father's mother or any of his paternal relatives, since he has little way to determine who his father is. He and his mom will be lifelong kin, but he won't have that much interest in her because she doesn't know how to be a successful male (being a female and all) and she isn't a good choice for mating.[12]

Now, consider a similar ape group but now with pair-bonding. Using the same "hangs around mom" rule, young apes can identify their likely father because he's now hanging around their mom, and father will be able to both identify his offspring and feel more confident that they are actually *his* offspring. Young males now may want to join their older brothers and start hanging around their father to learn about how to be a successful male. When this first begins to unfold, a father won't invest in his offspring, but he will at least tolerate them since by getting to hang around him and observe him, under his protection, they will have an edge later on. Through their association with their father, young males will now also be able to identify both their paternal half-brothers (their father may have other pair-bonds) and their father's brothers, since the father will hang around his brothers and will tolerate offspring from any of his pair-bonds. Because dad's mother knows her son, she can now locate her grandchildren. She can watch which females her son is around and which little ones he tolerates. Overall, pair-bonding holds the potential to transform a few scattered kin-based relationships into something that looks more like a family or a kin network, as the arrow L in Figure 16.1b indicates.

At some point in all this, females began to evolve what researchers call *concealed ovulation*, or *ovulatory crypsis*. In many primates, such as chimpanzees, female bodies unmistakably signal when they are sexually receptive and capable of getting pregnant, sometimes using shiny buttock swellings. This means that once a male has hung around a female long enough, he'll know her cycle, and thus know when it's safe to head off to find some more receptive females or build alliances among males. In humans, however, females are potentially sexually receptive all the time, and males cannot reliably predict when their mate can get pregnant. So, by concealing ovulation at least partially, males are forced to be around their mates more often than they would otherwise and end up engaging in a lot of reproductively unnecessary sex. As a by-product, this extra "hanging around" his mate will further solidify his relationships with any offspring who are hanging around their mother.[13]

Social learning, and later cultural learning, enhances the effects created by pair-bonding and kinship. Think about it like this. A social learner watches how mom responds to and behaves toward various other individuals. The offspring copies mom's actions, and potentially her feelings and emotions, toward these other individuals. If mom treats a new infant with care and feeding, then an older sister might copy mom and provide care and feeding to the infant as well. If mom expresses kindness toward dad and grooms him, then our young learner feels more positive toward this male as well. If nothing else, daughters may copy mom's practices of "hanging around dad," which will put her in contact with her brothers. Meanwhile, a male, once he has identified his father via mom, may copy his father's tendency to hang out with his brothers and his other offspring, thus effectively locating his uncles, brothers, and half-brothers. With more sophisticated cultural learning, a young male will copy his father's feelings toward his associates, which may put him in contact with his father's alliance partners. If his father invests in or protects mom and his sisters, then sons may copy these investing or protective acts or attitudes.[14] The arrow M captures these effects in figure 16.1a. Of course, as we discussed in chapter 4, once natural selection begins to improve and hone cultural learning, learners will update their cultural beliefs, preferences, and strategies from more successful and prestigious individuals.

Expanding one's recognized network of kin also means that learners will now have more opportunities for social learning (arrow T). This is because the affection, or even merely social tolerance, fostered by recognized kin relations allows learners to comfortably hang around older, more knowledgeable individuals and watch what they do. At the same time, these older individuals now have incentives to make what they know available to these youngsters—that is, teaching and the use of communicative cues in cultural transmission (see chapter 13). Increasing opportunities for social learning will increase the strength of natural selection for social learning, which can generate more useful, learnable, knowledge.

Help for Mother and the Division of Information

A combination of pair-bonding and a greater reliance on social learning will create both an expanded circle of kinship, with more enduring relationships, and greater *alloparental care* for offspring. Alloparental care refers to situations in which individuals besides the mother help care for offspring. In nonhumans, alloparents include relatives who have kin-based incentives to help and nonrelatives who are under coercive threat or have no other better option for survival. Among hunter-gatherers, alloparents include an offspring's father (or fathers, see chapter 9), sisters, brothers, aunts, and grandparents from both sides, as well as the mother's friends. Detailed studies of alloparental care in eight small-scale societies show that mothers do only about half of the direct child care. Of the remaining 50%, siblings and grandparents cover about half of that, with fathers, aunts, and everyone else picking up the final quarter. By contrast, other ape mothers do nearly 100% of the direct care. In some human foragers, though not all, aunts, grandmothers, and some others even do 10% to 25% of the breast-feeding.[15]

Pair-bonding means that many relatives who previously would have had little or no recognition of their relatedness can now identify and build relationships with each other. Synergistically, social learning means that anyone who copies mom, like her daughters or younger sisters, will care for the infant in the way mom does. For fathers, knowledge of these kin relationships also means they will now have reason to protect the mother and her offspring. Sons, seeking to learn how to be a successful male, will apprentice with their fathers, or later with other males, and learn to protect and care for family members in adaptive ways. This all contributes to the N-Q path shown in figure 16.1b.

Greater social learning also means that young parents and alloparents can rapidly tap into the know-how of prior generations for finding and extracting valuable food sources, such as meat, insects, nuts, water, honey, and underground tubers (reinforcing arrow Q) to help supply the extra energy needed for the lengthened period of a child's development.

Moreover, sophisticated social learning means that various females have incentives to help mom with her offspring (arrow P); less experienced females gain opportunities for observation and practice with infants, which they can later put to use in caring for their own infants. And mom will be much more inclined to allow others to stay around and learn if they are pitching in. Once social learning become selective, the more successful and prestigious mothers will get more help, as learners seek them out as models and care for their children as a form of deference (see chapter 8). Meanwhile, grandmothers may be inclined both to provide direct care to their grandchildren and to demonstrate to the mother how it's done (teaching). As noted in earlier chapters, grandparents have amassed a wealth of experience and knowledge, and once social learning improves, the door opens, and they have a means to bequeath this cultural inheritance to their children and grandchildren. This happens provided that they can locate those grandkids, which is the evolutionary door that pair-bonding opens.

In thinking about what inexperienced girls and young women might learn from mothers and grandmothers, it bears underlining just how dependent humans are on cultural learning for what should be basic mammalian functions. Human mothers, for example, have to be taught how to breast-feed properly, by latching the infant in a way that prevents the infant from biting through the nipple. Similarly, without cultural transmission, many mothers are inclined to toss away the viscous, yellowish fluid that secretes from their breasts after birth, before the milk arrives. This valuable fluid—colostrum—plays a number of important biological roles, including helping improve an infant's immune system. Nevertheless, many humans intuitively perceive colostrum as "sour milk," which should not be given to infants.[16] Other species do not make this serious mistake.

Of course, later in human evolution, alloparenting would have become increasingly influenced by social norms. For example, among Hadza hunter-gatherers, the evolutionary anthropologist Alyssa Crittenden tells of a young girl who was repeatedly scolded because she refused to assist a mother with her baby. Not only was the young girl scolded by the mother, to whom she wasn't related, but she was ostracized by the other children until she complied with requests for help with infant care.[17] Thus, social norms help explain why more distantly related and unrelated people supply 20% to 30% of direct child care in small-scale societies.

The convergence of cultural learning and alloparenting gives rise to a synergy between the psychological abilities for caring for others and for cultural transmission—both are enhanced by mentalizing or theory of mind (see chapter 4) and by greater prosociality. Alloparents will be more effective when they can assess the desires, beliefs, and goals of those in their care and are motivated to meet those needs. Recent experimental work across fourteen primate species confirms this by showing that those species with more intensive alloparenting are more proactive in helping their group mates. Like alloparents, cultural learners need to infer the goals, desires, beliefs, and strategies of those they are learning from. And teachers, who are often alloparents engaged in cultural transmission, will be aided by the ability to assess what learners know and what they don't, as well as by the inclination to instruct them. Thus, the more intensive alloparenting became, the better equipped a species would be to culturally learn, and vice versa; the evolution of better cultural learning capacities would have enhanced parenting and alloparenting (arrow P), by sharpening mind-reading abilities and shaping prosocial motivations.[18]

The combination of (1) an increasing reliance on social learning and (2) the emergence of pair-bonding and alloparenting may explain the emergence of the division of labor between men and women. However, the key to understanding the division of labor is to recognize that it's rooted in a *division of information*. As cultural information begins to accumulate such that no one individual can know everything, pair-bonded couples can specialize in complementary bodies of culturally acquired skills, practices, and knowledge. Female hunter-gatherers, by virtue of birthing, lactating, and providing primary care for babies, need to focus on learning about infant handling, nursing, weaning, food preparation and processing, and any foraging skills that guarantee a steady flow of calories. Males, by contrast, can specialize in know-how about toolmaking, defense, weapons, hunting, and tracking.

This division of information can be observed in many small-scale societies. Among the Hadza, for example, the anthropologist Frank Marlowe tells of a time he went with three adolescent males to dig up some tubers, a task usually done by women. The trio gathered the tubers and then they, plus Frank, ate them raw. Soon they all got sick. It turned out that the boys either misidentified the tubers or didn't realize that those particular tubers required cooking first. Apparently, this never happens when Hadza women are on the job.[19] Similarly, Fijian men don't know all the fish taboos that protect women and infants against ciguatera poisoning during pregnancy and nursing, and my male professor pals usually don't know about colostrum.

Our biases in social learning mean that at least initially these specializations will be acquired by focusing on and learning from either more successful males or more successful females—the "same-sex" bias we discussed in chapter 4. However, culture-gene coevolution may eventually have endowed males and females with different content biases, making them differentially more or less interested in learning about distinct topics. For example, girls might tend to be more interested in infants, whereas boys might tend to be more interested in projectiles. Supporting this possibility, male infants between six and nine months readily imitate the gentle propulsive actions on a balloon (by patting it around) by same-sex models more than female infants (who don't seem equally fascinated by propulsive movements).[20]

The Beginning of Tribes

Pair-bonds can also socially connect different groups, thereby opening the flow of cultural information and increasing the size and complexity of toolkits by expanding the collective brain (pathway R–S in figure 16.1b). This effect occurs because pair-bonds can establish enduring relationships between females and their brothers, fathers, and uncles. If females depart their home groups when they start looking for mates, they will often end up in nearby groups. As we saw in chapter 10, chimpanzee troops have hostile relations with their neighbors. Such persistent hostile relations inhibit adaptive cultural evolution by constraining the flow of cultural know-how among groups—it constrains the size of collective brains. Now, however, when two groups with pair-bonding encounter each other, sisters and brothers, or fathers and daughters, may recognize each other, and this affiliation may reduce the stress in such intergroup encounters. Moreover, those sisters and daughters may have pair-bonded with a male and had offspring. Fathers and brothers can then identify their nephews and grandchildren, as well as the fathers of those nephews and grandkids, who protect and care for them (and it's probably best not to kill that guy). These social connections may permit groups to more comfortably intermix, potentially sharing things like waterholes or fruit groves.[21]

Practices, tools, and techniques can flow between groups through the social channels created by family relationships. These newly expanded networks will increase the size and complexity of the repertoires of tools, skills, practices, and know-how. So, at this point, there will be more cultural complexity in the world, which means that larger brains can more readily pay for themselves because there's more to learn.

Later, once our ancestors began acquiring packages of social norms that prescribe, extend, and reinforce behavioral patterns, pair-bonds transform into marriages, and fathers into dads (see chapter 9). Cultural evolution has even repeatedly enriched the relationship between a son and his mother's brother, which anthropologists call the avunculate. The avunculate is a special relationship that often helps link different communities. Nephews get special privileges to visit and obtain gifts from these particular uncles (and learn stuff). To emphasize an earlier point, social norms mean that, although these relationships are underpinned by some basic aspects of primate kin psychology, they are now also monitored and enforced by the community, by third parties.[22]

Thus, by intersecting with the know-how pathway, the sociality-care pathway contributes to crossing the Rubicon in several ways. First, by expanding the kinship network, a combination of pair-bonding and social learning both creates more opportunities for learning from others and fosters alloparental care from newly located relatives. Kin bonds increase social tolerance, which, by creating more opportunities for learning from others, fuels the cultural evolution of larger and more complex suites of tools and know-how. Such suites put an even greater premium on equipping individuals with the psychological abilities and motivations to be able to learn from others; eventually, as skills get hard to learn, this situation opens the door to at least some simple forms of teaching, since we now have experienced individuals who recognize their relatives and have various incentives to invest in them. If nothing else, an offspring's kin network won't, and often can't, easily hide valuable know-how, skills, or food-extraction techniques. Second, the care and investment provided by newly bonded relatives and those seeking to learn about infants allows mothers to more quickly get pregnant again while at the same time allowing her offspring the time and opportunities they need to tune into the expanding body of practices, tools, techniques, and expertise that are becoming increasingly available. The result will be larger brains, and the longer childhoods needed to fill those brains with know-how gleaned and honed through a combination of observation and play or practice. Farther down the road, of course, cultural evolution will shape social norms and deploy social technologies like rituals to establish and maintain expanded networks that link groups and individuals across large populations.

Why Living Apes Haven't Crossed the Rubicon

Understanding the importance of group size, social interaction, and pair-bonding for cultural accumulation helps us understand why living apes haven't crossed the threshold into a regime of cumulative cultural evolution. Gorillas, for example, do pair-bond but live in single-family groups with only one male and several females. These groups are too small for cumulative cultural evolution. Similarly, orangutans are rather solitary and don't pair bond, which means that young orangutans often grow up with only their mother to learn from. With little access to others for social learning, it's very hard to generate aggregations of cultural know-how. Chimpanzees are more group oriented but have a fission-fusion form of social organization that still means that young chimps mostly hang around their moms. Studies of how infant and juvenile chimpanzees learn to use stick tools to fish for ants or termites show that young chimpanzees really only have access to mom as a model (90% of the time), though when given the chance, they will watch others, usually older, female relatives.[23] Thus, the narrow evolutionary bridge across the Rubicon I've constructed begins with a large ground-dwelling ape who is forced to live in larger groups (by predation) in which at least some members of both sexes have evolutionary incentives to pair-bond.

In summary, the key to explaining why our lineage crossed the Rubicon and so many other species have not is to understand how we solved the start-up problem: bigger brains calibrated to rely on learning from others can't pay for themselves unless there is already a lot to learn out there in the minds of others. So, assuming we are starting with a creature with good individual learning abilities (e.g., big apes), we first consider the conditions that would favor increasing the amount of know-how that one could potentially learn without increasing brain size. Terrestriality gives a boost by providing better opportunities for individual learning and more chances for social learning. Predation, by forcing primates to live in larger groups for protection, increases the size and interconnectedness of groups. Both terrestriality and predation will thereby increase the size of groups' cultural repertoires and may begin to push a species past the threshold.

Nevertheless, expanding brains to store more cultural information still face a crucial obstacle: the high cost to mothers of investing in offspring who take a long time to grow up and learn what they need to know to survive. Even when this pays in the short run for mothers, it can result in extinction in the long run when famines, droughts, and floods hit. Circumventing this constraint, the larger groups of social learners created by predation can also lead to the spread of pair-bonding strategies, which eventually expand kinship circles, increase social learning opportunities, and favor greater alloparental care. Combined with the increased production of high-quality foods made possible by more cultural know-how (due to more social learning opportunities), the alloparental help delivered by relatives and others to mom and her offspring lower her costs in the short and long run, which permits her to produce children more rapidly. This opens the door for a massive expansion of brains, increasingly selected for their cultural learning abilities.

CHAPTER 17

A NEW KIND OF ANIMAL

The case I've presented in this book suggests that humans are undergoing what biologists call a *major transition*. Such transitions occur when less complex forms of life combine in some way to give rise to more complex forms. Examples include the transition from independently replicating molecules to replicating packages called chromosomes or, the transition from different kinds of simple cells to more complex cells in which these once-distinct simple cell types came to perform critical functions and become entirely mutually interdependent, such as the nucleus and mitochondria in our own cells. Our species' dependence on cumulative culture for survival, on living in cooperative groups, on alloparenting and a division of labor and information, and on our communicative repertoires mean that humans have begun to satisfy all the requirements for a major biological transition. Thus, we are literally the beginnings of a new kind of animal.[1]

By contrast, the wrong way to understand humans is to think that we are just a really smart, though somewhat less hairy, chimpanzee. This view is surprisingly common.

Understanding how this major transition is occurring alters how we think about the origins of our species, about the reasons for our immense ecological success, and about the uniqueness of our place in nature. The insights generated alter our understandings of intelligence, faith, innovation, intergroup competition, cooperation, institutions, rituals, and the psychological differences between populations. Recog-

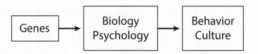

Figure 17.1. Canonical evolutionary approaches to human biology, psychology, behavior, and culture.

nizing that we are a cultural species means that, even in the short run (when genes don't have enough time to change), institutions, technologies, and languages are coevolving with psychological biases, cognitive abilities, emotional responses, and preferences. In the longer run, genes are evolving to adapt to these culturally constructed worlds, and this has been, and is now, the primary driver of human genetic evolution.

This is quite different from the usual story. The standard picture of human evolution proposes a long, fairly boring period of genetic evolution that culminated in an explosion of innovation and creativity sometime after 100,000; 50,000; or 10,000 years ago, depending on whom you are reading. After this, genetic evolution seems to stop and cultural evolution takes over. Culture is then divorced from brains and biology, as well as from genetics, and the rest is history.

More recent efforts to apply evolutionary thinking to human brains, biology, and behavior have made substantial progress, but these approaches still often portray a one-way causal avenue like the one shown in figure 17.1.

To the degree that these older evolutionary approaches recognize culture or cultural evolution, they are seen as relatively recent phenomena, which at most scratch at the surface of a large core of human nature that emerged through purely genetic evolutionary processes.[2] In twenty-first-century textbooks on evolutionary psychology, for example, culture is often politely acknowledged and then waved off as being mostly relevant for explaining phenomena like the "hula-hoop craze, changes in clothing styles or fashion, beliefs about alien beings, and jokes."[3]

While a big advance on prior nonevolutionary views, where culture existed in some ethereal sphere disconnected from genes and biology, these older evolutionary approaches are still off the mark. As we've seen, phenomena like Hula-Hoops could be replaced by Inuit snow houses, adaptive food taboos, composite bows, divination rituals that improve hunting, and the know-how to locate water in the Australian desert; that is, things that directly influence our species ability to survive as hunter-gatherers. It's not just that these older approaches fail to consider some minor influences of culture on biology or some recent and rare feedback loops showing how cultural practices, like drinking cow's milk, have shaped genetic changes; these now outdated evolutionary views fail to recognize that the *central force* driving human genetic evolution for hundreds of thousands of years, or longer, has been cultural evolution. The consequences of this run deep and wide (see table 5.1):

- Many aspects of our physiology and anatomy make sense only as genetically evolved responses to selective pressures created by the cultural evolution of things like fire, cooking, cutting tools, projectile weapons, water containers, artifacts, tracking know-how, and communicative repertoires. Among our numerous features, these help explain our small teeth, short colons, shrunken stomachs, poor plant-detoxification abilities, accurate throwing capabilities, nuchal ligaments (head stabilizer for running), numerous eccrine sweat glands, long postreproductive lives, lowered larynxes, dexterous tongues, whitened sclera, and enlarged brains (see chapters 5 and 13).
- Many of our cognitive abilities and biases make sense only as genetically evolved adaptations to the presence of valuable cultural information (chapters 4, 5, and 7). These evolved mechanisms include our well-honed cultural learning abilities, "over-imitative" tendencies, and folkbiological capacities for organizing and enriching what we learn about plants and animals, among many others.
- Much of our species' status psychology, including our deferential motivations, patterns of mimicry and imitation, facets of pride, cooperative tendencies, and bodily displays, appear to be genetically evolved adaptations to a world in which valuable cultural information was unevenly distributed across the minds of other members of our social groups (chapter 8).
- Our social psychology appears designed for navigating a world with social rules and reputations, where learning and complying with these rules is paramount and where different groups possess quite different norms (chapters 9–11). We internalize costly norms as goals in themselves, usually via cultural learning, and are particularly good at spotting norm violators, even when those violations have nothing to do with cooperation. To make sure we learn

the best norms for our own groups and avoid the dangers of miscoordinating with others, we preferentially use marker traits like dialect and language to distinguish potential models and then preferentially target our cultural learning and social interactions toward those who share our marker traits.

My point is that trying to understand the evolution of human anatomy, physiology, and psychology without considering culture-gene coevolution would be like studying the evolution of fish while ignoring the fact that fish live, and evolved, underwater.[4]

To bring the ideas together that I've developed in this book, let me now pose a series of key questions and lay out my answers.

Why Are Humans Unique?

Of course, humans are different from other animals in myriad ways that are related to our physiology, anatomy, and psychology. We humans are impressive endurance runners, throwers, trackers, communicators (with either spoken or gestural languages), food sharers, teachers, toolmakers, cooks, causal-model builders, mind-readers, and ritual performers, just to name a few. However, rather than picking a particular product, like language, cooperation, or toolmaking, and then backing out an evolutionary story, I began this book by focusing on a type of evolutionary process—culture-gene coevolution—and then sought to track down its implications for our species.

The answer to why humans are different is that we crossed the Rubicon. Cultural evolution became cumulative, and then both this accumulating body of information and its cultural products, like fire and food-sharing norms, developed as the central driving forces in human genetic evolution. We seem so unique because no other living animal has gone down this road, and those that did, such as Neanderthals, were replaced during one of our species' many expansions. In earlier chapters, I've sought to explain how culture-gene coevolution gave rise to that impressive list above. Thus, the key to understanding our uniqueness lies in understanding the process, not in highlighting particular products of that process, like languages, cooperation, or tools.

Having crossed the Rubicon, we can't go back. The impact of this transition is underlined by the fact that, despite our long evolutionary history as foragers, we generally can't survive by hunting and gathering when we have been stripped of that relevant culturally acquired know-how. We saw big-brained explorers repeatedly flounder in environments ranging from the Arctic to the Australian outback. As our heroes sought to confront the recurrent challenges faced by our Paleolithic ancestors, like finding food and water, they struggled. No foraging modules fired up and no fire-making instincts kicked in. Mostly, they just fell ill and died as a result of blunders that any local, indigenous adolescent equipped with cultural know-how inherited from earlier generations would have easily avoided. It's not merely that people in modern societies need culture to survive. Hunter-gatherers, as well as other small-scale societies studied by anthropologists, are massively dependent on large bodies of culturally acquired know-how, related to tracking, food processing, hunting, and tool manufacture. This expertise is often complex, well-adapted to local challenges, and not causally well understood by most practitioners; remember cassava processing to remove the cyanide, corn-ash mixtures to prevent pellagra, and Fuegian arrow manufacturing. All humans societies, whether they live as hunter-gatherers or not, are entirely dependent on culture.

As I noted at the outset of this chapter, humans are at the beginning of a major biological transition, the formation of a new kind of animal. In our species, the extent and sophistication of our technical repertoire—and of our ecological dominance—depends on the size and interconnectedness of our collective brains. In turn, our collective brains depend heavily on the packages of social norms and institutions that weave together our communities, create interdependence, foster solidarity, and subdivide our cultural information and labor. These social norms, which were gradually selected by intergroup competition over eons, have domesticated us to be better rule followers, as well as more attentive parents, loyal mates, good friends (reciprocators), and upstanding community members. Like the cells in our bodies, all human societies possess a division of labor and information, with different subgroups specializing in different tasks and cultural knowledge. The extent of this division, and the interdependence it creates, continues to expand. Similarly, most human societies possess institutions that make group-level decisions—decisions that strongly influence the long-term survival and reproduction of their members—and even sacrifice some members in pursuit of group goals. It appears that as with the emergence of individual organisms from collections of cells, culture-driven genetic evolution is gradually shaping our societies into superorganisms of a sort.

Why Are Humans So Cooperative Compared to Other Animals?

Once cultural learning evolved to the point where people could acquire social behaviors, motivations and rules for judging others' behavior, norms emerged spontaneously. This makes the shared recognition of how people should behave possible and permits the flow of reputational information among individuals who share those standards. At this point, genes had to survive in a dynamic social landscape in which different groups were doing different things, and failure to do those things properly (performing a ritual or sharing food) meant facing reputational damage, diminished mating prospects, ostracism, and, in the extreme, execution by the group. Natural selection shaped our psychology to make us docile, ashamed at norm violations, and adept at acquiring and internalizing social norms. This is the process of self-domestication.

The differences among groups created by cultural evolution and social norms would have created intergroup competition if it hadn't already existed. Various forms of intergroup competition, only one of which involves violence, increasingly favored social norms that fostered success in this competition, which would have commonly included norms that increase group size, solidarity, social interconnectivity, cooperation, economic production, internal harmony, and risk sharing, among many other domains. This process meant that genes would increasingly find themselves fighting to survive in a world of prosocial norms, where narrowly self-interested norm violations were punished. This would have favored genes for a prosocial psychology, prepared for navigating a world where norms about harm and fairness toward fellow community members were likely to be important. In chapter 11, I described hints of these effects from experiments with babies younger than nine months old.

To most effectively construct prosocial norms, cultural evolution often harnesses, extends, or sometimes suppresses aspects of our innate psychology. In chapter 9, we saw that cultural evolution has repeatedly constructed norms that harness our kin psychology, pair-bonding instincts, and incest aversion to weave an expanded kinship network that includes affinal relatives, classificatory brothers and sisters, and uncles whom we call "father." We also saw that cultural evolution can reinforce our pair-bonding psychology to make better fathers, or suppress that psychology and eliminate any social role for a child's genetic father.

The threat of reputational damage and punishment created by social norms for kinship and reciprocity favors genes that further enhanced our evolved kin and reciprocity psychologies. This is how we became more cooperative than other species at both the level of family and friends, as well as at the levels of communities and tribes. Of course, this also creates an enduring tension between these levels that continues to pervade modern life and institutions, and establishes one of the major challenges to well-functioning organizations, governments, and states.

Thus, human societies vary so much in the scale and intensity of their cooperation because different societies have culturally evolved different social norms. Such norms have profound psychological effects, often by tapping into innate mechanisms, that modify our motivations, hormones, judgments, and perceptions, making us more or less cooperative in different contexts.

Overall, this culture-gene coevolutionary process meets the key challenges to explaining the particular nature of human cooperation. It not only accounts for why our species is so much more cooperative than others, but it also explains why human cooperation (1) varies so much across societies and behavioral domains (e.g., food sharing, community defense, ritual participation, etc.), (2) has increased so dramatically in the last 10,000 years, (3) is so readily influenced by cultural learning, (4) relies on the same reputational enforcement mechanisms operating in many noncooperative domains like ritual practices and food taboos, and (5) is sustained by quite different systems of incentives across societies, variously involving rewards, punishments, signals of quality, and selective exploitation of norm violators.[5]

Why Do we Seem So Smart Compared to Other Animals?

The first thing to realize is that you are much smarter than you would otherwise be because you've tapped into and downloaded an immense repository of mental apps from a vast pool of culturally inherited know-how and practices. The full list of things I've discussed is rather long and is scattered over many chapters. Here's a brief list of a few of the mental tools you have at your disposal, which you never would have figured out on your own: base-10 counting, fractions, temporal partitions (minutes, hours, days, etc.), pulleys, fire, wheels, levers, eleven basic color terms, wind power, writing, elastic energy, multiplication, reading, kites, writing, a variety of knots, 3D spatial reference systems, and subordinating conjunctions. And, as I noted in chapter 16, while some of these are installed in your brain as firmware, some are now hardware. Since you are reading this, we can bet that your corpus callosum—the information highway connecting your brain hemispheres—is thicker than it might otherwise be.

Now, if we were to begin moving back in time, century by century, picking adults at random from all living humans and seeing what culturally acquired mental tools they had in their heads, we'd find that the ones now in your head would gradually disappear. Often, we'd find other very cool mental tools and abilities, such as the mental abacus, improved underwater vision, tons of subtle farming heuristics, mental maps for spotting animal tracks, and an acute sense of smell. As I've noted, many of these mental tools come with some costs. For example, acquiring eleven basic color terms improves our ability to distinguish between colors with different verbal labels, although it also degrades our ability to distinguish shades with the same label. And although biological categories can highlight important relationships (penguins are a type of bird, so they lay eggs), these categories can also conceal other relationships (penguins are aquatic, so they have solid bones, unlike most other birds). Nevertheless, it's still the case that past peoples would be less good at doing many of the things we now think are "smart." For example, if we measured IQ scores and scaled them using today's scale, we'd find that at our second stop, in in 1815, the average American would have an IQ below 70.[6]

Recall that we saw just how well less enculturated humans—that is, children—performed against apes, both young and old. If humans were equipped with more powerful genetically installed hardware, we'd expect that these kids, who have much bigger brains than the apes they were competing against, would have mopped up the floor with their hairy brethren. Instead, they mostly tied in a wide range of cognitive domains. Where the kids did excel was in the domain of social learning, as we saw in both chapter 2 and again in the cumulative cultural transmission experiments in chapter 12. Of course, as the kids get older and have time to download those aforementioned apps, they will rapidly surpass the apes in all the cognitive domains. The young apes, by contrast, will get older but no better at these cognitive tasks.

This of course raises the question of where all these elaborate psychological tools, complex artifacts, and training regimes that make us smarter come from, if not from the genius of individual minds. The answer is that they arose via cumulative cultural evolution from our collective brains, often without anyone realizing what was going on. Because of the combination of our powerful cultural learning and our sociality, information in the form of ideas, tools, practices, insights, and mental models can flow among individuals, recombine with other such information, and gradually improve as the selective filters of human learning, reproductive success, and intergroup competition toss some stuff on the trash heap of history and push other stuff onward to the next generation. Unique feats of genius are rare, since once the tectonic forces of cumulative cultural evolution have closed the knowledge gap to a point where one person can step across, intellectual histories show that multiple people often independently manage to take that step.[7]

I illustrated the importance of the size and social interconnectedness of collective brains in Tasmania, Northern Greenland, and Oceania as well as in the laboratory. Not only do larger collective brains generate more and faster cumulative cultural evolution, but if the size or interconnectivity of a group suddenly shrinks, that group can collectively begin losing cultural know-how over generations.

This all means that the power of a group's collective brain depends on its social norms and institutions. This is why, in chapters 9 and 10, I emphasized the importance of affinal relatives (in-laws) and both exchange and ritual ties in hunter-gatherers, because it's these culturally constructed relationships—not genealogical relatedness—that nourish and enlarge their groups' collective brains. Thus, these kinds of social institutions are part of the reason why expanding populations, such as speakers of the Pama-Nyungan, Inuit, and Numic languages, had and maintained a technological advantage over those they replaced. In humans, sociality and technological know-how are intimately interwoven.

Of course, we humans do have superior abilities to build causal models of how the world works. But we should ask why. My view is that *first* cultural evolution began producing increasingly complex practices and technologies, which involved things like chemical reactions (burnt sea shells in maize), compressed air (blowguns), aerodynamics (straight, smooth spears), extended moment arms (spear-throwers), and elastic energy (bows), just to name a few. To be able to more effectively learn and retransmit these valuable bits of culture, our species needed to be able to "back out," or reverse engineer, what I've called mini causal models. These help the learner, in adapting to a variety of circumstances, make sure he or she is achieving the required goals or subgoals for the task. For example, straightening a spear shaft often requires a complex process involving soaking, heating, rubbing, and polishing. A learner has to understand that the procedure is in the service of straightening, balancing, and smoothing the spear, because straighter, smoother, and more balanced spears can be reliably thrown with greater accuracy. With this in mind, someone crafting a spear can periodically examine the shaft for smoothness, balance, and straightness and also test its flight for accuracy. If it's not accurate, she knows what to do: more rubbing and polishing. That's the beginning of a causal model because it realizes that this protocol *causes* (or is supposed to cause) straightness, smoothness, and balance, and these characteristics are then supposed to *cause* predictable flight patterns that favor greater accuracy.

The point is that being able to construct or acquire these mini-models of causality evolved genetically because it improved cultural transmission. The selective pressures that drove this evolutionary process were produced by the emergence of increasingly complex tools, practices, and technologies. By this view, the ability to construct mini causal models didn't *cause* fancy tools and practices. The cultural evolution of increasingly sophisticated tools and practices first drove the emergence of this cognitive ability, and then the two entered into a culture-gene coevolutionary duet. Consistent with this, we saw that observing a demonstrator use an artifact fired up our causal inference machinery more readily than the world simply presenting us with the same causal information.

Overall, we are smarter than other animals for a bunch of reasons, both cultural and genetic, but the ultimate causal explanation is that our species found a bridge across the Rubicon, and cumulative cultural evolution finally got some traction on the other side.

So, yes, we are smart, but not because we stand on the shoulders of giants or are giants ourselves. We stand on the shoulders of a very large pyramid of hobbits. The hobbits do get a bit taller as the pyramid ascends, but it's still the number of hobbits, not the height of particular hobbits, that's allowing us to see farther.[8]

Is All This Still Going On?

Yes, all of this is still going on: Cumulative cultural evolution, intergroup competition, and culture-gene coevolution still continue and have only accelerated in the last ten millennia.[9] With the stabilization of global climates 10,000 years ago and the increased ease of food production, intergroup competition intensified to foster new institutional forms, leading to larger and larger societies with more and more people. This competition eventually favored the rise and diffusion of new social norms favoring trust, fairness, and cooperation with strangers, which were sustained by a diversity of increasingly complex political, religious, and social institutions.[10] Politically, these involved institutions like laws, courts, judges, and police, which backstopped the usual community-level monitoring and punishment that had dominated small-scale human societies for eons. In the domain of religion, new packages of supernatural beliefs, rituals, and norms emerged, spread, and recombined. Over time, these processes gave rise to novel "high gods," who were morally concerned about behavior among coreligionists, even strangers, and increasingly better equipped to monitor (omniscience) and punish (e.g., hell) norm violators. New communal rituals developed and spread that combined cues of prestige and conformity with credibility-enhancing displays (CREDs) to deepen faith in these new gods and to build larger communities of belief that extended beyond the local community or the tribe.[11] The consequence is that modern religions, just like our political institutions, are quite unlike the religions and rituals that existed for most of our species' evolutionary history, although they were all created by the same cultural evolutionary processes.

For social institutions, some ancient societies even began to develop a package of social norms that fostered and enforced rules that limited the number of wives a man, even a rich one, could have to just one (at a time).[12] This is odd, given that 85% of human societies permit men to marry multiple wives. By exploiting various aspects of human psychology, normative monogamous marriage may have spread because it suppressed male-male competition within societies, which reduces crime, violence, rape, and murder while increasing infant health and survival, in part by increasing male investment in children (more on this below).

What is striking about institutions in the modern world, such as those related to marriage and religion, is that most people still don't understand how or why these institutions work or how they harness aspects of our innate psychology and alter our brains and biology. Not much has changed on this front.

Technologically, of course, these larger societies, which were often interconnected by trade and migration especially along latitudinal lines, had larger collective brains and so continued to accumulate increasingly complex tools, technologies, practices, and bodies of know-how. This can be seen by comparing the most complex technological suites found on each continent around 1500 CE. By far, Eurasia had the most complex tools and accumulation of know-how, which arose in part from the synergistic cultural exchanges between places in the Middle East, China, India, and Europe. At the other end were Australia and New Guinea. Australia is the smallest continent, a situation further accentuated by its arid interior deserts, and New Guinea is a mini-continent, or really, a very big island. In the middle were North and South America, which are large but have a north-south orientation that is further exacerbated by the extreme narrowing at the Darien gap in Panama, which impedes land travel even today, and by various mountains and deserts. Until international trade fully opened the oceans and made them highways, our collective brains were constrained by the sizes and geography of our continents.[13]

It bears emphasizing that once the body of know-how becomes sufficiently complex, cultural evolution will often favor an increasingly complex division of labor (really, a division of information). In this new world, the size of the collective brain will be influenced by the size and interconnectedness of people at the *knowledge frontier*, the place at which individuals know enough to have any chance of making improvements on existing forms. For example, suppose I wanted to improve my iPhone, which is now running slow. I could probably figure out how to pry it open, though that's not obvious, and I could start monkeying around with stuff inside. I'd probably try spraying WD-40 multipurpose lubricant inside, to see if that helps (it seems to work on my lawn mower). This approach, however, is unlikely to be successful. The point is that once things get complex, it's pretty hard for the uninformed to improve complex technologies, even with luck. Thus, in more complex societies, technological accumulations will depend heavily on the size and interconnectedness of the subpopulation at the knowledge frontier—that is, the number of people who know enough to potentially take the next baby step or to recognize a lucky mistake. Getting lots of people to the knowledge frontier will depend on the particular cultural transmission institutions (that is, the educational institutions) of the society.

Once we understand the importance of collective brains, we begin to see why modern societies vary in their innovativeness. It's not the smartness of individuals or the formal incentives. It's the willingness and ability of large numbers of individuals at the knowledge frontier to freely interact, exchange views, disagree, learn from each other, build collaborations, trust strangers, and be wrong. Innovation does not take a genius or a village; it takes a big network of freely interacting minds. Achieving this depends on people's psychology, which arises from a package of social norms and beliefs, along with the formal institutions they foster or permit.[14]

With the spread of the Internet, our collective brains have the potential to expand dramatically, although differences in languages will still prevent a truly global collective brain. The other challenge to expanding our collective brains on the Internet is the same one that we've always faced: the cooperative dilemma of sharing information. Without social norms or some sort of institutions, self-interest will favor individuals who cream off all the good ideas and insights from the Web without posting their own good ideas and novel recombinations for others to use. Right now, there seem to be sufficient incentives, often based on acquiring prestige, but that may change as new strategies spread that allow people to get the informational benefits without paying the costs. A key issue will be the degree to which prosocial norms for information sharing can be sustained on the Internet over the long run.

Finally, my undergraduates sometimes ask me if human evolution has stopped or reversed. Their intuition is that natural selection operates on our genes to better adapt individuals to survive in their environments. Since the products of recent cultural evolution now permit us to cure once-deadly infectious diseases, fix once-disabling torn knee ligaments, and artificially inseminate once-infertile couples, natural selection somehow isn't acting to better adapt us to the "natural world," the world without culture. The answer is, of course, that natural selection certainly has not stopped; it has merely changed direction. This, however, is not something new.

To throw this question into stark relief, realize that Early Homo or *Homo erectus* could have asked the same question. These apes were gradually losing their big teeth, powerful jaw muscles, and large digestive systems while they were becoming addicted to high-quality foods and food-processing technologies, which required the know-how to find tubers and involved manufactured cutting tools and, at some point, fire and cooking. Some precocious *erectus*, with the same confusion as some of my students, might have worried that natural selection had stopped or reversed because stone tools and fire were now doing the work that their powerful jaw muscles and big teeth once did. The effect of culture on human genetic evolution is not new. Cultural evolution is just, once again, sending our species' genetic evolution off in some new direction, down a coevolutionary alley that no other species has yet experienced.

How Does This Change How We Study History, Psychology, Economics, and Anthropology?

All this means that to understand human behavior, we need to untangle the ways in which psychology, biology, culture, genes, and history are intertwined. Perhaps the first step is to recognize that people who experience very different institutions, technologies, languages, and religions, just to name a few key domains, will be both psychologically and biologically different, even if they are not genetically different. Ample experimental evidence already shows us how growing up in different culturally constructed environments alters our visual perception, fairness motivations, patience, responses to honor threats, analytic thinking, tendency to cheat, frame dependence, overconfidence, and endowment biases. And all these psychological differences are in some way biological differences.[15]

Thus, to explain much of *human* psychology, and certainly much of the content of current textbooks of psychology, one needs to establish the causal interconnections between the various products of cultural evolution, like institutions (e.g., monogamous marriage) and technologies (e.g., reading), and features of our brains, biology, genes, and psychology. For example, as I mentioned above, getting married in a monogamous society lowers a man's testosterone, reduces his probability of committing a crime, increases his aversion to risk, and may strengthen his ability to defer gratification. In polygamous societies, many poor men cannot get married, because the high-status men attract most of the women as first, second, and third wives, so the crime rates for these unmarried poor men go up, not down. Meanwhile, married men in polygamous societies probably don't have the drop in testosterone because, unlike married men in monogamous societies, they are still openly and actively on the marriage market and testosterone is linked to the pursuit of female romantic partners. This means that monogamous marriage may act as a kind of society-wide testosterone suppression system. As I mentioned above, the psychological effects created by this unusual package of marriage norms may be precisely the reason for its successful global spread in the last few hundred years.[16]

On the technology side, as we discussed in chapter 16, highly literate people, who were rare until relatively recently in human history, have some neuronal rewiring, longer verbal memories, greater brain activation for spoken language, and some losses in the domain of face recognition. The details of writing systems clearly evolved to fit the genetically specified architecture of our brains, but it's worth asking what the implications were when reading—and its related neurological changes—first became widespread, due to the religious convictions that rapidly spread with Protestantism and the printing press, in the run up to the industrial revolution. This changed the brains of many people while opening the cultural transmission pathways among the numerous writers and readers in Europe for the first time. The result was a sudden expansion of collective brains.

Can This Help Us Build Stuff?

When politicians, CEOs, generals, or economists design new laws, organizations, counterterrorism plans, or policies they inevitably import implicit assumptions about human nature. These assumptions often come from a combination of their own life experiences, personal introspection, and our cultural folk beliefs, which often seem rooted in the dreams of some Enlightenment philosopher. Such assumptions have a big impact. Let's look at some examples. In Iraq, in the aftermath of the American military victory in 2003, many assumed that once freed from the dictatorial oppression of Saddam Hussein and presented with new state-of-the-art political and economic institutions imported from the United States and Europe, Iraqis would rapidly take to these institutions and start acting like people in Ohio.[17] That did not happen, probably in part because new formal institutions and organizations have to fit with people's social norms, informal institutions, and cultural psychology.

Aiming to reduce deadly diarrhea, malaria, or sexually transmitted infections, public health specialists have long emphasized the need for "education." Many in public health are (or were) committed to the idea that if people just have the facts, they will behave in sensible ways, meaning they will use toilets, wash their hands, sleep under bed nets, and use condoms. Empirically, however, what surely doesn't work in case after case is providing people with "the facts" or "education," in part because we are selective cultural learners who evolved to acquire practices and respond to social norms. The framing of the message and the messenger are crucial, but the mini causal models (the "facts") are merely secondary—only necessary to support any acquired practices or social norms.[18]

In Haifa, day care centers wanted to get parents to pick up their children on time. At six of these centers, fines were imposed on late parents, following typical prescriptions from economics. If people respond to incentives, then fewer parents should come late if they will be fined. In response, the rate of late parents doubled. Twelve weeks later, when the fine was revoked, parents still kept picking up their children just as late and did not return to the pre-fine levels. That is, the fine made things a lot worse. Adding a fine had apparently changed the implicit social norm, making arriving late merely something that could be purchased with a fee instead of a violation of an interpersonal social obligation that induces shame, embarrassment, or empathy for the staff. I suspect a better approach would have been to go the other way: reinforcing the interpersonal obligation with an explicit social norm and enriching the relationship between the parents and day-care staff.[19]

Underlying these failures is the assumption that we, as humans, all perceive the world similarly, want the same things, pursue these things based on our beliefs (the "facts" about the world), and process new information and experience in the same way. We already know all these assumptions are wrong. In chapter 14, we saw that East Asians' perceptual judgments were different from those of Euro-Americans and that each group struggled neurologically in perceptual tasks that the other did not. In chapter 8, I discussed how and why Angelina Jolie's op-ed flooded medical systems from the United Kingdom to New Zealand with women seeking genetic tests, even though it did not increase anyone's knowledge about either breast cancer or genetic testing. We also saw that exactly the same hallway "bump" fired up men raised in the Deep South, readying them for aggression and violence, while Northerners blew it off. It seems likely that the Northerners wouldn't have predicted the behavior of their classmates from the South, unless they had spent substantial time there visiting. In chapter 11, we saw that social norms become automatic and intuitive motivations, literally wired into our brains.

Once we understand humans as a cultural species, the toolbox for designing new organizations, policies, and institutions begins to look quite different. Here are eight insights drawn from this book.

1. Humans are adaptive cultural learners who acquire ideas, beliefs, values, social norms, motivations, and worldviews from others in their communities. To focus our cultural learning, we use cues of prestige, success, sex, dialect, and ethnicity, among others, and especially attend to particular domains, such as those involving food, sex, danger, and norm violations. We do this especially under uncertainty, time pressure, and stress. If you doubt the power of

cultural learning, remember the celebrity copycat suicides in chapter 4.

2. However, we aren't suckers. To adopt costly practices or nonintuitive beliefs, such as eating a strange food or believing in life after death, we demand Credibility Enhancing Displays (CREDs). Our models must endure costs, such as extreme pain or big financial hits, that demonstrate their deep commitment to their expressed beliefs or practices. CREDs can turn pain into pleasure and make martyrs into the most powerful of cultural transmitters.

3. Humans are status seekers and are strongly influenced by prestige. But what's highly flexible is which behaviors or actions lead to high prestige. People will grant others great prestige for being fierce warriors or gentle nuns. Remember Saint Ambrose, who convinced rich Romans in late antiquity that they had to give their wealth to the poor. Only by giving generously would they prove themselves worthy of the kingdom of heaven. Of course, before Ambrose began this campaign, he gave away most of his substantial wealth (a CRED).

4. The social norms we acquire often come with internalized motivations and ways of viewing the world (guiding our attention and memory), as well as with standards for judging and punishing others. People's preferences and motivations are not fixed, and a well-designed program or policy can change what people find desirable, automatic, and intuitive.

5. Social norms are especially strong and enduring when they hook into our innate psychology. For example, social norms for fairness toward foreigners will be much harder to spread and sustain than those that demand mothers care for their children. Throughout this book, I've discussed norms that have locked into various innate aspects of our psychology, including our favoritism toward close kin, aversion to incest, preference for reciprocity, readiness to avoid meat, and desire for pair-bonding. As we saw, rituals have also evolved culturally to powerfully tap many innate aspects of our psychology.

6. Innovation depends on the expansion of our collective brains, which themselves depend on the ability of social norms, institutions, and the psychologies they create to encourage people to freely generate, share, and recombine novel ideas, beliefs, insights, and practices.

7. Different societies possess quite different social norms, institutions, languages, and technologies, and consequently they possess different ways of reasoning, mental heuristics, motivations, and emotional reactions. The imposition of new formal institutions, imported from elsewhere, on populations often create mismatches. The result is that such imposed formal institutions will work rather differently, and perhaps not at all.

8. Humans are bad at intentionally designing effective institutions and organizations, though I'm hoping that as we get deeper insights into human nature and cultural evolution this can improve. Until then, we should take a page from cultural evolution's playbook and design "variation and selection systems" that will allow alternative institutions or organizational forms to compete. We can dump the losers, keep the winners, and hopefully gain some general insights during the process.

To move forward in our quest to better understand human life, we need to embrace a new kind of evolutionary science, one that focuses on the rich interaction and coevolution of psychology, culture, biology, history, and genes. This scientific road is largely untraveled, and no doubt many obstacles and pitfalls lie ahead, but it promises an exciting journey into unexplored intellectual territories, as we seek to understand a new kind of animal.

NOTES

Chapter 1. A Puzzling Primate

1. This introduction draws on Chudek, Muthukrishna, and Henrich, forthcoming.

Chapter 2. It's Not Our Intelligence

1. See Vitousek et al. 1997 and Smil 2002, 2011. Also, see http://www.newstates man.com/node/147330. Thanks to Kim Hill for pointing me toward this information.
2. By "success" I mean the ecological success of our species in diverse global environments in terms of energy capture.
3. There remains much debate about the forces contributing to these extinctions, including the hypothesis that humans transmitted diseases to megafauna. In total, however, it seems likely that humans made substantial contributions to many of these extinctions, through both direct hunting and indirect effects such as fire (expansive burning in Australia) and other ecological disturbances (competition with other top carnivores). See Surovell 2008 and Lorenzen et al. 2011.
4. Of course, the magnitude and speed of the impacts created by industrialized societies on the earth is unparalleled in the history of our species, or any other (Smil 2011).
5. The ants material is drawn from Hölldobler and Wilson 1990.
6. See discussion in Boyd and Silk 2012. Because our global expansion occurred relatively recently in evolutionary terms, there hasn't been much time for genetic differentiation.
7. See Dugatkin 1999 and Dunbar 1998.
8. The notion that the key to our species' success is our "intelligence" is ubiquitous (Bingham 1999). Most recently, however, this idea has emerged in the work of evolutionary psychologists, such as Barrett, Tooby, and Cosmides (2007) and Pinker (2010). The "on the fly" quotation comes from Pinker, the term "improvisational intelligence" from Barrett et. al. See Boyd et. al. 2011a for further discussion.
9. Pinker 1997: 184.

10. This view is widespread but most recently can be seen in the works of E. O. Wilson (2012) and D. S. Wilson (2005).

11. I'll generally use the word "intelligence" in its common-sense way, unless otherwise stated. Intelligence is a feature of individuals that allows them to figure out new and better solutions to difficult problems. The more intelligent a person is, the better she is able to devise solutions to problems or challenges on her own, including to previously unrecognized problems. We typically do not include the ability to copy from (or imitate) others as part of our "intelligence." For example, children taking IQ tests, or almost any kind of test in school, are prohibited from using their preferred cultural learning strategies (see chapter 4)—copy answers from the smartest kid in the room. Similarly, groups can have *group intelligence*, which aims to measure the group's ability to solve problems. This need not in any direct way reflect the intelligence of its individual members (Woolley et al. 2010). Group intelligence also does not include copying solutions from other groups. Thus, trying to include cultural learning strategies as a type of intelligence violates common usage.

12. These findings are drawn from Herrmann et. al. 2007, 2010. In drawing out this data, I've focused narrowly on the key results relevant to my point and have not, for example, presented the findings related to communication or mentalizing abilities. These would only further underline the overall argument in this book.

13. Actually, in the space subset, there is a tiny increase in performance with age. Older animals perform slightly better (Ester Herrmann, pers. comm., 2013).

14. We should consider three concerns with this study (De Waal et al. 2008). First, the apes may have been at a disadvantage in learning socially because the demonstrators in these tasks were always humans, regardless of the participant's species. However, work by Dean et. al. (2012) shows that using same-species demonstrators does not close the human-chimpanzee gap in social learning. Second, the ape participants were not fully wild but were wild-born orphans who were moved to ape sanctuaries where they were incorporated into mixed-age social groups. This means that (1) they have had extensive exposure to humans and (2) don't face food shortages or serious predation threats. Though this is a real concern, prior work suggests that, if anything, exposure to humans and greater security improves cognitive abilities, especially social learning (van Schaik and Burkart 2011, Henrich and Tennie, forthcoming). Moreover, these sanctuaries provide social groups with access to primary tropical forest, where they spend much of their time. Third, perhaps the apes (who didn't bring their moms with them) were shyer or more uncertain, leading to less impressive results. Herrmann et. al. measured "inhibition" and "temperament," which aimed to capture this shyness. These results indicate that not only were humans more (not less) inhibited than the apes (who were eager), but measures of temperament and inhibition are not associated with performance on social learning. It is also not clear why such differences would only operate in the social learning tasks but not in all the other tests.

15. See Fry and Hale 1996 and Kail 2007.

16. See Inoue and Matsuzawa 2007.

17. See Silberberg and Kearns 2009 and Cook and Wilson 2010.

18. The humans would no doubt counter by complaining that although the chim-

panzees were rewarded with snacks for each correct sequence, the students received no snacks (and may thus have been lacking a key glucose boost). The humans would also argue that Ayumu is clearly a ringer who figured out some secret way of winning that none of his fellow chimps have replicated. Humphrey (2012) provides an interesting discussion of potential issues with this research.

19. See Byrne and Whiten 1992, Dunbar 1998, and Humphrey 1976.
20. See Martin et al. 2014. The average deviation from the Nash equilibrium target was 0.02 for the chimps but 0.14 for the humans.
21. See Cook et al. 2012, Belot, Crawford, and Heyes 2013, and Naber, Pashkam, and Nakayama 2013.
22. On heuristics and biases from psychology and economics, see Gilovich, Griffin, and Kahneman 2002, Kahneman 2011, Kahneman, Slovic, and Tversky 1982, Camerer 1989, Gilovich, Vallone, and Tversky 1985, and Camerer 1995. For asking the question of how we are so well adapted given our apparent irrationality, see Henrich 2002 and Henrich et al. 2001a. For work in nonhumans, see Real 1991, Kagel, McDonald, and Battalio 1990, Stanovich 2013, and Herbranson and Schroeder 2010.

Chapter 3. Lost European Explorers

1. My discussion of the Franklin expedition draws on material from various sources (Lambert 2009, Cookman 2000, Mowat 1960, Woodman 1991, Boyd, Richerson, and Henrich 2011a). Lambert refers to the "Apollo program."
2. The Franklin expedition has been the subject of sustained and intense scholarly interest. Research has suggested that both lead and food poisoning, both associated with the use of the newly canned foods, may have contributed to the expedition's problems. While the lead hypothesis has stood up in tests of the human remains from the expedition, the onset of lead poisoning can only be a relatively small contributor. The food-poisoning hypothesis was not been well supported, though it's not implausible. Neither of these can-related concerns nor the incidence of scurvy would have been a problem if the expedition's men had adopted Inuit lifestyles. The crews of both Ross and Amundsen did well with Inuit foods supplementing their diets.
3. Drawn from Boyd, Richerson and Henrich 2011a.
4. Ibid..
5. It is worth noting that the west side of King William Island is known by the Netsilik to be less productive than elsewhere on the island or nearby regions (Balikci 1989). However, all three of our explorers—Franklin, Ross and Amundsen—ended up in roughly the same area, and Franklin's men located a cairn left by Ross. Moreover, Inuit testimony and archeological remains suggest that Franklin's men eventually broke up into multiple parties and wandered around both sides of the island (Woodman 1991).
6. The quotations and statements on clothes, sledges, and snow houses come, respectively, from Amundsen 1908: 149, 156, and 142.
7. This label was coined by Rob Boyd.
8. This material is drawn from a variety of Burke and Wills resources, including

Phoenix 2003, Henrich and McElreath 2003, and Wills, Wills, and Farmer 1863 and two valuable websites: burkeandwills.slv.vic.gov.au and www.burkeandwills.net.au.

9. Drawn from direct transcriptions of Will's posthumously published journal, the first part of the quotation comes from the entry on June 20, 1861, and the second is from the final entry, which was dated June 26 but may be as late as June 28, 1861. See www.burkeandwills.net.au/Journals/Wills_Journals/Wills_Journal_of_a_trip.htm. Interestingly, the first entry does not appear in its complete form in the version of Will's journal published by his father in 1863. The second part does appear in full on p. 302 (Wills, Wills, and Farmer 1863).

10. Assembled from several sources (Earl and Mccleary 1994, Mccleary and Chick 1977, Earl 1996) and Phoenix's views at http://burkeandwills.slv.vic.gov.au/ask-an-expert/did-burke-and-wills-die-because-they-ate-nardoo.

11. My account is drawn almost entirely from Goodwin's (2008) recent book, with supplement material on the Karankawa from other sources, including www.tshaonline.org/handbook/online/articles/bmk05.

12. Sadly, this heroine arrived at the mission in Santa Barbara only to find herself still isolated, because no one could speak her language. All of her fellow Nicoleños had died of disease or disappeared. She lasted only a few weeks herself, despite much care and attention. My account is drawn from several sources (Hardacre 1880, Hudson 1981, Morgan 1979, Kroeber 1925). This true story was the basis for the well-known novel by Scott O'Dell, *The Island of the Blue Dolphins*. The quoted phrases come from Hardacre's 1880 article in *Schribner's Monthly*.

Chapter 4. How to Make a Cultural Species

1. Boyd and Richerson (1985) built on the pioneering efforts of Luca Luigi Cavalli-Sforza and Marc Feldman (1981), who led the way in modeling cultural evolution as a separate process from genetic evolution. Other key contributors to this early work on this issue include Durham (1982), Sperber (1996), Campbell (1965), Lumsden and Wilson (1981), and Pulliam and Dunford (1980). Intellectual threads can be traced back to James Mark Baldwin (1896). For valuable and insightful overviews see Hoppitt and Laland 2013, Brown et al. 2011, Rendell et al. 2011.

2. Most of the items in this list are covered at some point in this book. For those that are not, see judgment heuristics (Rosenthal and Zimmerman 1978), standards of punishment (Salali, Juda, and Henrich, 2015), and gods/germs (Harris et al. 2006).

3. See Bandura and Kupers 1964.

4. See Henrich and Broesch 2011.

5. For these hunting examples, see Henrich and Gil-White 2001.

6. At the University of British Columbia, I was a faculty member in the Department of Economics and the Vancouver School of Economics for nearly a decade. I've also taught MBAs in NYU's Stern School of Business and been a visiting professor at the Business School at the University of Michigan. Consequently, I'm familiar with both MBAs and economists.

7. See Kroll and Levy 1992.

8. See Henrich and Gil-White 2001, Rogers 1995b, Henrich and Broesch 2011, and N. Henrich and Henrich 2007: chapter 2.
9. Evolutionary models predict that cultural learning should dominate when individual learning is difficult or costly and when learners are uncertain (Hoppitt and Laland 2013, Laland, Atton, and Webster 2011, Laland 2004, Boyd and Richerson 1988, Nakahashi, Wakano, and Henrich 2012, Wakano and Aoki 2006, Wakano, Aoki, and Feldman 2004).
10. Thanks to Michael Muthukrishna for pointing this out. See www.forbes.com/sites/moneybuilder/2013/11/14/investing-with-billionaires-the-ibillionaire-index/.
11. See Pingle 1995, Pingle and Day 1996, Selten and Apesteguia 2005, N. Henrich and Henrich 2007, Fowler and Christakis 2010, Apesteguia, Huck, and Oechssler 2007, Offerman, Potters, and Sonnemans 2002, Offerman and Sonnemans 1998, Rogers 1995a, Conley and Udry 2010, and Morgan et al. 2012.
12. Work on cultural learning has a long history in psychology (Rosenbaum and Tucker 1962, Baron 1970, Kelman 1958, Mausner 1954, Mausner and Bloch 1957, Greenfield and Kuznicki 1975, Chalmers, Horne, and Rosenbaum 1963, Miller and Dollard 1941, Bandura 1977). See Henrich and Gil-White 2001 for discussion and review.
13. Mesoudi and O Brien 2008, Atkisson, O'Brien, and Mesoudi 2012, Mesoudi 2011a.
14. This experiment comes from Kim and Kwak (2011). One might worry that in this particular experiment the stranger was more active than mom, which might have biased the infants' referencing toward the stranger. However, related work with Swedish (Stenberg 2009) and American (Walden and Kim 2005) infants allays these concerns.
15. This is from Zmyj et. al. 2010, but also see Poulin-Dubois, Brooker, and Polonia 2011, and Chow, Poulin-Dubois, and Lewis 2008.
16. In one now paradigmatic experiment, Kathleen Corriveau and Paul Harris (2009b) exposed three- and four-year-olds to two potential models (adults) who gave their opinions regarding the names (linguistic labels) of four common things, like ducks and spoons, with which the children would already be familiar. One of the models accurately named all the items, while the other gave incorrect labels. Then the young subjects saw the potential models name a novel object that was unfamiliar to the children. Having heard both models give different object labels, the kids were asked to label the object. Whom should they believe? It turns out that not only do kids track who is a competent labeler of things in providing their label for the object, but they also remember this information for at least a week: when the same children were retested a week later, without hearing the models label familiar objects again, the children still copied the labels used by the previously more accurate person. The reader should also see Koenig and Harris 2005, Corriveau, Meints, and Harris 2009, Scofield and Behrend 2008, and Harris and Corriveau 2011 on word learning, and Birch, Vauthier, and Bloom 2008 for artifact-function learning. Young children also prefer to learn from more confident models (Birch, Akmal, and Frampton 2010, Jaswal and Malone 2007, Sabbagh and Baldwin 2001).
17. See Henrich and Gil-White 2001 for a review.

18. This experiment is drawn from Chudek et. al. 2012. On adults, see Atkisson, O'Brien, and Mesoudi 2012.

19. A sampling of evidence for same-sex cultural learning biases comes from Bussey and Bandura 1984, Bussey and Perry 1982, Perry and Bussey 1979, Basow and Howe 1980, Rosekrans 1967, Shutts, Banaji, and Spelke 2010, Wolf 1973, 1975, Bandura 1977, Bradbard et al. 1986, Bradbard and Endsley 1983, Martin and Little 1990, and Martin, Eisenbud, and Rose 1995. For recent work in six- to nine-month-olds, see Benenson, Tennyson, and Wrangham 2011.

20. For research on language and dialect cues, see Kinzler et al. 2009, Kinzler, Dupoux, and Spelke 2007, Shutts et al. 2009, and Kinzler, Corriveau, and Harris 2011. Children (Gottfried and Katz, 1977) and adults (e.g., Hilmert, Kulik, and Christenfeld 2006) also seem particularly disposed to learn from those who share their existing beliefs. See Buttelmann et al. 2012 for evidence of selective imitation in infant using ethnic cues (language).

21. Drawn from Hoffmann and Oreopoulos 2009 and Fairlie, Hoffmann, and Oreopoulos 2011, but also see Nixon and Robinson 1999, Bettinger and Long 2005, and Dee 2005.

22. For experimental work in children supporting the effects of age and age vs. competence tradeoffs see Jaswal and Neely 2006 and Brody and Stoneman 1981, 1985. Children can use age in sophisticated ways, sometimes using it as cue to competence, and other times deploying it as a self-similarity cue (VanderBorght and Jaswal 2009, Hilmert, Kulik, and Christenfeld 2006). On the acquisition of food preferences, see Birch 1980 and Duncker 1938. For infants, 14- to 18-month-olds more closely imitate the actions of models that are nearer to them in age (Ryalls, Gul, and Ryalls 2000).

23. Research in small-scale societies examining how senior status influences cultural transmission is only just beginning—see work in the Bolivian Amazon by Reyes-García and colleagues (2008, 2009) as well as my work with James Broesch in Fiji (Henrich and Broesch 2011). However, anthropological ethnography across diverse societies reveals a clear association between age and prestige, and prestige has potent effects on cultural learning. Chapter 8 explains how the rate of change in society influences the link between age and prestige, which explains why the elderly are not particularly prestigious in our own societies.

24. Morgan et. al. (2012) and Muthukrishna et al. (forthcoming) provide the best current evidence of conformist transmission in humans, though also see Efferson et al. 2008, McElreath et al. 2005, 2008, Rendell et al. 2011, and Morgan and Laland 2012. On conformist transmission in fish, see Pike and Laland 2010. For an entry into the theoretical modeling literature, see Nakahashi, Wakano, and Henrich 2012 and Perreault, Moya, and Boyd 2012. This modeling work suggests we should find conformist transmission in many species reliant on social learning.

25. For cross-national data, see: on United States, Stack 1990; on Germany, Jonas 1992; and on Japan, Stack 1996. For evidence of prestige and self-similarity effects as well as copying methods, see Stack 1987, 1990, 1992, 1996, Wasserman, Stack, and Reeves 1994, Kessler and Stipp 1984, and Kessler, Downey, and Stipp 1988.

26. For an overview, see Rubinstein 1983. For evidence from the United States of adolescent epidemics, see Bearman 2004.

27. See Chudek et al., n.d., Birch and Bloom 2002, Barrett et al. 2013, Scott et al. 2010, Hamlin, forthcoming, Tomasello, Strosberg, and Akhtar 1996, Harris and Corriveau 2011, Corriveau and Harris 2009a, Koenig and Harris 2005, Buttelmann, Carpenter, and Tomasello 2009, and Hamlin, Hallinan, and Woodward 2008.
28. See Byrne and Whiten 1988 and Humphrey 1976.
29. Humphrey (1976) sketched both the Machiavellian intelligence hypothesis (Byrne and Whiten 1992) and the cultural intelligence hypothesis (Herrmann et al. 2007, Whiten and van Schaik 2007).
30. See Schmelz, Call, and Tomasello 2011, 2013, Hare et al. 2000, and Hare and Tomasello 2004.
31. See Heyes 2012a. Of course, showing that something can be influenced by experience doesn't tell you much at all about whether its development has been fostered or shaped by natural selection.
32. See Heyes 2012b.
33. See Whiten and van Schaik 2007 and van Schaik and Burkart 2011.
34. One of the debates in this literature involves opposing "innate" and "learned" in explaining our abilities and behaviors. As we'll see, much behavior is both 100% innate and 100% learned. For example, humans have clearly evolved to walk on two legs, and it's one of our species' behavioral signatures. Yet we also clearly *learn* to walk. From natural selection's point of view, it only cares that the phenotype it "wants" emerges when it needs it. To get there, it will use learning, attention biases, motivational changes, anatomical adjustments, inferential biases, and pain responses to make sure that the required developmental processes run to completion, on schedule. Thus, showing that something is learned only tells us about the developmental process but not about whether it was favored by natural selection acting on genes. For example, many people throughout history have had to figure out sexual intercourse on the fly, with no information from other people, so they clearly had to learn about it for themselves. Yet, to suggest that it has not been shaped by natural selection seems unlikely, despite the importance of learning in the process. To canalize learning about intercourse, natural selection made some things "feel right" and other things not so much. Consequently, most couples can eventually figure out what to put where and for how long, at least well enough for natural selection's purposes. Despite the importance of learning for both walking and sexual intercourse, there are no remote societies studied that only hop or crawl or that don't make babies. For evidence of differences among human populations in cultural learning, see Mesoudi et al. 2014.

Chapter 5. What Are Big Brains For?
Or, How Culture Stole Our Guts

1. See Tomasello 1999. Other important recent efforts to understand the role of cultural evolution and its influence on genetic evolution are Sterelny 2012a and Pagel 2012.
2. See Roth and Dicke 2005, Lee and Wolpoff 2002, and Striedter 2004.
3. The data in figure 5.2 are drawn from Miller et. al. 2012. I averaged the fractions

for each of the brain regions in their table S2. I have two concerns about these data. First, the samples are small. Second, it is not clear to what degree these differences may be due to the relatively enriched environments that humans experience compared to chimpanzees.

4. See Sterelny 2012a.

5. See Campbell 2011, Thompson and Nelson 2011, Kaplan et al. 2000, Bogin 2009, and Nielsen 2012.

6. Clancy, Darlington, and Finlay (2001) compared the timing of 95 neurological events across nine species to show the developmentally advanced state of human baby brains at birth. Hamlin (2013a) showed that eight-month-olds use intentions in judging others.

7. The material on food processing and cooking is drawn principally from Wrangham 2009, Wrangham, Machanda, and McCarthy 2005, and Wrangham and Conklin-Brittain 2003.

8. On fire starting among the Tasmanians, Sirionó, and Andaman Islanders, see Radcliffe-Brown 1964, Holmberg 1950, and Gott 2002. Kim Hill told me about the Northern Aché.

9. See Aldeias et al. 2012 and Sandgathe et al. 2011a. Of course, this claim about Neanderthals is controversial (Sandgathe et al. 2011b, Shimelmitz et al. 2014). From my point of view, however, much work in paleoarcheology is plagued by the assumption that once some tool or technology appears in the material record, then our lineage forever has it. As you will see in chapter 12, this assumption is dubious and arises from thinking about tools and technologies as the product of individual cognitive abilities rather than as products of cultural evolution.

10. This schooling in fire control was particularly discouraging since I had been an Eagle Scout and thought I knew something about campfires. Feeling like a dim-witted child turned out to be a recurrent experience in my anthropological fieldwork.

11. On polar bear livers, see Rodahl and Moore 1943. The same may be true of marine mammals.

12. Other species also have a taste for cooked food (Felix Warneken, pers. comm., 2012), and this probably served as a kind of preadaptation that paved the way for cooking (Wrangham 2009). Like other animals, we generally prefer foods that are easier to digest.

13. See Fessler 2006.

14. For evidence on the effects of food processing with stone tools, see Zink, Lieberman, and Lucas 2014.

15. See Noell and Himber 1979.

16. See Leonard et al. 2003 and Leonard, Snodgrass, and Robertson 2007.

17. Also, avoid any kind of jumping contests with members of the genus *Pan* (Scholz et al. 2006), which includes chimpanzees and bonobos.

18. See Striedter 2004 on brains and dexterity. On throwing, see Roach and Lieberman 2012, 2013, and Bingham 1999.

19. See Gelman 2003, Greif et al. 2006, and Meltzoff, Waismeyer, and Gopnik 2012.

20. Horses are tough to beat, though they can be beaten, as shown by the 22-mile

Man versus Horse Marathon, which is held annually in Wales. See http://news
.bbc.co.uk/2/hi/uk_news/wales/mid_/6737619.stm.

21. The material on endurance running draws on Bramble and Lieberman 2004,
Lieberman et al. 2009, 2010, Carrier 1984, Heinrich 2002, and Liebenberg 1990,
2006. For a breezy introduction, see McDougall 2009.

22. This does nothing for walking but cuts the metabolic costs for running in half.

23. From Liebenberg 2006, Heinrich 2002, and Falk 1990.

24. See Carrier 1984 and Newman 1970.

25. See Liebenberg 1990 and Gregor 1977. It is also likely that foragers can identify
humans by their tracks. Numerous ethnographies, as well as my own experience
in the South Pacific, attest to people's abilities to identify individuals by their
tracks. When the tracking expert and field ethnographer Louis Liebenberg asked
!Xo foragers in the Kalahari desert if they could identify individual animals by
their tracks, the group of hunters laughed at such a stupid question. They wondered how someone could *not* be able to distinguish individuals by their tracks.
In the South Pacific, during many years of walking along the beaches of Yasawa
Island, I've also noticed that many villagers have an uncanny ability to predict
who we will see when we round into the next cove based entirely on footprints.
I've even formally tested villagers by asking them to identify a set of footprints
that I had a villager make in secret (and not tell anyone). I ran ten random adult
villagers through my one question test and all ten got the answer correct.

26. See Heinrich 2002 and Carrier 1984.

27. See Liebenberg 1990, 2006. For a video clip of persistence hunting, see http://
www.youtube.com/watch?v=826HMLoiE_o.

28. From conversations and correspondence with Dan Lieberman (2013–14).

29. See Atran and Medin 2008, Atran, Medin, and Ross 2005, Lopez et al. 1997,
Atran 1993, 1998, and Medin and Atran 1999.

30. See Atran, Medin, and Ross 2004 and Atran et al. 2001.

31. See Gelman 2003, Lopez et al. 1997, Coley, Medin, and Atran 1997, Atran et al.
2001, 2002, Wolff, Medin, and Pankratz 1999, Medin and Atran 2004, and Atran,
Medin, and Ross 2005.

32. See Wertz and Wynn 2014a, 2014b.

33. The cognitive system for learning about animals also possesses other adaptive
content biases, which focus learners on certain kinds of information and away
from particular kinds of mistakes. Clark Barrett, James Broesch, and I have explored this topic using a teaching-and-recall task with children and adults from
Fiji, the Ecuadorian Amazon, and Los Angeles. We gave children and adults information about animals that they'd never encountered before, using pictures as
visual aids. We then tested what they remembered both immediately after the
teaching phase and then a week later. Our results suggest that often children
preferentially recalled the information about dangerousness over other kinds of
information, such as that pertaining to the species' habitat or diet. Moreover,
when our participants made mistakes of recall about an animals' dangerousness,
they tended to more frequently recall an animal as dangerous (when it was not)
than mistakenly recall it as safe (when it was actually dangerous). Thus, our recall system is adaptively biased to avoid the more costly mistake of thinking a

dangerous animal is safe rather than vice versa (Barrett and Broesch 2012, Broesch, Henrich, and Barrett 2014). Similarly, in the domain of food, Dan Fessler has argued we have an evolved readiness to acquire avoidances for animal foods (e.g., beef) due to the threats posed by pathogens over our evolutionary history. This may explain why taboos on animal foods are so common cross-culturally, and why vegetarians are relatively common but few people taboo vegetables (Fessler 2002, 2003, Fessler et al. 2003).

Chapter 6. Why Some People Have Blue Eyes

1. Thanks to Matt Ridley for pointing me to this case. See Kayser et al. 2008.
2. See Jablonski and Chaplin 2000, 2010.
3. Eiberg et al. 2008, Sturm et al. 2008, Kayser et al. 2008. Also see http://essays .backintyme.biz/item/4. The genetic variant for blue or green eyes could have been favored by natural selection directly or by sexual selection indirectly. Via sexual selection under the conditions described, any preferences for mating with those with blue or green eyes would have made it more likely for those with such preferences to have children who were better able to synthesize vitamin D from the sun. Such preferences could evolve either culturally or genetically, or both.
4. See Carrigan et al. 2014.
5. See Tolstrup et al. 2008, Edenberg et al. 2006, Danenberg and Edenberg 2005, Edenberg 2000, Gizer et al. 2011, Meyers et al. 2013, and Luczak, Glatt, and Wall 2006. For dating, see Peng et al. 2010, though see Li et al. 2011 for a more recent date.
6. See Borinskaya et al. 2009 and Peng et al. 2010.
7. See Peng et al. 2010.
8. See McGovern et al. 2004.
9. The milk of some marine mammals contains little or no lactose (Lomer, Parkes, and Sanderson 2008). Estimates of global lactose persistence range from 30% to 40% (Gerbault et al. 2013, Lomer, Parkes, and Sanderson 2008, Bloom and Sherman 2005). My 68% estimate in the main text is from Gerbault et al. 2013. For overview and context, see O'Brien and Laland 2012. The presence of the symptoms of lactose intolerance seem to depend on particular microbiota in the colon. Somali nomads, for example, possess a gut flora that allows them to drink milk to obtain the calcium and water without being lactose persistent (so, they can't get most of the calories).
10. See Ingram, Mulcare, et al. 2009, O'Brien and Laland 2012, Bloom and Sherman 2005, Gerbault et al. 2009, 2011, 2013, and Leonardi et al. 2012. This more recent work builds on early and important work (Simoons 1970, Aoki 1986, Durham 1991).
11. See Gerbault et al. 2011, 2013, Leonardi et al. 2012, Itan et al. 2010, Ingram, Raga, et al., 2009, and Ingram, Mulcare, et al. 2009. The extraction of DNA from Mesolithic European hunter-gatherers and early Neolithic farmers shows that these populations have lactase persistence genes at very low frequencies (Gerbault et al. 2013), making it clear that cultural evolution drove the spread of the lactase persistence gene. Prior to this evidence, it was possible to argue that the cultural

practices (herding and milking) may have spread into populations in which the gene was already at high frequencies.

12. The earliest medical findings date to a journal article in *The Lancet* in 1965 (Cuatreca, Lockwood, and Caldwell 1965), where a difference was noted between African- and European-descent Americans in milk processing. Interestingly, behavioral differences in milk drinking between people of Asian, African, and European descent have been noted by researchers since at least 1931. These differences were widely attributed to differences in education or income (Paige, Bayless, and Graham 1972). A failure to understand the origins of this behavioral variation permitted the U.S. government to promote milk drinking for all, for decades. For the Got Milk? material, see Wiley 2004. The lesson here is not that income and education are unimportant (they are very important), but rather that what policy makers need is good behavioral science.

13. See Laland, Odling-Smee, and Myles 2010, Richerson, Boyd, and Henrich 2010, and Fisher and Ridley 2013.

14. See Perry et al. 2007.

15. See Oota et al. 2001.

16. See Cavalli-Sforza and Feldman 2003 and Brown and Armelagos 2001.

17. For a textbook treatment of this aspect, see Boyd and Silk 2012.

18. Recently, the journalist Nicholas Wade (2014) has sought to argue that continental races do indeed capture behaviorally important genetic variation in humans. Wade combines three lines of evidence: (1) analyses of global genetic variation, (2) specific cases of natural selection favoring locally or regionally adaptive traits, as discussed in this chapter, and (3) phenotypic differences in behavior, psychology, or biology (IQ, Aggression, etc.). His first line of evidence uses recent analyses of global samples to establish a genetic reality for classical continental races. And, yes, there is continental-level genetic variation, but as I'll explain, that doesn't imply natural selection is operating to differentiate these continental populations. Then Wade points to the local cases in which natural selection can be more or less isolated as the cause of particular genetic changes. At this point, he leads the reader to infer that if natural selection caused these local or regional genetic changes, then it's probably also responsible for the continental-level genetic variation. Further, he argues, if natural selection explains the continental-level variation, then perhaps it also explains the prevalent psychological, behavioral, and biological variation observed across continents.

Both of Wade's inferential moves between these lines of evidence are fraught with problems. To understand the issues with the first inference, we must realize that the genetic variation among different continental populations traces to the spread of humans out of Africa, which occurred relatively recently. These migrations gave rise to evolutionary genetic drift and founder effects, as small samplings (groups) of much larger populations set off to become founding populations on new continents. These migrations created genetic variation, but not functional variation due to natural selection. The genetic variation most suitable for studies of such ancient migration is specifically neutral (not under selection). DNA frequently undergoes mutations that do not operationally change the organism, either because a particular sequence is nonfunctional or because DNA bases can flip without altering the coding of proteins. Thus, finding continental-level genetic variation is precisely what one should expect after such ancient migrations, but this does not imply that any important functional variation exists. Moreover, since no continental-level selection pressures have been identified, there's no evidence that much of this variation is due to natural selection. Then, when Wade points to local or regional examples of natural selection acting on specific genes, he fails to realize that this actually works *against his idea* of continental races. As I explained in the main text, these local processes often make continental races less genetically similar while at the same time increasing the similarity between different continental populations. Thus, natural selection may often operate to reduce the variation between distant populations.

Finally, Wade's inference from behavior and psychology back to genes conflates genes with biology and thereby reveals a lack of understanding of modern cultural evolutionary theory. He dismisses culture as an alternative explanation for continental-level behavioral, psychological, and biological variation without seriously considering what we now know about human learning, development, motivation, and cultural neuroscience. For example, he casually points to the fact that after the American invasion in 2003, Iraqis did not immediately adopt American political institutions as an argument against culture as an explanation. Clearly, he argues, if it was culture, the Iraqis would have immediately adopted American institutions, so it must be their tribal genes. In the coming chapters, as I develop a proper theory of cultural evolution grounded in evolutionary biology, neuroscience, psychology, and anthropology, you'll see how profoundly off-target such an argument is. Culture, social norms, and institutions all shape our brains, biology, and hormones, as well as our perceptions, motivations, and judgments. We can't pick our underlying cultural perceptions and motivations any more than we can suddenly speak a new language.

19. See Kinzler and Dautel 2012, Esteban, Mayoral, and Ray 2012a, Gil-White 2001, Moya, Boyd, and Henrich forthcoming, Astuti, Solomon, and Carey 2004, Dunham, Baron, and Banaji 2008, and Baron and Banaji 2006.

Chapter 7. On the Origin of Faith

1. For a review of the health effects, see Nhassico et al. 2008.
2. See Dufour 1994, Wilson and Dufour 2002, Jackson and Jackson 1990, and Dufour 1988a, 1988b. Varieties of manioc respond to drought by massively increasing their cyanogenic output. Bitter manioc supplies 70% of the Tukanoans' calories.
3. See Dufour 1984, 1985.
4. This appears to have happened in the Democratic Republic of Congo (Tylleskar et al. 1991, 1992).
5. The emergence of specific negative health impacts is complex and depends on other factors such as the presence of sulfur in the diet (Jackson and Jackson 1990, Tylleskar et al. 1992, 1993 Peterson, Legue, et al. 1995, Peterson, Rosling, et al. 1995). Jackson and Jackson discuss a processing technique that actually increases cyanogenic content. See Padmaja 1995 for a review of processing techniques.
6. I replaced all personal names from my ethnographic work with pseudonyms.

7. See Henrich and Henrich 2010. Also see Henrich and Broesch 2011.
8. See Henrich 2002.
9. We also elicited descriptions of any actual cases of fish poisoning that women might have heard. Almost everyone relayed the same few cases. This means that the repertoire of taboos cannot be composed of "case knowledge," with individual women assembling their taboos from stories—most of the tabooed species appear in zero of the reported cases.
10. See Katz, Hediger, and Valleroy 1974 and Mcdonough et al. 1987.
11. See Bollet 1992 and Roe 1973.
12. This is from Bollet 1992. The quotations "constitutionally resistant" and "absurd" are on p. 217. See Jobling and Petersen 1916.
13. See Whiting 1963, Beck 1992, and Mann 2012.
14. On the Gambler's Fallacy and our problems with randomness, see Kahneman 2011 and Gilovich, Griffin, and Kahneman 2002.
15. Beaver hips were used for hunting beavers and fish jaws for locating fish.
16. See Moore 1957.
17. Statistical data shows that rainfall patterns and floods are random, without distinguishable cycles or streaks.
18. See Dove 1993 and Henrich 2002. For a similar case, see Lawless 1975.
19. On arrow making, see Lothrop 1928. For an extended discussion and more examples, see Henrich 2008.
20. McGuigan 2012, McGuigan, Makinson, and Whiten 2011, McGuigan et al. 2007, Horner and Whiten 2005.
21. See Lyons, Young, and Keil 2007.
22. This assumes that the relative competence, age, and skill of the model is appropriately adjusted. Adults won't overimitate three year olds ... much.
23. See Nielsen and Tomaselli 2010, McGuigan, Gladstone, and Cook 2012, McGuigan 2012, 2013, and McGuigan, Makinson, and Whiten 2011.
24. See Horner and Whiten 2005.
25. For a detailed discussion of chimpanzee culture see Henrich and Tennie, forthcoming.
26. For important lines on work on all these angles, see Herrmann et al. 2013, Over and Carpenter 2012, 2013, and Kenward 2012.
27. See Billing and Sherman 1998, Sherman and Billing 1999, Sherman and Flaxman 2001, and Sherman and Hash 2001.
28. The evidence for this is only suggestive at this point (Billing and Sherman 1998, Sherman and Billing 1999, Sherman and Flaxman 2001, Sherman and Hash 2001).
29. It's worth noting that cultural learning has overcome other aversions that are probably innate. For example, we likely have innate aversions to eating feces, but Inuit foragers will eat deer poop like berries (apparently, they are good in soup: Wrangham 2009), and Hadza hunter-gatherers enjoy picking the partially digested nuts from baboon poop (Marlowe 2010).
30. See Rozin, Gruss, and Berk 1979, Rozin and Schiller 1980, Rozin, Mark, and Schiller 1981, Rozin, Ebert, and Schull 1982, and Rozin and Kennel 1983. There is some evidence of a weak desensitization to the pain-inducing effects of capsicum after high levels of capsicum consumption (Rozin and Schiller 1980, Rozin,

Mark, and Schiller 1981). However, this doesn't account for the clear enjoyment of the burning sensation and for preferences for chili peppers. Efforts to train rats to like capsicum have failed (Rozin, Gruss, and Berk 1979), though they can be trained to selectively eat capsicum-containing food if the unpleasant burning sensation is correlated with future desirable states (less pain). In Mexico, dogs and pigs, who can only survive by eating chili-laden-food garbage, come to be indifferent to capsicum (which is a big step, since otherwise it's aversive). The only nonhuman evidence for acquiring a taste for capsicum is from two juvenile, human-reared chimpanzees and three pet dogs. Rozin and Kennel (1983) argue that it's the experience of human environments during ontogeny that sets the stage for such taste acquisitions. This will be important when we consider our species' likely evolutionary pathway in chapter 16.

31. See Williams 1987 and Basalla 1988.
32. See Meltzoff, Waismeyer, and Gopnik 2012.
33. See Buss et al. 1998 and Pinker and Bloom 1990.
34. See Boyd, Richerson, and Henrich 2013.
35. See Boyd, Richerson, and Henrich 2011a.

Chapter 8. Prestige, Dominance, and Menopause

1. Krakauer 1997: 78.
2. Radcliffe-Brown 1964: 45. In Northern Canada, the ethnographer Robert Paine summarized work among Arctic hunter-gatherers by writing: "Acknowledged expertise attracts, though perhaps only temporarily, what we may term a following of dependent persons. These persons will be welcomed as a principal source of prestige—as a capital benefit of the hunter's expertise. Nor is this expertise necessarily reduced or dissipated through having to share it with other persons attached to him" (1971: 165).
3. Among egalitarian hunter-gatherers in the Kalahari desert in Africa, the ethnographer Richard Lee (1979: 343) observed that particularly skilled orators, arguers, ritual specialists, and hunters "may speak out more than others, may be deferred to by other discussants, and one gets the feeling that their opinions hold a bit more weight than the opinions of other discussants." Clear descriptions of prestige in Amazonia can also be found in Goldman 1979 and Krackle 1978.
4. This theory was developed by Henrich and Gil-White (2001).
5. For the learner, this prestige information can then be integrated with their own direct observations of success and skill. Early on, their choice of models may be dominated by prestige information—the patterns of deference they observe. Later, as learners accumulate their own skill, know-how, and abilities to assess excellence, their judgments of whom to learn from will often shift from relying primarily on observations of others' deference patterns to their own direct observations.
6. See Henrich and Broesch 2011, Henrich and Gil-White 2001, and Chudek et al. 2012.
7. See James et al. 2013.
8. See Boyd and Silk 2012, Fessler 1999, Henrich and Gil-White 2001, and Eibl-Eibesfeldt 2007.

9. To make the claim that dominance and prestige are genetically evolved forms of social status in humans, it's crucial that both link to higher reproductive success in small-scale societies. However, in the modern world, the links between status and reproductive fitness are more complicated due to the demographic transition. In Europe in the mid-nineteenth century, women began substantially reducing their fertility (number of children). The pattern has since spread to many countries. The most educated and richest women have reduced their numbers of children most dramatically. Thus, in the modern world, achieving high status may actually be associated with having fewer children, not more, perhaps because having fewer children allows one to achieve greater prestige in a world with meritocratic institutions (Richerson and Boyd 2005).

10. See von Rueden, Gurven, and Kaplan 2008, 2011. I do worry about how prestige was operationalized in this study, since dominance may contribute to "community influence." However, Chris tried to remove this effect by controlling for fighting ability when he looked at the relationships between community influence and his various fitness proxies.

11. Here I use these terms in a specific theoretical sense as developed in this literature. Consequently, their usage may not completely correspond with the intuitions of every speaker of English.

12. Many evolutionary researchers now distinguish prestige and dominance as different types of human status (Cheng et al. 2013, Chudek et al. 2012, Atkisson, O'Brien, and Mesoudi 2012, von Rueden, Gurven, and Kaplan 2011, Horner et al. 2010, Hill and Kintigh 2009, Reyes-Garcia et al. 2008, 2009, Snyder, Kirkpatrick, and Barrett 2008, Johnson, Burk, and Kirkpatrick 2007). Prestige hierarchies arise when deference is freely conferred by others, out of a positive desire to interact with the higher-ups, whereas dominance hierarchies emerge because others have been compelled to defer by force, or threat of it, to accept the status-quo, at least temporarily. Besides valuable cultural information, like hunting know-how, high-status individuals might possess "goods" that might be traded for deference benefits. For example, a beautiful woman might attract many male suitors—more than she can handle or wants. Other, less-attractive, women might want to be around her in order to hang out where the men are (Pinker 1997). Or the son or daughter of a former president or prime minister might receive deference, not because they are coercive or because of their knowledge or skill, but because of their valuable (inherited) social connections. Merely being able to hang around them could result in making many important friends and contacts. From this perspective, the informational goods we discuss above might be merely one type of "good" that can be acquired by paying deference and no different from accessing mates, allies, or social contacts.

Understanding these kinds of situations and the importance of noninformational goods are certainly features of many diverse human societies and strongly suggest that a person's prestige status may be influenced by factors besides the possession of skills and knowledge related to success. However, the idea that the evolutionary pressure to acquire mates, allies, or social contacts (and not information) drove the evolution of prestige does not account for many real-world aspects of prestige. First, by contrast with dominant individuals, prestigious individuals are truly persuasive, meaning that their subordinates actually shift their views closer to those of the prestigious individual. It's not clear why exchanging deference for mates or allies should result in actual opinion shifts, rather than merely causing shifts in superficially expressed agreement. Similarly, people preferentially imitate prestigious individuals, using a variety of cues to figure out from whom to copy. People copy everything from food preferences to charitable giving from prestigious people, even when there's no chance they will ever actually meet the prestigious person. Again, it's not clear why a deference-for-allies or deference-for-mate-access exchange should result in such biased imitation. Experimental evidence suggests, for example, that women will preferentially copy mate-choice preferences from more attractive women. They learn both which men are attractive and which elements (clothing styles, conversational choices) attract men (see chapter 14). Second, the suite of emotions—awe and admiration—and the ethological patterns found in subordinates in a prestige hierarchy are well suited to situations in which learners need to seek out and hang around a model for the purposes of learning. They are less readily explained by noninformational exchanges. Finally, since only humans as far as we know have fully developed systems of prestige status, alternative theories would have to explain why prestige has not strongly emerged in nonhuman primates, who have to seek mates, build alliances, and rely on specific social partners. By contrast, approaching prestige from an "information-goods" viewpoint readily explains this, since only our evolutionary lineage has crossed the Rubicon of cumulative cultural evolution to enter a regime of culture-gene coevolution.

Of course, it's possible that there is a third kind of status that no one has yet isolated, effectively characterized, and studied. It would be a kind of prestige, but without all the information-gathering components. Some scholars, for example, have suggested that wealth, income, or education might be a kind of status. Famously, the sociologist Max Weber distinguished three types of status, two of which correspond roughly to dominance and prestige (see Henrich and Gil-White 2001 for discussion). His third type was based on wealth. However, wealth, like income and education, merely acts as a cue to either (1) skill, knowledge, and success (prestige), or to (2) control over costs and benefits (dominance). Moreover, from any evolutionary perspective, wealth can only accumulate in the hands of one individual because of a set of social norms and institutions that enforce property rights. In a primate world without social norms, if you have 100 bananas and we have zero bananas, you are only going to be able to keep as many of those bananas as you can defend by force.

13. The measures of prestige and dominance, based on peer reports, were developed and validated by Cheng Tracy, and Henrich (2010). The Lost on the Moon findings are from Cheng et al. 2013.

14. Yes, these results all apply to women too (Cheng et al. 2013, Cheng, Tracy, and Henrich 2010).

15. For discussions of the effect, see Gregory, Webster, and Huang 1993, Gregory, Dagan, and Webster 1997, and Chartrand and Bargh 1999.

16. See Gregory and Webster 1996.

17. I'm updating Neal Gabler's "Zsa Zsa factor," since some readers probably haven't a clue who Zsa Zsa Gabor is. Other relevant terms include "famesque" and "celebutante." See *Wikipedia* under any of these entries.

18. See Watts 2011.
19. For infant research, see Thomsen et al. 2011. Similar work by Mascaro and Csibra (2012) indicates that infants readily infer that dominance relationships are stable across contexts but do not assume transitivity.
20. Since prestige evolved long after dominance in our lineage, it's not surprising that it has exapted some of the emotions and displays of dominance (Henrich and Gil-White 2001). Like many human facial and bodily expressions, these displays are part of complex feedback systems in which internal motivations and contextual cues lead individuals to display; but, at the same time, making the displays also causes psychological and physiological feedback. For example, when students are placed into either expansive or diminutive postures, they subsequently show behavior shifts consistent with high or low status, taking more risks and showing greater pain tolerance in experiments. Dominant and submissive postures may even create predictable hormonal shifts in testosterone and cortisol (Bohns and Wiltermuth 2012, Carney, Cuddy, and Yap 2010), though these findings need further testing.
21. Pride and shame displays are recognized across diverse societies (Tracy and Matsumoto 2008, Tracy and Robins 2008, Fessler 1999) and among young children (Tracy, Robins, and Lagattuta 2005), and they automatically and unconsciously communicate status information to others, or at least the displayers own beliefs about their status (Tracy et al. 2013).
22. The link between these facets of pride and prestige and dominance was established by Cheng, Tracy, and Henrich (2010). See Johnson, Burk, and Kirkpatrick 2007 on hormonal evidence.
23. See Fessler 1999 and Eibl-Eibesfeldt 2007.
24. The management of *proximity* by subordinates also highlights the contrast between dominance and prestige. In dominance situations, subordinates try to keep their distance from the dominant, since dominants are prone to erratic fits of anger that may have evolved both to remind subordinates and observers of who the boss is and potentially to create chronic stress on subordinates as a means to lower their fitness by damaging their health and cognitive abilities (Silk 2002). In contrast, subordinates in a prestige hierarchy seek proximity, looking to engage with prestigious individuals. This is why prestigious individuals have a "following."
25. See Brown 2012.
26. The material on Astor comes from the *New York Times* (Mar. 30, 2002, by Alex Kuczynski); also see Potters, Sefton, and Vesterlund 2005. Similarly, when asked why the university requests permission from donors to announce their contributions, the chairman of the Johns Hopkins trustees explained, "Fundamentally we are all followers. If I can get somebody to be a leader, others will follow. I can leverage that gift many times over" (Kumru and Vesterlund 2010).
27. Of course, the generosity of high-status people is a complex phenomenon with numerous causes and contributions. In many small-scale societies, for example, successful individuals give generously because if they do not, they will be envied, and envy is often believed to cause negative consequences for its targets, like sickness, injury, and death. I suspect that envy is most likely to occur when a person's success is perceived to be disproportionate to their worth, effort, or tal-

ent. Nevertheless, in some places, nearly all success is assumed to be dispro-portionate.

28. See Kumru and Vesterlund 2010. For related work, see Potters, Sefton, and Vesterlund 2005, 2007, Guth et al. 2007, Gillet, Cartwright, and Van Vugt 2009, Ball et al. 2001, Eckel and Wilson 2000, and Eckel, Fatas, and Wilson 2010.

29. From Birdsell 1979.

30. See Henrich and Gil-White 2001, Simmons 1945, and Silverman and Maxwell 1978. Restricting certain forms of political leadership to older individuals may take advantage of their superior reasoning about social conflicts (Grossmann et al. 2010).

31. For the quotation, see Simmons 1945: 79.

32. This "information grandparent hypothesis" was developed in the context of culture-gene coevolutionary theory in the electronic supplementary informa-tion of Henrich and Henrich 2010, although it is closely related to theoretical and empirical lines of research developed by Hill, Kaplan, Gurven, and their collaborators (Kaplan et al. 2000, 2010, Gurven et al. 2012, Gurven and Kaplan 2007). See Kaplan et. al. 2010 for data and discussion on the long postreproduc-tive period in humans and Alberts et. al. 2013 for a comparison to nonhuman primates.

33. See Sear and Mace 2008.

34. One of the tricky parts to studying this empirically is that the benefits created by cultural transmission to the children and grandchildren of the older person may be indirect. For example, in the Fijian communities I've studied, older people readily dispense their knowledge and wisdom to essentially anyone from the village (at least), which gives their relatives no particular informational advan-tage (though it does benefit their community). However, as a consequence, grandparents accrue prestige and obtain the associated deference that accompa-nies it. This may then convert back to benefits shared more narrowly by the grandparents' own families.

35. For a review of culture in whales and dolphins see Rendell and Whitehead 2001, including a discussion of menopause (also see McAuliffe and Whitehead 2005). For the experimental work with killer whales, see Abramson et al. 2013. For the demographic study, see Foster et al. 2012. For general information, see Baird 2000.

36. See Foley, Pettorelli, and Foley 2008.

37. For field experiments on age and the recognition of male lions, fellow elephants, and dangerous humans, see McComb et al. 2001, 2011, 2014, and Mutinda, Poole, and Moss 2011. From this work, I think there's probably a case to be made that elephants have a type of prestige status. Note, however, there is debate about whether elephants have true menopause or just a rapid decline in fertility. This is interesting but not really an issue for my point here since the road to the evo-lution of true menopause would begin with declining fertility at older ages. It may be that elephant females are more similar to human males in their fertility decline than to human females.

38. Even in the Sanhedrin, the speaking order was reversed when discussing issues of purity; see Schnall and Greenberg 2012 and Hoenig 1953. Also, see chapter 4

of Tractate Sanhedrin at http://www.come-and-hear.com/sanhedrin/sanhedrin _32.html#chapter_iv.

Chapter 9. In-Laws, Incest Taboos, and Rituals

1. In Yasawa, unlike other places in Fiji, one's closest cross-cousins should be avoided for sex and marriage. Also, "Kula" is a pseudonym.
2. See Pinker 1997 and Dawkins 1976, 2006. Other purely genetic evolutionary mechanisms, such as those based on partner choice or "biological markets" (Baumard, Andre, and Sperber 2013), similarly cannot account for human cooperation (Chudek, Zhao, and Henrich 2013, Chudek and Henrich 2010), because they don't address the five challenges (see chapter 10).
3. For a more detailed explanation of these and other aspects of cooperation, see *Why Humans Cooperate* (N. Henrich and Henrich 2007).
4. For an introduction, see chapters 4 and 5 of *Why Humans Cooperate* (N.Henrich and Henrich 2007).
5. You may have read the opposite, that foraging bands are mostly close relatives. Though this is widely repeated, this claim has little or no real evidentiary support. The best available evidence on relatedness in foraging bands is presented later in this chapter.
6. To enter this literature, begin with Chudek and Henrich's (2010) review.
7. See Edgerton 1992 and Durham 1991.
8. See Henrich, Boyd, and Richerson 2012, especially the supplement to that article.
9. On pair-bonding, see Chapais 2008. On paternal care and paternity certainty, see Buchan et al. 2003, Neff 2003.
10. Nevertheless, primates do seem to have some mechanisms to figure out who their paternal relatives are, but these kin identification mechanisms are rather weak (Langergraber 2012).
11. Male siamangs, a smaller, arboreal ape from Southeast Asia, help a bit more, by carrying their mate's infants. As expected, monogamously paired males do much more infant carrying than males who share a single female mate (Lappan 2008).
12. See Lee 1986, Draper and Haney 2005, and Marshall 1976.
13. This maternal bias has been observed in several modern societies (Gaulin, Mc-Burney, and Brakeman-Wartell 1997, Pashos 2000, Euler and Weitzel 1996). Crucially, this effect vanishes in societies with explicit social norms and beliefs that reduce paternity uncertainty and favor both males and the male side of descent (Pashos 2000).
14. See Garner 2005; this is part of a large literature on name effects, including studies showing that people like products with brand names similar to their own (Brendl et al. 2005). On the use of similarity in appearance to assess relatedness, see DeBruine 2002.
15. See Hill and Hurtado 1996 and Lee and Daly 1999.
16. Hua 2001.
17. See Beckerman and Valentine 2002a, 2002b, Beckerman et al. 2002, Crocker 2002, Hill and Hurtado 1996, and Walker, Flinn, and Hill 2010.

18. It's not clear why additional fathers beyond two reduces child survival from its peak at two fathers. My suspicion is that it may create a diffusion of responsibility. That is, if the primary father dies or is injured, and there is only one father left, the responsibility clearly falls to him. However, if two or more fathers remain, it's not clear who should do what, or who should step up. Among Westerners, psychologists have documented this diffusion of responsibility phenomenon and call it the "bystander effect" (Fischer et al. 2011).

19. See Lieberman, Fessler, and Smith 2011, Chapais 2008, Sepher 1983, Wolf 1995, and Hill et al. 2011.

20. See Fessler and Navarrete 2004 and Lieberman, Tooby, and Cosmides 2003.

21. See Henrich 2014, Henrich, Boyd, and Richerson 2012, and Talhelm et al. 2014.

22. See Fiske 1992 and N. Henrich and Henrich 2007.

23. See Richerson and Boyd 1998, Simon 1990, Richerson and Henrich 2012.

24. Several background points are important here. First, hunter-gatherer societies are, in fact, extraordinarily diverse. Many ethnographically and historically known hunter-gatherers were sedentary and had complex divisions of labor, accumulated wealth, hereditary leaders, and social classes, including slaves. In contrast to the standard view, I suspect that some of this complexity existed for at least periods during the Paleolithic, prior to the earliest signs of agriculture (Price and Brown 1988). While important for thinking about human evolution broadly, it is enough for me to show here that even mobile foraging populations rely on culture for cooperation. Second, I'm not using these populations as representatives of the Paleolithic or as "primitive." That would be stupid. The few remaining hunter-gatherer societies are not relics from the Paleolithic, and they have been heavily shaped by their own histories, innovations, and interactions with other groups. In chapter 10, you'll see that I take advantage of this very fact and use it to illustrate some key points. However, at the same time, studies of diverse small-scale societies, including foraging populations, provide a valuable tool to understand what human social life can be like—in all its diversity—in subsistence societies governed by kinship systems, away from modern states, taxes, police, hospitals, and industrial technology. When combined with evidence from paleoanthropology (the stones and bones of ancient populations), primatology, and genetics, the full spectrum of insights from diverse small-scale societies immensely enriches our understanding of life in the distant past (Flannery and Marcus 2012), as well as what it means to be human.

25. See Hill et al. 2011.

26. See Lee (1986), who notes that the Ju/'hoansi do have a term for "friend" or "equal" that is only used when two nonrelatives cannot be distinguished by age (which prevents the older-younger terminology from being applied).

27. For a summary of paleoanthropology, see Boyd and Silk 2012. On the importance of hunting for prestige, see Henrich and Gil-White 2001.

28. Some have argued that hunters have to share because they cannot store the meat. However, among the Hadza, we know this is untrue, since hunters who own the meat know how to dry and store it. It's social norms about distribution, and a sense of entitlement from other Hadza, that prevents storage, not a lack of know-how (Woodburn 1982, Marlowe 2010).

29. There is a large literature on food sharing in foragers (e.g., Gurven 2004a, 2004b, Marlowe 2004). Early efforts to explain sharing from an evolutionary perspective focused on genealogical relatedness and reciprocity. While these clearly play a role for many kinds of foods, the band-wide distribution of large game could not be readily explained. The patterns of meat sharing behavior call for an evolutionary approach that includes social norms (Hill and Hurtado 2009, Hill 2002).

30. See Lee 1979 and Wiessner 2002.

31. See Wiessner 1982, 2002.

32. See Schapera 1930. It would be difficult to secretly violate the taboos, since all large game must be brought back to camp and tasted by the headman before portions are allocated. Portions are then cooked and consumed publically, but the different categories of consumers cook at separate hearths.

33. See meat taboos among the Aché (Kim Hill, pers. comm., 2012, 2013), Mbuti (Ichikawa 1987), and Hadza (Woodburn 1982, 1998, Marlowe 2010) and on the island of Lembata in Indonesia (Barnes 1996, Alvard 2003).

34. See Fessler et al. 2003 and Fessler and Navarrete 2003.

35. The Hadza attribute illness to eating the ritually tabooed *Epeme* meat (God's meat: Woodburn 1998). Another reason is that anyone who does figure out that taboo violations don't cause bad things will be forced to eat tabooed portions in secret, to avoid reputational damage, so learners won't be able to copy him, thus inhibiting the spread of taboo violations.

36. See Marshall 1976, Wiessner 2002, Altman and Peterson 1988, Endicott 1988, Heinz 1994, Myers 1988, and Woodburn 1982.

37. One telling feature of cooperation and sharing among mobile hunter-gatherers involves what happens when a group encounters a novel situation for which they lack sharing norms. The evolutionary researcher Nicholas Blurton Jones tells a story of when he wanted to reward a group of Hadza men for assisting him. Blurton Jones first attempted to pay the group in a lump sum of tobacco, which he assumed they'd readily share, just as they routinely did with meat and honey. The men, however, absolutely did not want to take the payment in a lump sum and asked Blurton Jones to please create individual shares. They feared that if they themselves had to divide it up, fights would break out and relationships might be damaged (Blurton Jones, pers. comm.).

38. See Wade 2009: chapter 5, Marshall 1976: 63–90, and Biesele 1978.

39. Biesele 1978: 169.

40. I'm drawing together recent work on ritual (Whitehouse 2004, Fischer et al. 2014, Xygalatas et al. 2013, Konvalinka et al. 2011, Atran and Henrich 2010, Soler 2010, Alcorta, Sosis, and Finkel 2008, Sosis, Kress, and Boster 2007, Alcorta and Sosis 2005, McNeill 1995, Ehrenreich 2007, Whitehouse and Lanman 2014). For Ibn Khaldûn's work, see Khaldûn 2005. Of course, the linking of ritual and sociality was most famously made by social theorists such as Durkheim ([1915] 1965) and Frazer (1996).

41. See Wiltermuth and Heath 2009. For other relevant work on synchrony, see Hove and Risen 2009, Valdesolo and DeSteno 2011, Valdesolo, Ouyang, and DeSteno 2010, and Paladino et al. 2010.

42. For work with children, see Kirschner and Tomasello 2009, 2010.

43. See Spencer and Gillen 1968.
44. See Birdsell 1979 and Elkin 1964.
45. Spencer and Gillen 1968: 271.
46. See Whitehouse et al. 2014, Whitehouse 1996, and Whitehouse and Lanman 2014.; these authors use the term "rites of terror."
47. See Chapais 2008, Apicella et al. 2012, and Wiessner 1982, 2002.
48. See Hill et al., 2014.
49. On the *Epeme* ritual, see Woodburn 1998.
50. See Wiessner 1982, 2002.

Chapter 10. Intergroup Competition Shapes Cultural Evolution

1. See Mitani, Watts, and Amsler 2010.
2. For early work on this idea, see Darwin 1981, Boyd and Richerson 1985, 1990 and Hayek and Bartley 1988.
3. For an overview of this way of thinking, see Henrich 2004a.
4. See Choi and Bowles 2007, Bowles 2006, Boyd, Richerson, and Henrich 2011b, Boyd et al. 2003, and Wrangham and Glowacki 2012.
5. See Smaldino, Schank, and McElreath 2013. To show that rates of intergroup violence were low, and thus argue that intergroup competition was unimportant, fails to recognize that *only* one type of intergroup competition involves physical violence.
6. See Boyd and Richerson 2009 for a theoretical model. See Knauft 1985 and Tuzin 2001, 1976 on the effects of differential migration into more successful groups in small-scale societies.
7. See Richerson and Boyd 2005, and for a review of the work on religion and fertility, see Blume 2009, Norenzayan 2013, and Slingerland, Henrich, and Norenzayan, 2013.
8. See Boyd and Richerson 2002 and Henrich 2004a.
9. See Wrangham and Glowacki 2012, Wilson et al. 2012, and Wilson and Wrangham 2003. For a review on the use of chimpanzees as a model for the common ancestor for chimps and humans see Muller, Wrangham, and Pilbeam, forthcoming. Notably, on current knowledge, our other closest primate relative, the bonobo, does not engage in intergroup violence, so the inference that our common ancestor with chimpanzees and bonobos had intergroup conflicts is far from automatic. However, bonobos are clearly an unusual ape in several ways, so there's a good case to support the use of chimpanzees as providing relatively more insights about our last common ancestor than bonobos (Muller, Wrangham, and Pilbeam, forthcoming).
10. For a theoretically driven comparison of culture in chimpanzees and humans, see Henrich and Tennie, forthcoming.
11. See Pinker 2011 and Morris 2014.
12. I'm drawing mostly on the supplement to Bowles 2006, as well as Keeley 1997, P. Lambert 1997, Ember 1978, 2013, and Ember and Ember 1992. It's not relevant to my concerns here whether some groups also experienced enduring

peace, especially since violent conflicts represent only one form of intergroup competition.

13. By "war" and "warfare." I'm including all sort of violent inter-group interactions, including raiding and ambushes.

14. See Ember, Adem, and Skoggard 2013, Ember and Ember 1992, and Lambert 1997.

15. See Boyd 2001.

16. Wiessner and Tumu 1998: 195–196.

17. See Tuzin 1976, 2001. Tuzin argues that Ilahita also acquired their elaborate garden technology for growing yams from the Abelam. He also points out that the transmission was one way, from more successful to less successful. The rich mythology and elaborate hunting magic of the Ilahita Arapesh did not transmit to the Abelam or anyone else (Tuzin 1976: 79).

18. See Sosis, Kress, and Boster 2007.

19. The ethnographic cases of the effect of intergroup competition on cultural evolution are plentiful (Currie and Mace 2009). For example, Atran et. al. (2002) have shown how conservation-oriented ecological beliefs spread from locally prestigious Itza Maya to Ladinos in Guatemala, and how highland Q'eqchi Maya, with tightly bound cooperative institutions and commercially oriented economic production, are spreading at the expense of both Itza and Ladinos. In New Guinea, using quantitative data gleaned from ethnographies, Soltis Boyd, and Richerson (1995) have shown that even the slowest forms of cultural group selection (conquest) can occur on time scales of 500 to 1,000 years. In Africa, using ethnohistorical data, Kelly (1985) has demonstrated how differences in culturally acquired beliefs about bride-price fueled the Nuer expansion over the Dinka, and how different social institutions, underpinned by cultural beliefs about segmentary lineages, provided the decisive competitive advantage (Richerson and Boyd 2005). Sahlins (1961) has also argued that cultural beliefs in segmentary lineages facilitated both the Nuer and Tiv expansions. Using archeological data, anthropologists are increasingly arguing for the importance of intergroup competition on cultural evolution and political complexity during prehistory (Flannery and Marcus 2000, Spencer and Redmond 2001).

20. See Evans and McConvell 1998, Bowern and Atkinson 2012, McConvell 1985, 1996, and Evans 2005, 2012.

21. See Evans and McConvell 1998. Thanks to Nick Evans for very helpful email correspondence.

22. See Elkin 1964: 32–35 and McConvell 1985, 1996.

23. See McConvell 1996.

24. See Maxwell 1984, Hayes, Coltrain, and O'Rourke 2003, and McGhee 1984.

25. See Spencer 1984, McGhee 1984, Johnson and Earle 2000, Anderson 1984, and Briggs 1970.

26. See Burch 2007.

27. See Maxwell 1984, McGhee 1984, Anderson 1984, and Sturtevant 1978.

28. Cultural transmission did go both ways, with the Inuit acquiring a harpoon design, and perhaps their use of soapstone for lamps and building of snow-houses, from the Dorset (Maxwell 1984: 368).

29. See Bettinger and Baumhoff 1982, Young and Bettinger 1992, and Bettinger 1994. For evidence from oral traditions, see Sutton 1986, 1993. It's worth noting that some evidence indicates that Numic-speaking peoples are part of the Uto-Aztecan expansion, which moved north out of Mexico. This expansion began with farmers, so our Numic foragers are probably the cultural, if not genetic, descendants of farmers (see Jane Hill's piece at lingweb.eva.mpg.de/Hunter-GathererWorkshop2006/Hill.pdf).
30. See Hämäläinen 2008. We know something of Comanche life because the Comanche would kidnap the children of white settlers and eventually adopt them into the tribe. Once liberated, sometimes against their will after many years, these former captives told their stories (Zesch 2004).
31. On shifting Paleolithic environments, see Richerson, Boyd, and Bettinger 2001.
32. Triangulating from other species and small-scale societies, the best inference is that these cases arose from violent intergroup interactions This is opposed to killing and eating one's own community members or peacefully consuming dead relatives at mortuary feasts (Stringer 2012). On "backed blades," see Ambrose 2001. On bows and arrows as part of the out of Africa expansion, see Shea 2006 and Shea and Sisk 2010. For overall context, see Klein 2009 and Boyd and Silk 2012.
33. On the rise of complex societies, see Ensminger and Henrich 2014: chapter 2 and Turchin 2010. For Diamond's view, see Diamond 1997 and Diamond and Bellwood 2003. Following Diamond, Ian Morris (2014) further substantiates the historical case for the importance of war in driving the cultural evolution of complex societies. Unfortunately, Ian focuses too narrowly on war and fails to realize that war is merely one type of intergroup competition. He also oddly opposes his explanation to "cultural explanations," not realizing that war is in fact influencing cultural evolution (the explanations are not at odds).
34. While most evolutionary researchers agree that intergroup competition, especially in the form of violent conflict, was probably part of life in ancient hunter-gatherer societies, there are two main alternative views regarding how this competition influenced our genetic evolution. The canonical view, staunchly defended by the psychologist Steve Pinker, is that intergroup competition plays no role in shaping either genetic or cultural evolution. An alternative view, recently reenergized, is that intergroup competition shaped not cultural evolution—as I argue—but genetic evolution. Under this view, warfare and differential extinction drove genetic evolution and shaped human nature directly (Haidt 2012, Wilson 2012, Wilson and Wilson 2007, Bowles 2006). The first view is contradicted by evidence showing that intergroup competition leads to the differential spread of certain cultural traits, including both social norms and technologies. Intergroup competition also helps account for the intricate and subtle institutions we commonly observe across diverse societies that expand and sustain cooperation. Adherents to the canonical view are stuck arguing that yes, intergroup competition was common, but no, it somehow never influenced which social norms or practices survived, were copied, and spread. You've now only seen the tip of an iceberg of the evidence showing the importance of intergroup competition for cultural evolution (Richerson et al., forth-

coming). For Pinker's view on group selection, see http://edge.org/conversation /the-false-allure-of-group-selection. However, be sure to read my commentary on Pinker's piece at the same site.

The other view, that intergroup competition has directly shaped human genetic evolution, need not conflict with what I'm focused on here. However, there are a couple of reasons to suspect that the direct role of intergroup competition on genetic evolution is, at least, secondary to the processes I'm describing, and possibly trivial. Here's the key: for intergroup competition to have any effect on evolutionary processes, be they cultural or genetic, groups have to remain relatively distinct along whatever dimensions are providing some groups with competitive advantages over others. For social norms, this is easy to see. If I move into your group from another group, my kids and I have to adopt your kinship and marriage norms. If we don't, my kids either won't have any relationships (which govern helping, food sharing, sex, trade, etc.) or they will be doing all the wrong things (norm violations). They might, for example, repeatedly make Kula's mistake and violate an incest taboo by sitting near the wrong girl or boy, which will get them sanctioned in some way. However, for genes, if people from different groups have sex, the relevant genetic differences between the groups will quickly go away. Either the initially advantaged groups will get the "bad" genes from the disadvantaged groups, or the disadvantaged groups will get the "good" genes. This genetic mixing means the groups will become increasingly indistinct. The point is, cultural evolution can sustain differences between groups in a manner that genetic evolution cannot. Exacerbating this genetic mixing is the fact that human-style intergroup competition often increases the flow of genes between groups. Victorious groups in warfare frequently take the younger women and girls from the defeated groups as "wives"—in fact, access to "wives" is often the explicit reason why men from one group attack another. This creates a big inflow of genes from the losers to the winners. Or, in the absence of violence, it's still the case that men from more successful groups look for, and often find, their future wives (or short-term mates) in less successful groups. This again causes genes to flow rapidly into the more successful groups—which will wipe out the genetic differences between the groups. The couple's children might adopt all of their father's social norms, by living in his community, but no matter what, they retain half of their mother's genes. This, and other forms of differential migration, deplete genetic differences among groups while not reducing cultural differences. Data on genes and culture from the modern world confirm these stark differences, with many genetically indistinguishable groups remaining culturally quite different; see Bell, Richerson, and McElreath 2009 for analyses of genetic versus cultural variation. More generally, see Henrich 2004a, N. Henrich and Henrich 2007, and Boyd, Richerson, and Henrich 2011b.

Beyond this, our species' capacity for large-scale cooperation is tightly hinged to the presence of culturally evolved reputational and sanctioning systems, and on internalized social norms. Thus, the psychological evidence regarding human sociality and morality is most consistent with innate mechanisms adapted to a culturally constructed world (see chapter 11). It's difficult to square this empirical evidence with either of the alternative views described above.

Chapter 11. Self-Domestication

1. See Schmidt, Rakoczy, and Tomasello 2012, Schmidt and Tomasello 2012, Rakoczy et al. 2009, and Rakoczy, Wameken, and Tomasello 2008.

2. These insights are gleaned from across many ethnographies, see, for example, Boehm 1993, Bowles et al. 2012, Mathew and Boyd 2011, and Wiessner 2005.

3. The parallels between humans and domesticated animals has long been recognized and discussed (Leach 2003). I don't mean to imply that humans intentionally domesticated dogs, any more than human communities intentionally domesticated their members.

4. Those who exploit a norm violator are able to remain anonymous because the rest of the community is unmotivated to figure out who committed the crime. When someone with a good reputation is harmed, the community energetically engages and the sharing of gossip, which often reveals the culprit (Henrich and Henrich 2014). For a formal model of this evolutionary mechanism, see Chudek and Henrich n.d.

5. See Engelmann et al. 2013, Engelmann, Herrmann, and Tomasello 2012, Cummins 1996b, 1996a, and Nunez and Harris 1998 on reputation and norm-violation detection in children.

6. On norm psychology, see Chudek, Zhao, and Henrich 2013, and Chudek and Henrich 2010. For more discussion on why human evolved to internalize preferences, see Ensminger and Henrich 2014.

7. See Bryan 1971, Bryan, Redfield, and Mader 1971, Bryan and Test 1967, Bryan and Walbek 1970a, 1970b, Grusec 1971, M. Harris 1971, 1970, Elliot and Vasta 1970, Rice and Grusec 1975, Presbie and Coiteux 1971, Rushton and Campbell 1977, Rushton 1975, and Midlarsky and Bryan 1972.

8. On the enduring effects of modeling, see Mischel and Liebert 1966.

9. Of course, none of this is limited to children. In natural settings, providing models to demonstrate social norms has been shown to increase (1) volunteering in experiments, (2) helping stranded motorists, (3) donating to a Salvation Army kettle, and (4) giving blood. Modeling treatments often increase helping rates by 100% (Bryan and Test 1967, Rosenbaum and Blake 1955, Schachter and Hall 1952, Rushton and Campbell 1977).

10. For cross-cultural experiments, see Ensminger and Henrich 2014, Henrich et al., *Foundations*, 2004, Gowdy, Iorgulescu, and Onyeiwu 2003, and Paciotti and Hadley 2003. For primate work, see Silk and House 2011, Silk et al. 2005, Cronin et al. 2009, Jensen, Call, and Tomasello 2007a, 2007b, 2013, Jensen et al. 2006, de Waal, Leimgruber, and Greenberg 2008, and Burkart et al. 2007. Of course, some have tried to argue that nonhuman primates behave like humans in these experiments (Burkart et al. 2007, Proctor et al. 2013, Brosnan and de Waal 2003). Despite their broad coverage in the popular media, these claims fail for a number of methodological reasons, most notably that they don't randomly pair strangers or don't pair strangers at all (Henrich and Silk 2013, Henrich 2004c, Jensen, Call, and Tomasello 2013).

11. Educated westerners over about age 25 typically give half in the Dictator Game. However, many experiments run with students reveal lower offers in the Dictator Game, a fact that has caused great confusion among researchers. This is be-

cause Dictator Game offers keep increasing with age, toward one-half, until roughly the mid-twenties (N. Henrich and Henrich 2007, Henrich, Heine, and Norenzayan 2010b). This result suggests that it takes a long-time to fully internalize such a motivation for equality toward strangers. Other anomalies emerge from focusing on students, such as the much-discussed effect of double-blind conditions on Dictator Game offers (Cherry, Frykblom, and Shogren 2002, Lesorogol and Ensminger 2013). For more on why experimental games measure social norms, see Chudek, Zhao, and Henrich 2013, Chudek and Henrich 2010, Henrich et al., "Overview," 2004, and Henrich and Henrich 2014.

12. See Henrich 2000, Henrich, Boyd, et al. 2001, Henrich et al., *Foundations*, 2004, Henrich et al. 2005, 2006, 2010, Silk et al. 2005, Vonk et al. 2008, Brosnan et al. 2009, House, Silk, et al. 2013, House, Henrich, et al. 2013, and Ensminger and Henrich 2014.

13. See Henrich and Smith 2004 and Ledyard 1995.

14. For Rand et. al.'s studies, see Rand, Greene, and Nowak 2012, 2013 and Rand et al. 2014.

15. Ultimatum Game proposers also make more equal offers when under time pressure (Crockett et al. 2008, 2010, Cappelletti, Guth, and Ploner 2011, van't Wout et al. 2006).

16. See Kimbrough and Vostroknutov 2013.

17. See de Quervain et al. 2004, Fehr and Camerer 2007, Rilling et al. 2004, Sanfey et al. 2003, Tabibnia, Satpute, and Lieberman 2008, and Harbaugh, Mayr, and Burghart 2007. This is true of simple norm-complying choices (Zaki and Mitchell, forthcoming). More complicated situations that trade off "goods" (like fairness vs. getting money) activate both these quick, intuitive-value areas and those associated with reflective control and strategic thinking. It's also the case that giving to charity seems to activate both norm-complying reward areas (the mesolimbic system) as well as the affiliative centers associated with social attachment, which are quintessentially associated with empathic concerns (Zahn et al. 2009, Moll et al. 2006).

18. See Baumgartner et al. 2009 and Greene et al. 2004.

19. See Cummins 1996b for this specific example. More broadly, on this point see Cummins 1996a, Harris and Nunez 1996, Harris, Nunez, and Brett 2001, Nunez and Harris 1998, and Cummins 2013. For similar experimental work in adults, see Cosmides, Barrett, and Tooby 2010 and Cosmides and Tooby 1989. Cosmides and her collaborators pioneered this interesting line of work, although they see these findings as due to a psychology for reciprocal altruism. The problem with this reasoning is that it doesn't explain why it seems to work with any costly norm or why the rules can be culturally transmitted (N. Henrich and Henrich 2007).

20. See Fessler 1999, 2004. For research showing the universality of shame displays, see Tracy and Matsumoto 2008, and for work examining the automatic and unconscious signals communicated by shame and pride in diverse societies, see Tracy et al. 2013.

21. See Hamlin et al. 2011, 2013, Hamlin, forthcoming, 2013b, Hamlin and Wynn 2011, Hamlin, Wynn, and Bloom 2007, and Sloane, Baillargeon, and Premack 2012. For work on fairness in infants and toddlers, see Sloane, Baillargeon, and

Premack 2012. I've noted the ages in months that these papers cite, but there's no reason to suspect these maturational patterns are the same across diverse societies. On the operation of these reputational logics in small-scale societies, see Henrich and Henrich 2014 and Mathew n.d.. For models predicting these patterns, see Panchanathan and Boyd 2004, Henrich and Boyd 2001, Chudek and Henrich n.d., Boyd and Richerson 1992, and Axelrod 1986.

22. Anthropologists have long argued that talking of "tribes" suggests that everyone belongs to a single discrete, bounded, and hermetically sealed group that never changes and endures through all time. Since this book is all about the dynamics of cultural evolution, I hope my usage here won't be misinterpreted as to suggest any of this outdated baggage.

23. See Diamond 1997.

24. See McElreath, Boyd, and Richerson 2003, Boyd and Richerson 1987, and N. Henrich and Henrich 2007.

25. See Shutts, Kinzler, and DeJesus 2013, Kinzler, Dupoux, and Spelke 2007, Kinzler, Shutts, and Spelke 2012, and Kinzler et al. 2009.

26. Consistent with the hard-to-fake features of ethnic markers, simply speaking the language in some way is not enough. You need to speak it "properly," without an accent (from the learners' point of view). The data here also make it clear that it's not the children's comprehension that is driving these preferences. Also, as we saw in chapter 8, people of all ages are keenly attuned to prestige differences and prefer to interact and learn from more prestigious individuals. Thus, when an individual's language or dialect cues prestige for the observer, this too can influence decisions to interact and learn (Kinzler, Shutts, and Spelke 2012).

27. See our book on this research (N. Henrich and Henrich 2007). Clearly, Chaldeans resemble many other successful immigrant populations, such as Jews, Koreans, and Armenians. Some Chaldeans explicitly pointed to Jews as worthy of emulation on these counts.

28. See Gerszten and Gerszten 1995 and Tubbs, Salter, and Oakes 2006.

29. See Kanovsky 2007, Gil-White 2001, Hirschfeld 1996, Moya, Boyd, and Henrich, forthcoming, Baron et al. 2014, and Dunham, Baron, and Banaji 2008.

30. Of course, it's possible to get children and adults to react to the norm violations of out-group people as well, simply by increasing the degree of the violation. The point is that there is an asymmetric response, which goes to the detriment of fellow in-group members (Schmidt, Rakoczy, and Tomasello 2012). For cross-cultural work with adults, see Bernhard, Fischbacher, and Fehr 2006, and Gil-White 2004.

31. Further work on this question is needed, though see Esteban, Mayoral, Ray 2012a, 2012b, and Fearon 2008. To be clear, the claim here is not that more ethnically heterogeneous places will have more civil wars; rather, when civil wars do occur, the divisions tend to fall along ethnic or religious lines.

32. Dominated by Americans, psychology has been missing the key fault lines in human psychology by focusing either on laboratory-friendly arbitrary groups (people who all like a certain painting) or on the peculiar differences between blacks and whites in the United States.

33. See Kinzler et al. 2009 and Pietraszewski and Schwartz 2014a, 2014b. Where there's antagonism among social categories marked by features like skin color,

people may also use such features to mark alliances (Pietraszewski, Cosmides, and Tooby 2014).

34. For the idea, see Mathew, Boyd, and van Veelen 2013 and N. Henrich and Henrich 2007. For the Gebusi example, see Knauft 1985.

35. See Gilligan, Benjamin, and Samii 2011.

36. To be clear, this process doesn't require genetic group selection. Intergroup competition will favor cultural practices that sanction people for norm violations, including those who fail to sanction norm violators (if necessary) even more harshly when groups are under threat. Intragroup sanctioning mechanisms, like a loss of mating opportunities, can be sufficient to favor the relevant genes. If, however, there were sufficiently stable intergroup genetic variation, it could augment the culture-gene coevolutionary process.

37. See Bauer et al. 2013.

38. See Voors et al. 2012, Gneezy and Fessler 2011, Bellows and Miguel 2009, and Blattman 2009. Cassar, Grosjean, and Whitt's (2013) evidence might appear contradictory, but it is actually supportive of the theory presented here because this civil war pitted neighbor against neighbor, so there was no in-group or local community to bond with. From the psychology laboratory, converging evidence from controlled experiments with undergraduates demonstrates that even the perception of intergroup competition instantly increases cooperation in Public Goods Games (Puurtinen and Mappes 2009, Saaksvuori, Mappes, and Puurtinen 2011, Bornstein and Erev 1994, Bornstein and Benyossef 1994). Similarly, induced threats of uncertainty or death in the laboratory motivate both greater norm compliance and greater willingness to punish norm violators (Heine, Proulx, and Vohs 2006, Hogg and Adelman 2013, Grant and Hogg 2012, Smith et al. 2007).

39. See Bauer et al. 2013.

Chapter 12. Our Collective Brains

1. For the case of the Polar Inuit, see Boyd, Richerson, and Henrich 2011a, Rasmussen, Herring, and Moltke 1908, and Gilberg 1984. I mention the importance of the readoption of the technologies because some researchers are inclined to argue that all foraging people are always behaving optimally, ergo, there must be some minor ecological differences that make all the lost technologies inefficient.

2. For theoretical work on these processes, see Shennan 2001, Powell, Shennan, and Thomas 2009, Henrich 2004b, 2009b, Kobayashi and Aoki 2012, Lehmann, Aoki, and Feldman 2011, and van Schaik and Pradhan 2003.

3. For a full exposition of this model, see Henrich 2009.

4. Muthukrishna et al. 2014.

5. See Derex et al. 2013.

6. Kline and Boyd (2010) analyzed a large number of ecological and environmental variables but found that few showed any relationship with the size of the toolkit or the sophistication of the technology, and none substantially diminished the magnitude of the relationship with population size.

7. See Collard, Ruttle, et al. 2013.

8. In contrast, the number of food-related techniques is not readily explained by free time or for the necessity of additional food sources (van Schaik et al. 2003; also see Jaeggi et al. 2010. Among chimpanzees, a similar relationship does emerge, but the dataset is small (Lind and Lindenfors 2010).

9. W.H.R. River's essay is at http://en.wikisource.org/wiki/The_Disappearance_of _Useful_Arts#cite_note-1.

10. See Henrich 2004b, 2006, Jones 1974, 1976, 1977c, 1977b, and Diamond 1978. On fire: the claim is that Tasmanians lost the ability to make fire. They still had fire. While losing the ability to make fire is not unheard of (Holmberg 1950, Radcliffe-Brown 1964), this particular claim is controversial (Gott 2002).

11. Tasmanian technology is also simple compared to that of other foraging populations at the same southern latitudes, like the Fuegians at southern tip of South America and the inhabitants of both southern New Zealand and the Chatham islands (Henrich 2004b, 2006).

12. On Paleolithic evidence, see McBrearty and Brooks 2000, Boyd and Silk 2012, and Klein 2009; pelagic fishing, O'Connor, Ono, and Clarkson 2011; bone tools, Yellen et al. 1995; stone tools, Jones 1977a, 1977b; arrivals, Boyd and Silk 2012; hafted stone points, Wilkins et al. 2012.

13. See Jones 1974, 1976, 1977a, 1977b, 1977c, 1990, 1995, Colley and Jones 1988, and Diamond 1978. I've reviewed the Tasmania case and consider various objections in Henrich 2004b and 2006.

14. There is one important set of anomalous data relevant to this line of argument that I've not discussed. Analyzing tool-complexity data from human foragers, Collard and his colleagues have argued that they find no relationship between population size or interconnectedness and tool measures of complexity (Collard et al. 2011, 2012, Collard, Kemery, and Banks 2005). Instead, they argue that eco-logical risk favors more individual investment in *risk-relevant* technological complexity. While their point is important to keep in mind, there are two problems with this effort. First, it's well established that foragers respond to risk by building far-reaching networks of social relationships, which they can turn to when disaster strikes (Wiessner 1982, 1998, 2002). Thus, finding a positive relationship between ecological risk and technological complexity is largely also supportive of the view presented in this chapter, since greater risk will cause individuals to develop broader networks using their cultural technologies (e.g., rituals, naming, gift giving), which will result in more-complex tools as a product (and potentially a by-product). Evidence for this view is found in Collard et al. 2013, where ecological risk is positively related to possessing all kinds of technologies, including those that have nothing to do with managing risk. Second, the nature of foraging groups makes determining the population size relevant for information flow about technology tricky and unreliable. In many places, foraging bands are loosely interconnected networks without clear boundaries. This is unlike many populations of farmers and herders, who often control and defend territory. Consequently, negative results aren't unexpected for foragers, given the challenge of isolating the relevant pool of learners. This also explains why we see the predicted relationships for farmers and herders, but not for foragers, in Collard's various analyses.

15. Chimpanzees and capuchins are both interesting because they have relatively

large brains, and field studies indicate some simple patterns of cultural variation. For this excellent study, see Dean et al. 2012.

16. See Henrich and Tennie, forthcoming.
17. See Stringer 2012, Klein 2009, and Pearce, Stringer, and Dunbar 2013.
18. See Deaner et al. 2007.
19. On the importance of rich coastal environments, see Jerardino and Marean 2010.
20. For an earlier version of this point, see Henrich 2004b. For evidence of differences between the life length of Neanderthals and subsequent Upper Paleolithic peoples, see Caspari and Lee 2004, 2006, and Bocquet-Appel and Degioanni 2013. On population size estimates and variation, see Klein 2009 and Mellars and French 2011. On projectiles for the expanding African populations, see Shea and Sisk 2010. For a discussion of the differences in trade networks between Upper Paleolithic peoples and Neanderthals, which suggest a difference social network interconnectedness see Ridley 2010.
21. See Henrich 2004b.
22. See McBrearty and Brooks 2000. For information on Paleolithic bows and arrows, see Shea and Sisk 2010, Shea 2006, and Lombard 2011. On the loss of bows and arrows, boats, and pottery, see Rivers 1931.
23. See Powell, Shennan, and Thomas 2009.
24. See van Schaik and Burkart 2011.
25. Gruber et al. 2009, 2011.
26. On Australian technology, see Testart 1988. On the wheel, see Diamond 1997. Mayan toys had wheels, which profoundly underlines my point.
27. See Frank and Barner 2012. Stripped of their calculators, calculator users typically flounder. Thanks to Yarrow Dunham for pointing me toward this work.

Chapter 13. Communicative Tools with Rules

1. In the nineteenth century, discussions of the origins of language invited so much undisciplined speculation that in 1866 the influential Société de Linguistique de Paris banned the topic (Deutscher 2005, Bickerton 2009).
2. For a discussion of this point, see Deutscher 2005.
3. See Tomasello 2010, Kuhl 2000, and Fitch 2000.
4. See Webb 1959, Kendon 1988, Mallery 2001 (1881), Tomkins 1936, and Kroeber 1958. The Plains Indians had also developed specialized signs for military purposes and long-distance communication. These inspired a host of developments in the U.S. military, including the Signal Corp.
5. See Kendon 1988.
6. See Busnel and Classe 1976 and Meyer 2004. Busnel and Classe argue that the length of human ear canal is actually better calibrated to pick the frequency of whistles than spoken language. To see people conversing in a whistled language, see https://www.youtube.com/watch?v=P0aoguO_tvI. Or, see https://www.youtube.com/watch?v=C0CIRCjoICA. For a discussion of drums, horns, etc., see Stern 1957.
7. See Munroe, Fought, and Macaulay 2009 and Fought et al. 2004.
8. See Ember and Ember 2007 and Nettle 2007.

9. See Nettle 2007.
10. For word counts, see Bloom 2000 and Deutscher 2010. It's important to keep in mind that people in small-scale societies are usually multilingual, so the total number of words known might be quite large. Nevertheless, the number of words available in any particular language are fewer than in the languages of larger-scale societies.
11. From W.H.R. Rivers as described in Deutscher 2010.
12. See Kay and Regier 2006, Webster and Kay 2005, Kay 2005, Berlin and Kay 1991, and D'Andrade 1995.
13. See Deutscher 2010. Following a long history of research on the topic, Brent Berlin and Paul Kay (1969) have argued that basic color terms emerge in response to the cultural evolution of technologies for separating color from its host objects—when color could be picked for clothing and other cultural products. Cultural evolution builds color term inventories in many ways, for example by extracting terms from the objects they previously infused. Words for "green" often derive from the words for "unripe" or "immature" (as in fruit); "violet" comes from words for flowers. Or, terms are simply borrowed through language contact.
14. There remains some debate on this, see Kay and Regier 2006, Xu, Dowman, and Griffiths 2013, Franklin et al. 2005, and Baronchelli et al. 2010. One worry about this work is that it assumes that the perceptual contours in human color perception are fixed and universal. This is noteworthy, since the color-term maps for some languages do deviate from the predictions of this approach. An outstanding question is what other routes cultural evolution might take to build a color-naming system. The dominant world pattern may be only one among multiple possible routes. I also predict that population size and interconnectedness will predict the degree to which a language is optimized to exploit the contours of our visual system.
15. Vowels sounds may arise through an analogous process (Lindblom 1986).
16. See Franklin et al. 2005, D'Andrade 1995, Goldstein, Davidoff, and Roberson 2009, and Kwok et al. 2011.
17. Gordon 2005, Dehaene 1997. On counting systems in New Guinea, see http://www.uog.ac.pg/glec/thesis/thesis.htm.
18. See Pitchford and Mullen 2002.
19. See Flynn 2012.
20. See Tomasello 2000a, 2000b.
21. See Deutscher 2010 and Everett 2005.
22. Not surprisingly, there remains much controversy about this, see Hay and Bauer 2007, Moran, McCloy, and Wright 2012, Atkinson 2011, and Wichmann, Rama, and Holman 2011.
23. See Nettle 2012 and Wichmann, Rama, and Holman 2011.
24. See Goldin-Meadow et al. 2008. Cultural learning provides humans with a final principle: communicative economy. Statements are made with the shared goal of communication, so contextually obvious information need not be included and basic information should not be repeated. The listener has to assume that the speaker is trying to communicate something and taking what the listener knows into account (though see Pawley 1987). How do humans come to share

each others' intentions and goals? They copy them—it's a product of cultural learning. If you have a goal of honest communication and I think you are a great model, that is, well worth copying, then I will tend to copy the goal of honest communication. Now you and I have *shared intentionality* (Tomasello 1999), at least about communication.

25. See Christiansen and Kirby 2003, Heine and Kuteva 2002a, 2002b, 2007, and Deutscher 2005.
26. See Fedzechkina, Jaeger, and Newport 2012.
27. See Deutscher 2005.
28. See Wray and Grace 2007, Kalmar 1985, Newmeyer 2002, Pawley 1987, and Mithun 1984. Kalmar argues that a Canadian Inuit language is in the process of evolving a full-blown subordination tool due to the spread of literacy and writing.
29. See Lupyan and Dale 2010.
30. See Deacon 1997 and Kirby 1999.
31. Cultural evolution can even explain the emergence of elements as basic as words—or, what linguists call compositionality. Communicative systems may have started without individual words that could be recombined in myriad ways. Instead, sounds or sound combinations might have mapped onto what we can think of as multiword combinations or phrases. The word *bamakuba* might have meant "cook the meat more," without separate words for "cook," "meat," and "more." With a separate sound or set of sounds for every phrase or sentence, vocabularies can quickly explode and become unmanageable. However, cultural transmission with limited memories favors breaking things down—compositionality—in ways that are easy to remember (Brighton, Kirby, and Smith 2005). Maybe the process was something like how "-gate" in "Watergate Hotel" got broken off and redeployed to mean "scandal," as in Monicagate and Climategate. Individuals who begin breaking things down in ways that are easily learnable for others will be more successful and be more likely to be imitated. Languages with compositionality (words) will persist and outcompete noncompositional languages.
32. See Kirby, Christiansen, and Chater 2013, Smith and Kirby 2008, Kirby, Cornish, and Smith 2008, and Christiansen and Chater 2008.
33. See Striedter 2004.
34. See Striedter 2004 and Fitch 2000.
35. Any account of this early culture-gene coevolutionary process is highly speculative. However, this "gesture first" account of our culturally transmitted and evolving communicative system is consistent with several empirical facts. First, to the degree that other apes can learn communicative elements, they learn manual gestures (sign systems), not verbalizations or facial expressions. Efforts to teach apes to speak have failed. Apes do communicate with vocalizations, but these vocalizations are a fixed repertoire of sounds that don't vary among groups, unlike their gestures. This suggests that human ancestors would have been much more susceptible to culturally transmitted gestures than to vocalization (Tomasello 2010). Second, as we've seen, gestures are still part of our communicative systems, and many hunter-gathers have both spoken languages and gestural sign languages. Third, infants are just as good, and maybe better, at learning gestural

signs for communication compared to speech. Learning a sign language doesn't appear to be any more difficult for children than learning a spoken language. Infants engage in gestural mimicry when learning to make speech sounds. They watch their models' mouths closely, and this influences their performance. Adults will confuse sounds like /b/ and /p/, as in "bat" and "pat," unless they can see the mouth of the speaker (Tomasello 2010, Kuhl 2000, Corballis 2003), so mouth gestures are part of speech processing. Fourth, tool use, gestures, and speech all share a substantial swath of neural circuitry.

36. See Fitch 2000.

37. See Csibra and Gergely 2009 and Kuhl 2000. Our team studied pedagogical cues across seven diverse societies and found the spontaneous use of at least some cues in all seven. However, the frequency of cueing varied substantially, as did which of a small set of specific cues were used. Only instructive pauses were found everywhere. For motherese, some have claimed it doesn't exist in some societies, though these claims were not based on systematic observational data collection and quantitative analysis. Working in my Fijian field site in the South Pacific, the developmental psychologist Tanya Broesch (2011) found much lower rates of motherese than among American mothers, but some motherese was still clearly present. In general, educated Westerners are at the high end of the distribution for both the use of pedagogical cues and motherese.

38. Bickerton 2009, Christiansen and Kirby 2003.

39. See Sterelny 2012b. Wadley (2010) discusses this recursion with regard to adhesive manufacturing dating back hundreds of thousands of years.

40. See Conway and Christiansen 2001. Another feature of languages that impresses philosophers is that we can use it to discuss the past and future, as well as people, things, and events that are not present. So, language is "stimulus independent." But as discussed for hierarchical constructions in language, having the capacity to think and plan independent of stimulus, and about the past and future, may not be that valuable for communication unless others can do it as well. Here, the culture-gene coevolution for toolmaking, learned skills, and social norms might again have led the way for languages. For example, difficult skills, like accurate spear-throwing, need to be practiced off-line (i.e., when not actually hunting or attacking) but in anticipation of future situations. Unlike language, there need not be anyone else with this capability (for spear-throwing, e.g.) to make it useful. And selection pressures for practicing will get stronger as tools and skills get more complex and others start practicing (Sterelny 2012b). Similarly, social norms, like many objects in language, are not visible, but as they emerge, individuals will need to anticipate what happens if they are violated. Thus, the evolution of both skills and norms will then select for minds better able to think about the past and future as well as things not physically manifest, like social norms.

41. See Conway and Christiansen 2001 and Reali and Christiansen 2009.

42. See Reali and Christiansen 2009, Tomblin, Mainela-Arnold, and Zhang 2007, and Enard et al. 2009. For a more general review of *FOXP2*, see Enard 2011.

43. See Stout and Chaminade 2012, Stout et al. 2008, Stout and Chaminade 2007, and Calvin 1993.

44. See Dediu and Ladd 2007.

45. For modeling work on deception and language, see Lachmann and Bergstrom 2004.
46. See N. Henrich and Henrich 2007 and Boyd and Mathew n.d..
47. For an introduction, see Henrich 2009a. For experiments on the transmission of counterintuitive beliefs, see Willard et. al., n.d. For related work, see Sperber et al. 2010.

Chapter 14. Enculturated Brains and Honorable Hormones

1. More technically, the area is usually called the visual word form area (Coltheart 2014). I'm drawing "letterbox" from Dehaene 2009.
2. The exact location of the letterbox does vary depending on the writing system. Japanese readers, for example, appear to have separate letterboxes for their syllabic Kana script and their logographic Kanji script. The important point is that location is constrained by the demands of the task and the innate neurogeography of human brains (Coltheart 2014, Dehaene 2014).
3. See Dehaene 2009, Ventura et al. 2013, Szwed et al. 2012, and Dehaene et al. 2010.
4. See Dehaene 2009, Ventura et al. 2013, Szwed et al. 2012, Dehaene et al. 2010, Carreiras et al. 2009, and Castro-Caldas et al. 1999.
5. See Ventura et al. 2013 and Dehaene et al. 2010. This means that the apparent brain asymmetry in face processing in "humans" was due to relying on highly literate participants in experiments.
6. See Coltheart 2014 and Dehaene 2014.
7. See Downey 2014.
8. See Little et al. 2008, 2011, Jones et al. 2007, Bowers et al. 2012, and Place et al. 2010. See Buss 2007 for an overview of the evolutionary psychology of mating.
9. See Zaki, Schirmer, and Mitchell 2011 and Klucharev et al. 2009.
10. See Plassmann et al. 2008.
11. On blind tasting, see Goldstein et al. 2008.
12. See Woollett and Maguire 2009, 2011, Woollett, Spiers, and Maguire 2009, Maguire, Woollett, and Spiers 2006, and Draganski and May 2008.
13. See Hedden et al. 2008.
14. See Nisbett 2003.
15. For the immigration studies, see Algan and Cahuc 2010, Fernandez and Fogli 2006, 2009, Guiso, Sapienza, and Zingales 2006, 2009, Giuliano and Alesina 2010, and Almond and Edlund 2008.
16. See Nisbett and Cohen 1996.
17. See Grosjean 2011.
18. See Benedetti and Amanzio 2011, 2013, Benedetti, Carlino, and Pollo 2011, Finniss et al. 2010, Price, Finniss, and Benedetti 2008, Benedetti 2008, 2009, and Guess 2002.
19. See Moerman 2000, 2002.
20. For reviews, see Finniss et al. 2010, Price, Finniss, and Benedetti 2008, and Benedetti 2008. For a study showing how tactile stimuli is turned into pain via the nocebo, see Colloca, Sigaudo, and Benedetti 2008.

21. Of course, being good at self-regulation may have caused people to do other things correctly besides take the medications. See Horwitz et al. 1990 for an analysis that seeks to address these issues.
22. See Benedetti et al. 2013.
23. This experiment is from Craig and Prkachin 1978. Also, see Goubert et al. 2011 and Craig 1986. For a nice, more recent experiment showing how powerful observational learning is compared to verbal suggestions or conditioning, see Colloca and Benedetti 2009.
24. See Finniss et al. 2010, Price, Finniss, and Benedetti 2008, Benedetti 2008, Kong et al. 2008, and Scott et al. 2008.
25. See Phillips, Ruth, and Wagner 1993.
26. Results like these provide a major challenge to economic theory. The models that economists make of choices have to map people's choices onto their eventual outcomes. For example, if Herb chooses to take Drug Alpha for his back pain, an economist might assume that Drug Alpha will cure Herb with a probability p (say, 65% chance) and not cure Herb with a probability $1 - p$ (say, a 35% chance). This probability p is typically assumed to be a feature of the world. However, what I just explained in the main text is that in a real biological way p actually depends substantially on what Herb believes about Drug Alpha. There's a causal connection between Herb's beliefs about Alpha and the chances of various outcomes (cure or not) in the world. This is not some wacky special case, applicable only to a narrow range of drug choices. As we saw, it affects longevity in California and probably also influences whole industries of "traditional," New Age, and spiritual healing and health practices, not to mention fad diets and exercise routines. More important, in many places in the world, witchcraft is a widespread and remarkably stable set of beliefs. These beliefs may persist in part because they are reinforced by the operation of the nocebo effect, such that perceiving others as angry or envious in the context of witchcraft beliefs actually then causes biological effects that increase the chances of illness. So, in a sense, witchcraft works.

Chapter 15. When We Crossed the Rubicon

1. For an introduction, see Klein 2009 and Boyd and Silk 2012. For inferences to the last common ancestor, see Henrich and Tennie, forthcoming. For sundry dates on our divergence from our last common ancestor with chimpanzees, see Suwa et al. 2009, Scally et al. 2012, and Klein 2009. The reason I'm using chimpanzees to set an *upper limit* on the cultural abilities of this common ancestor is because one of the selection pressures—climatic variation—that has likely made our lineage more cultural since we split may have similarly influenced other species, including chimpanzees. Much of the mathematical theory underpinning culture-gene coevolution suggests that the kind of relatively rapid environmental variation that has occurred over the last 3 million years, up to ten millennia ago, should favor a greater reliance on social learning (Boyd and Richerson 1985, 1988, Wakano and Aoki 2006, Aoki and Feldman 2014), making many species more reliant on social learning in adapting to shifting environments. Consequently, since the lineages leading to modern apes experienced these vari-

ations as well, they too may have evolved an increased reliance on social learning during this epoch.

2. See Klein 2009.

3. On chimpanzee culture, see Henrich and Tennie, forthcoming; for the "brush heads," see Sanz and Morgan 2007, 2011. On the effects of overall brain size on social learning and other cognitive abilities, see Deaner et al. 2007, Reader, Hager, and Laland 2011, Klein 2009, and Boyd and Silk 2012. On comparative brain sizes, see Klein 2009 and Boyd and Silk 2012.

4. See McPherron et al. 2010.

5. See Hyde et al. 2009.

6. See Panger et al. 2002, but note that the pollicis longus tendon may have appeared earlier in *Ardipithecus ramidus* (White et al. 2009).

7. See Klein 2009 and Boyd and Silk 2012.

8. See Stout and Chaminade 2012. On Kanzi, see Schick et al. 1999 and Toth and Schick 2009.

9. See Klein 2009, Ambrose 2001, Wrangham and Carmody 2010, and Boyd and Silk 2012. On fish and turtles, see Stewart 1994 and Archer et al. 2014. Note that fish may have also been used by the hominins that preceded Early Homo.

10. See Stedman et al. 2003, 2004, McCollum et al. 2006, and Perry, Verrelli, and Stone 2005.

11. See Backwell and d'Errico 2003, d'Errico and Backwell 2003, and d'Errico, Backwell, and Berger 2001.

12. See Stout and Chaminade 2012, Stout 2011, Faisal et al. 2010, Stout et al. 2008, and Klein 2009.

13. See Morgan et al. 2015.

14. See Stout 2011, Faisal et al. 2010, Stout et al. 2010, Klein 2009, and Delagnes and Roche 2005.

15. On the anatomical changes related to food processing, see Wrangham and Carmody 2010 and Wrangham 2009. On the use of fire, see Goren-Inbar et al. 2004, Klein 2009, and Berna et al. 2012.

16. See Stout 2002, 2011, Beyene et al. 2013, and Perreault et al. 2013. This newer view is controversial, since many researchers have long claimed that this was a period of technological stasis. I think the reason researchers saw stasis is a lack of appreciation for the dynamics of cultural evolution, and specifically, the ways in which populations size, sociality, migration, and ecological shocks influence it.

17. See Roach et al. 2013. The evolution of throwing *Homo erectus* probably benefited from some anatomical preadaptation in Australopiths that were byproducts of the evolution of bipedal walking or running.

18. See chapter 5 for references.

19. My thanks to the KGA Research Project and Berhane Asfaw for permitting me to use this image (figure 15.1).

20. On change in early Acheulean tool tradition, see Beyene et al. 2013, and on various techniques that were added cumulatively, see Stout and Chaminade 2012 and Stout 2011.

21. See Stout et al. 2010.

22. Anatomically, I'm referring to the styloid process on the end of the third metacarpal (Ward et al. 2013, 2014).

23. For evidence and discussion of temporal accumulations in complexity, see Stout 2011 and Perreault et al. 2013. As noted, this view remains quite controversial, though even fairly conservative approaches distinguish less and more complex variants of both the Oldowan and Acheulean tool industries (Klein 2009).
24. See Alperson-Afil et al. 2009, Goren-Inbar et al. 2002, 2004, Rabinovich, Gaudzinski-Windheuser, and Goren-Inbar 2008, Goren-Inbar 2011, and Sharon, Alperson-Afil, and Goren-Inbar 2011.
25. See Wilkins et al. 2012, Wilkins and Chazan 2012, Klein 2009, Wadley 2010, Wadley, Hodgskiss, and Grant 2009, and McBrearty and Brooks 2000. On the ears and the auditory capacities of *Homo heidelbergensis*, see Martinez et al. 2013.

Chapter 16. Why Us?

1. See Reader, Hager, and Laland 2011, Reader and Laland 2002, van Schaik, Isler, and Burkart 2012, Pradhan, Tennie, and van Schaik 2012, van Schaik and Burkart 2011, and Whiten and van Schaik 2007.
2. See Boyd and Richerson 1996.
3. By "individual learning" I'm referring to a broad class of cognitive abilities that allow individuals through their direct experience with the environment to select, on average, more adaptive behaviors or to better achieve their goals or satisfy their preferences. This is also called asocial learning. These abilities need not be either entirely "domain general" or narrowly "domain specific." I expect them to be applicable to many problems, but not all problems.
4. See Meulman et al. 2012 for the basic argument here.
5. Admittedly, this leaves open the question of why bonobos and gorillas don't use tools more than orangutans.
6. See Suwa et al. 2009 and White et al. 2009 on *Ardipithecus ramidus*. Ground-dwelling apes may be much older than 5 million years, but these details are not relevant for my argument.
7. On predators, see Plummer 2004 and Klein 2009: 277.
8. See Boyd and Richerson 1985, 1988, and Aoki and Feldman 2014 for evolutionary models of social learning. For papers combining theoretical and empirical insights, see Richerson and Boyd 2000a, 2000b. If environments change too quickly, say every generation or every decade, then natural selection will favor individual learning. At the other extreme, if environments are changing very rapidly, say on an hour-by-hour basis, then neither individual nor social learning can help. Natural selection will go back to favoring genes that fix on the best trait appropriately averaged across the range of environments experienced.
9. See Isler and van Schaik 2009, 2012 and Isler et al. 2012.
10. See Langergraber, Mitani, and Vigilant 2007, 2009. It is important not to overstate the implications of these data. We do not have equivalent data from other chimpanzee communities, so we shouldn't be too confident that the pairing is being caused by the unusually large group size.
11. In addition to the effects of group size and social learning, predatory threats, especially in the form of raids from other groups, may further reduce the effectiveness of dominance competition. Too many fights and injuries from one's competitors could leave males weakened and ill-prepared to deal with predators.

12. See Chapais 2008: 205. In my discussion here, I'm assuming that we are starting with a large primate group in which the females depart and the males stay around, as in chimpanzees. However, this is not crucial. We could start with gorilla-like groups, who already have pair-bonding, and examine what happens when predatory threats force them into larger groups. Also note that primates do appear to have some limited ability to detect relatives, including paternal relatives, besides through the familiarity mechanisms I've described here (Langergraber 2012).

13. As pair-bonding spreads as a strategy, male-male competition may decline, resulting in a reduction in the large canines males use in fighting each other (Lovejoy 2009).

14. Part of this kin recognition package is an aversion to sex with these relatives, so we don't have to worry about a daughter's copying mom's sexual desires toward dad.

15. On alloparenting in hunter-gatherers and other small-scale societies, see Crittenden and Marlowe 2008, Hewlett and Winn 2012, Kramer 2010, and Kaplan et al. 2000.

16. See Morse, Jehle, and Gamble 1990 and Lozoff 1983. Mothers in many societies toss away the colostrum.

17. See Crittenden and Marlowe 2008 for the case of the girl who was scolded for not helping with an infant. This case was further enriched by Alyssa via email correspondence (2014). See Kramer 2010 for a review of data alloparental care.

18. See Hrdy 2009, Burkart, Hrdy, and Van Schaik 2009, and Burkart et al. 2014. Of course, the prosociality shown in these experiments is limited to very small groups of close relatives. It cannot begin to explain the patterns of prosociality discussed in chapter 11.

19. It's also possible that the boys lacked the required tuber expertise due to their relative youth and inexperience. Nevertheless, Frank assured me via email (2014) that Hadza women are more knowledgeable than men about tubers.

20. See Benenson, Tennyson, and Wrangham 2011.

21. See Chapais 2008 and Hill et al. 2011.

22. As noted in chapter 11, cultural learning contributes dramatically to the effectiveness of reciprocity-based cooperation. Consequently, as natural selection improves individuals' cultural learning abilities, reciprocity will become a more effective strategy for sustaining cooperation and establishing enduring social relationships. These relationships can only further fuel the opportunities for social learning from others and the possibilities for alloparental care from mom's friends (Crittenden and Marlowe 2008, Hewlett and Winn 2012).

23. On orangutans, see Jaeggi et al. 2010, and on chimpanzees, see Henrich and Tennie, forthcoming.

Chapter 17. A New Kind of Animal

1. See Maynard Smith and Szathmáry 1999.

2. See Buss 1999, Tooby and Cosmides 1992, Pinker 1997, 2002, and Smith and Winterhalder 1992.

3. Buss 2007: 419.

4. For a review of the evolution of the human body, see Lieberman 2013.
5. See Richerson and Henrich 2012.
6. See Flynn 2007, 2012.
7. See Basalla 1988, Mokyr 1990, Diamond 1997, and Henrich 2009b.
8. This metaphor builds on one often used by Rob Boyd.
9. See Cochran and Harpending 2009 on increasing rates of genetic evolution. On cultural evolutionary rates, see Mesoudi 2011b and Perreault 2012. On intergroup competition, see Turchin 2005, 2010.
10. For research on the spread of prosocial norms for dealing with strangers after the origins of agriculture, see Ensminger and Henrich 2014, chapters 2 and 4.
11. The older "rites of terror" were often filtered out by cultural evolution because, by binding smaller political units too tightly together, they threatened the integrity and stability of the new and larger political units encompassing them (Norenzayan et al., forthcoming).
12. For research on the evolution of religions with big, moralizing gods, see Norenzayan 2014, Atran and Henrich 2010, and Norenzayan et al., forthcoming.
13. See Diamond 1997. Much of Diamond's famous argument in *Guns, Germs, and Steel* only makes sense in the light of the evolutionary foundations I've developed here.
14. See Henrich 2009b and also my next book.
15. See Henrich, Heine, and Norenzayan 2010a, 2010b, Apicella et al., forthcoming, Muthukrishna et al., n.d., and S. Heine 2008.
16. See Henrich, Boyd, and Richerson 2012.
17. See Chandrasekaran 2006.
18. See World Bank Group 2015.
19. See Bowles 2008 and Gneezy and Rustichini 2000.

REFERENCES

Abramson, J. Z., V. Hernandez-Lloreda, J. Call, and F. Colmenares. 2013. "Experimental evidence for action imitation in killer whales (*Orcinus orca*)." *Animal Cognition* 16 (1):11–22.

Alberts, S. C., J. Altmann, D. K. Brockman, M. Cords, L. M. Fedigan, A. Pusey, T. S. Stoinski, K. B. Strier, W. F. Morris, and A. M. Bronikowski. 2013. "Reproductive aging patterns in primates reveal that humans are distinct." *Proceedings of the National Academy of Sciences, USA* 110 (33):13440–13445.

Alcorta, C. S., and R. Sosis. 2005. "Ritual, emotion, and sacred symbols—The evolution of religion as an adaptive complex." *Human Nature* 16 (4):323–359.

Alcorta, C. S., R. Sosis, and D. Finkel. 2008. "Ritual harmony: Toward an evolutionary theory of music." *Behavioral and Brain Sciences* 31 (5):576–577.

Aldeias, V., P. Goldberg, D. Sandgathe, F. Berna, H. L. Dibble, S. P. McPherron, A. Turq, and Z. Rezek. 2012. "Evidence for Neandertal use of fire at Roc de Marsal (France)." *Journal of Archaeological Science* 39 (7):2414–2423.

Algan, Y., and P. Cahuc. 2010. "Inherited trust and growth." *American Economic Review* 100 (5):2060–2092.

Almond, D., and L. Edlund. 2008. "Son-biased sex ratios in the 2000 United States Census." *Proceedings of the National Academy of Sciences, USA* 105 (15):5681–5682.

Alperson-Afil, N., G. Sharon, M. Kislev, Y. Melamed, I. Zohar, S. Ashkenazi, R. Rabinovich, et al. 2009. "Spatial organization of hominin activities at Gesher Benot Ya'aqov, Israel." *Science* 326 (5960):1677–1680.

Altman, J., and N. Peterson. 1988. "Rights to game and rights to cash among contemporary Australian hunter-gatherers." In *Hunters and Gatherers: Property, Power and Ideology*, edited by T. Ingold, D. Riches and J. Woodburn, 75–94. Berg: Oxford.

Alvard, M. 2003. "Kinship, lineage, and an evolutionary perspective on cooperative hunting groups in Indonesia." *Human Nature* 14 (2):129–163.

Ambrose, S. H. 2001. "Paleolithic technology and human evolution." *Science* 291 (5509):1748–1753.

Amundsen, R. 1908. *The North West Passage, Being the Record of a Voyage of Exploration of the Ship "Gyöa" 1903–1907*. London: Constable.

Anderson, D. D. 1984. "Prehistory of North Alaska." In *Arctic*, Vol. 5 of *Handbook of*

REFERENCES

North American Indians, edited by D. Damas, 80–93. Washington DC: Smithsonian Institution Press.

Aoki, K. 1986. "A stochastic-model of gene culture coevolution suggested by the culture historical hypothesis for the evolution of adult lactose absorption in humans." *Proceedings of the National Academy of Science, USA* 83 (9):2929–2933.

Aoki, K., and M. W. Feldman. 2014. "Evolution of learning strategies in temporally and spatially variable environments: A review of theory." *Theoretical Population Biology* 91:3–19.

Apesteguia, J., S. Huck, and J. Oechssler. 2007. "Imitation—theory and experimental evidence." *Journal of Economic Theory* 136 (1):217–235.

Apicella, C., E. A. Azevedo, J. A. Fowler, and N. A. Christakis. 2014. "Isolated hunter-gatherers do not exhibit the endowment effect bias." *American Economic Review* 104(6) 1793–1805.

Apicella, C. L., F. Marlowe, J. Fowler, and N. Christakis. 2012. "Social networks and cooperation in Hadza hunter-gatherers." *American Journal of Physical Anthropology* 147:85–85.

Archer, W., D. R. Braun, J. W. K. Harris, J. T. McCoy, and B. G. Richmond. 2014. "Early Pleistocene aquatic resource use in the Turkana Basin." *Journal of Human Evolution* 77:74–87. http://dx.doi.org/10.1016/j.jhevol.2014.02.012.

Astuti, R., G. E. A. Solomon, and S. Carey. 2004. "Constraints on conceptual development." *Monographs of the Society for Research in Child Development* 69 (3):vii–135.

Atkinson, Q. D. 2011. "Phonemic diversity supports a serial founder effect model of language expansion from Africa." *Science* 332 (6027):346–349.

Atkisson, C., M. J. O'Brien, and A. Mesoudi. 2012. "Adult learners in a novel environment use prestige-biased social learning." *Evolutionary Psychology* 10 (3):519–537.

Atran, S. 1993. "Ethnobiological classification—Principles of categorization of plants and animals in traditional societies—Berlin, B." *Current Anthropology* 34 (2):195–198.

———. 1998. "Folk biology and the anthropology of science: Cognitive universals and cultural particulars." *Behavioral and Brain Sciences* 21:547–609.

Atran, S., and J. Henrich. 2010. "The evolution of religion: How cognitive by-products, adaptive learning heuristics, ritual displays, and group competition generate deep commitments to prosocial religions." *Biological Theory* 5 (1):1–13.

Atran, S., and D. L. Medin. 2008. *The Native Mind and the Cultural Construction of Nature*. Cambridge, MA: MIT Press.

Atran, S., D. L. Medin, E. Lynch, V. Vapnarsky, E. E. Ucan, and P. Sousa. 2001. "Folkbiology doesn't come from folkpsychology: Evidence from Yukatek Maya in cross-cultural perspective." *Journal of Cognition and Culture* 1 (1):3–42.

Atran, S., D. L. Medin, and N. Ross. 2004. "Evolution and devolution of knowledge: A tale of two biologies." *Journal of the Royal Anthropological Institute* 10 (2):395–420.

———. 2005. "The cultural mind: Environmental decision making and cultural Modeling within and across Populations." *Psychological Review* 112 (4):744–776.

Atran, S., D. L. Medin, N. Ross, E. Lynch, V. Vapnarsky, E. U. Ek, J. D. Coley, C. Timura, and M. Baran. 2002. "Folkecology, cultural epidemiology, and the spirit of the commons—A garden experiment in the Maya lowlands, 1991–2001." *Current Anthropology* 43 (3):421–450.

Axelrod, R. 1986. "An evolutionary approach to norms." *American Political Science Review* 80 (4):1095–1111.

Backwell, L. R., and F. d'Errico. 2003. "Additional evidence on the early hominid bone tools from Swartkrans with reference to spatial distribution of lithic and organic artefacts." *South African Journal of Science* 99 (5–6):259–267.

Baird, R. W. 2000. "The killer whale: Foraging specializations and group hunting." In *Cetacean Societies: Field Studies of Dolphins and Whales*, edited by J. Mann, R. C. O'Connor, P. L. Tyack, and H. Whitehead, 127–153. Chicago: University of Chicago Press.

Baldwin, J. M. 1896. "Physical and Social Heredity." *American Naturalist* 30:422–428.

Balikci, A. 1989. *The Netsilik Eskimo*. Long Grove, IL: Waveland Press.

Ball, S., C. Eckel, P. Grossman, and W. Zame. 2001. "Status in markets." *Quarterly Journal of Economics* 155 (1):161–181.

Bandura, A. 1977. *Social Learning Theory*. Englewood Cliffs, NJ: Prentice Hall.

Bandura, A., and C. J. Kupers. 1964. "Transmission of patterns of self-reinforcement through modeling." *Journal of Abnormal & Social Psychology* 69 (1):1–9.

Barnes, R. H. 1996. *Sea Hunters of Indonesia: Fishers and Weavers of Lamalera*. Oxford Studies in Social and Cultural Anthropology. Oxford: Clarendon Press.

Baron, A. S., and M. R. Banaji. 2006. "The development of implicit attitudes—Evidence of race evaluations from ages 6 and 10 and adulthood." *Psychological Science* 17 (1):53–58.

Baron, A. S., Y. Dunham, M. Banaji, and S. Carey. 2014. "Constraints on the acquisition of social category concepts." *Journal of Cognition and Development* 15 (2):238–268.

Baron, R. 1970. "Attraction toward the model and model's competence as determinants of adult imitative behavior." *Journal of Personality and Social Psychology* 14:345–351.

Baronchelli, A., T. Gong, A. Puglisi, and V. Loreto. 2010. "Modeling the emergence of universality in color naming patterns." *Proceedings of the National Academy of Sciences, USA* 107 (6):2403–2407.

Barrett, H. C., and J. Broesch. 2012. "Prepared social learning about dangerous animals in children." *Evolution and Human Behavior* 33 (5):499–508.

Barrett, H. C., T. Broesch, R. M. Scott, Z. J. He, R. Baillargeon, D. Wu, M. Bolz, et al. 2013. "Early false-belief understanding in traditional non-Western societies." *Proceedings of the Royal Society B: Biological Sciences* 280 (1755).

Barrett, H. C., L. Cosmides, and J. Tooby. 2007. "The hominid entry into the cognitive niche." In *Evolution of the Mind: Fundamental Questions and Controversies*, edited by S. Gangestad and J. Simpson, 241–248. New York: Guilford Press.

Basalla, G. 1988. *The Evolution of Technology*. New York: Cambridge University Press.

Basow, S. A., and K. G. Howe. 1980. "Role-model influence—Effects of sex and sex-role attitude in college students." *Psychology of Women Quarterly* 4 (4):558–572.

Bauer, M., A. Cassar, J. Chytilová, and J. Henrich. 2013. "War's enduring effects on the development of egalitarian motivations and in-group biases." *Psychological Science* 25 (1): 47–57 doi:10.1177/0956797613493444.

REFERENCES

Baumard, N., J. B. Andre, and D. Sperber. 2013. "A mutualistic approach to morality: The evolution of fairness by partner choice." *Behavioral and Brain Sciences* 36 (1):59–78.

Baumgartner, T., U. Fischbacher, A. Feierabend, K. Lutz, and E. Fehr. 2009. "The neural circuitry of a broken promise." *Neuron* 64 (5):756–770.

Bearman, P. 2004. "Suicide and friendship among American adolescents." *American Journal of Public Health* 94 (1):89–95.

Beck, W. 1992. "Aboriginal preparation of cycad seeds in Australia." *Economic Botany* 46 (2):133–147.

Beckerman, S., R. Lizarralde, M. Lizarralde, J. Bai, C. Ballew, S. Schroeder, D. Dajani, et al. 2002. "The Bari Partible Paternity Project, Phase One." In *Cultures of Multiple Fathers: The Theory and Practice of Partible Paternity in Lowland South America*, edited by S. Beckerman and P. Valentine, 14–26. Gainesville: University Press of Florida.

Beckerman, S., and P. Valentine, eds. 2002a. *Cultures of Multiple Fathers: The Theory and Practice of Partible Paternity in Lowland South America.* Gainesville: University Press of Florida.

———. 2002b. "Introduction: The concept of partible paternity among native South Americans." In *Cultures of Multiple Fathers: The Theory and Practice of Partible Paternity in Lowland South America*, edited by S. Beckerman and P. Valentine, 1–13. Gainesville: University Press of Florida.

Bell, A. V., P. J. Richerson, and R. McElreath. 2009. "Culture rather than genes provides greater scope for the evolution of large-scale human prosociality." *Proceedings of the National Academy of Sciences, USA* 106 (42):17671–17674. doi:10.1073/pnas.0903232106.

Bellows, J., and E. Miguel. 2009. "War and local collective action in Sierra Leone." *Journal of Public Economics* 93 (11–12):1144–1157.

Belot, M., V. P. Crawford, and C. Heyes. 2013. "Players of Matching Pennies automatically imitate opponents' gestures against strong incentives." *Proceedings of the National Academy of Sciences, USA* 110 (8):2763–2768.

Benedetti, F. 2008. "Mechanisms of placebo and placebo-related effects across diseases." *Annual Review of Pharmacology and Toxicology* 48:33–60.

———. 2009. *Placebo Effects: Understanding the Mechanisms in Health and Disease.* Oxford: Oxford University Press.

Benedetti, F., and M. Amanzio. 2011. "The placebo response: How words and rituals change the patient's brain." *Patient Education and Counseling* 84 (3):413–419.

———. 2013. "Mechanisms of the placebo response." *Pulmonary Pharmacology & Therapeutics* 26 (5):520–523.

Benedetti, F., E. Carlino, and A. Pollo. 2011. "How placebos change the patient's brain." *Neuropsychopharmacology* 36 (1):339–354.

Benedetti, F., W. Thoen, C. Blanchard, S. Vighetti, and C. Arduino. 2013. "Pain as a reward: Changing the meaning of pain from negative to positive co-activates opioid and cannabinoid systems." *Pain* 154 (3):361–367.

Benenson, J. F., R. Tennyson, and R. W. Wrangham. 2011. "Male more than female infants imitate propulsive motion." *Cognition* 121 (2):262–267. http://dx.doi.org/10.1016/j.cognition.2011.07.006.

Berlin, B., and P. Kay. 1991. *Basic Color Terms: Their Universality and Evolution*. Berkeley: University of California Press.

Berna, F., P. Goldberg, L. K. Horwitz, J. Brink, S. Holt, M. Bamford, and M. Chazan. 2012. "Microstratigraphic evidence of in situ fire in the Acheulean strata of Wonderwerk Cave, Northern Cape province, South Africa." *Proceedings of the National Academy of Sciences, USA* 109 (20):E1215-E1220.

Bernhard, H., U. Fischbacher, and E. Fehr. 2006. "Parochial altruism in humans." *Nature* 442 (7105):912–915.

Bettinger, E. P., and B. T. Long. 2005. "Do faculty serve as role models? The impact of instructor gender on female students." *American Economic Review* 95 (2):152–157.

Bettinger, R. L. 1994. "How, when and why Numic spread." In *Across the West: Human Population Movement and the Expansion of the Numa*, edited by D. Madsen and D. Rhode, 44–55. Salt Lake: University of Utah.

Bettinger, R. L., and M. A. Baumhoff. 1982. "The Numic spread: Great Basin cultures in competition." *American Antiquity* 47 (3):485–503.

Beyene, Y., S. Katoh, G. WoldeGabriel, W. K. Hart, K. Uto, M. Sudo, M. Kondo, et al. 2013. "The characteristics and chronology of the earliest Acheulean at Konso, Ethiopia." *Proceedings of the National Academy of Sciences, USA* 110 (5):1584–1591.

Bickerton, D. 2009. *Adam's Tongue: How Humans Made Language, How Language Made Humans*. New York: Hill and Wang.

Biesele, M. 1978. "Religion and folklore." In *The Bushmen*, edited by P. V. Tobias, 162–172 Cape Town: Human & Rousseau.

Billing, J., and P. W. Sherman. 1998. "Antimicrobial functions of spices: Why some like it hot." *Quarterly Review of Biology* 73 (1):3–49.

Bingham, P. M. 1999. "Human uniqueness: A general theory." *Quarterly Review of Biology* 74 (2):133–169.

Birch, L. L. 1980. "Effects of peer model's food choices on eating behaviors on preschooler's food preferences." *Child Development* 51:489–496.

Birch, S. A. J., N. Akmal, and K. L. Frampton. 2010. "Two-year-olds are vigilant of others' non-verbal cues to credibility." *Developmental Science* 13 (2):363–369.

Birch, S. A. J., and P. Bloom. 2002. "Preschoolers are sensitive to the speaker's knowledge when learning proper names." *Child Development* 73 (2):434–444.

Birch, S. A. J., S. A. Vauthier, and P. Bloom. 2008. "Three- and four-year-olds spontaneously use others' past performance to guide their learning." *Cognition* 107 (3):1018–1034.

Birdsell, J. B. 1979. "Ecological Influences on Australian aboriginal social organization." In *Primate Ecology and Human Origins*, edited by I. S. Bernstein and E. O. Smith, 117–151. New York: Garland STPM Press.

Blattman, C. 2009. "From violence to voting: War and political participation in Uganda." *American Political Science Review* 103 (2):231–247.

Bloom, G., and P. W. Sherman. 2005. "Dairying barriers affect the distribution of lactose malabsorption." *Evolution and Human Behavior* 26 (4):301–312.

Bloom, P. 2000. *How Children Learn the Meaning of Words*. Cambridge: MIT Press.

Blume, M. 2009. "The reproductive benefits of religious affiliation." In *The Biological*

Evolution of Religious Mind and Behavior, edited by E. Voland and W. Schiefen-hovel, 117–126. Berlin: Springer-Verlag.

Bocquet-Appel, J.-P., and A. Degioanni. 2013. "Neanderthal demographic estimates." *Current Anthropology* 54 (S8):S202–S213. doi:10.1086/673725.

Boehm, C. 1993. "Egalitarian behavior and reverse dominance hierarchy." *Current Anthrropology* 34 (3):227–254.

Bogin, B. 2009. "Childhood, adolescence, and longevity: A multilevel model of the evolution of reserve capacity in human life history." *American Journal of Human Biology* 21 (4):567–577.

Bohns, V. K., and S. S. Wiltermuth. 2012. "It hurts when I do this (or you do that): Posture and pain tolerance." *Journal of Experimental Social Psychology* 48 (1):341–345.

Bollet, A. J. 1992. "Politics and pellagra—The epidemic of pellagra in the United States in the early 20th century." *Yale Journal of Biology and Medicine* 65 (3):211–221.

Borinskaya, S., N. Kal'ina, A. Marusin, G. Faskhutdinova, I. Morozova, I. Kutuev, V. Koshechkin, et al. 2009. "Distribution of the alcohol dehydrogenase *ADH1B*47His* allele in Eurasia." *American Journal of Human Genetics* 84 (1): 89–92.

Bornstein, G., and M. Benyossef. 1994. "Cooperation in intergroup and single-group social dilemmas." *Journal of Experimental Social Psychology* 30 (1):52–67.

Bornstein, G., and I. Erev. 1994. "The enhancing effect of intergroup competition on group performance." *International Journal of Conflict Management* 5 (3):271–283.

Bowern, C., and Q. Atkinson. 2012. "Computational phylogenetics and the internal structure of Pama-Nyungan." *Language* 88 (4):817–845.

Bowers, R. I., S. S. Place, P. M. Todd, L. Penke, and J. B. Asendorpf. 2012. "Generalization in mate-choice copying in humans." *Behavioral Ecology* 23 (1):112–124.

Bowles, S. 2006. "Group competition, reproductive leveling, and the evolution of human altruism." *Science* 314 (5805):1569–1572.

———. 2008. "Policies designed for self-interested citizens may undermine "the moral sentiments": Evidence from economic experiments." *Science* 320 (5883): 1605–1609.

Bowles, S., R. Boyd, S. Mathew, and P. J. Richerson. 2012. "The punishment that sustains cooperation is often coordinated and costly." *Behavioral and Brain Sciences* 35 (1):20–21.

Boyd, D. 2001. "Life without pigs: Recent subsistence changes among the Irakia Awa, Papua New Guinea." *Human Ecology* 29 (3):259–281.

Boyd, R., H. Gintis, S. Bowles, and P. J. Richerson. 2003. "The evolution of altruistic punishment." *Proceedings of the National Academy of Sciences, USA* 100 (6):3531–3535.

Boyd, R., and S. Mathew. n.d. "The evolution of language may require third-party monitoring and sanctions." Unpublished manuscript.

Boyd, R., and P. J. Richerson. 1985. *Culture and the Evolutionary Process*. Chicago: University of Chicago Press.

———. 1987. "The evolution of ethnic markers." *Cultural Anthropology* 2 (1):27–38.

———. 1988. "An evolutionary model of social learning: the effects of spatial and

REFERENCES

temporal variation." In *Social Learning: Psychological and Biological Perspectives*, edited by T. R. Zentall and B. G. Galef, 29–48. Hillsdale, NJ: Lawrence Erlbaum.

———. 1990. "Group selection among alternative evolutionarily stable strategies." *Journal of Theoretical Biology* 145:331–342.

———. 1992. "Punishment allows the evolution of cooperation (or anything else) in sizable groups." *Ethology & Sociobiology* 13 (3):171–195.

———. 1996. "Why culture is common, but cultural evolution is rare." *Proceedings of the British Academy* 88:77–93.

———. 2002. "Group beneficial norms can spread rapidly in a structured population." *Journal of Theoretical Biology* 215:287–296.

———. 2009. "Voting with your feet: Payoff biased migration and the evolution of group beneficial behavior." *Journal of Theoretical Biology* 257 (2):331–339.

Boyd, R., P. J. Richerson, and J. Henrich. 2011a. "The cultural niche: Why social learning is essential for human adaptation." *Proceedings of the National Academy of Sciences, USA* 108:10918–10925.

———. 2011b. "Rapid cultural adaptation can facilitate the evolution of large-scale cooperation." *Behavioral Ecology and Sociobiology* 65 (3):431–444.

———. 2013. "The cultural evolution of technology." In *Cultural Evolution: Society, Language, and Religion*, edited by P. J. Richerson and M. H. Christiansen, 119–142. Cambridge, MA: MIT Press.

Boyd, R., and J. B. Silk. 2012. *How Humans Evolved*. 6th ed. New York: W.W. Norton.

Bradbard, M. R., and R. C. Endsley. 1983. "The effects of sex-typed labeling on preschool children's information-seeking and retention." *Sex Roles* 9 (2):247–260.

Bradbard, M. R., C. L. Martin, R. C. Endsley, and C. F. Halverson. 1986. "Influence of sex stereotypes on children's exploration and memory—A competence versus performance distinction." *Developmental Psychology* 22 (4):481–486.

Bramble, D. M., and D. E. Lieberman. 2004. "Endurance running and the evolution of Homo." *Nature* 432 (7015):345–352.

Brendl, C. M., A. Chattopadhyay, B. W. Pelham, and M. Carvallo. 2005. "Name letter branding: Valence transfers when product specific needs are active." *Journal of Consumer Research* 32 (3):405–415. doi:10.1086/497552.

Briggs, J. L. 1970. *Never in Anger: Portrait of an Eskimo Family*. Cambridge, MA: Harvard University Press.

Brighton, H., K. N. Kirby, and K. Smith. 2005. Cultural selection for learnability: Three principles underlying the view that language adapts to be learnable. In *Language Origins: Perspectives on Evolution*, edited by M. Tallerman, 291–309. Oxford: Oxford University Press.

Brody, G. H., and Z. Stoneman. 1981. "Selective imitation of same-age, older, and younger peer models." *Child Development* 52 (2):717–720.

———. 1985. "Peer imitation: An examination of status and competence hypotheses." *Journal of Genetic Psychology* 146 (2):161–170.

Broesch, J., J. Henrich, and H. C. Barrett. 2014. "Adaptive content biases in learning about animals across the lifecourse." *Human Nature* 25:181–199.

Broesch, T. 2011. *Social Learning across Cultures: Universality and Cultural Variability*. PhD diss., Emory University.

Brosnan, S., and F.B.M. de Waal. 2003. "Monkeys reject unequal pay." *Nature* 425:297–299.

REFERENCES

Brosnan, S. F., J. B. Silk, J. Henrich, M. C. Mareno, S. P. Lambeth, and S. J. Schapiro. 2009. "Chimpanzees (*Pan troglodytes*) do not develop contingent reciprocity in an experimental task." *Animal Cognition* 12 (4):587–597.

Brown, G. R., T. E. Dickins, R. Sear, and K. N. Laland. 2011. "Evolutionary accounts of human behavioural diversity. Introduction." *Philosophical Transactions of the Royal Society B: Biological Sciences* 366 (1563):313–324.

Brown, P. 2012. *Through the Eye of a Needle: Wealth, the Fall of Rome, and the Making of Christianity in the West, 350–550 AD*. Princeton, NJ: Princeton University Press.

Brown, R., and G. Armelagos. 2001. "Apportionment of racial diversity: A review." *Evolutionary Anthropology* 10:34–40.

Bryan, J. H. 1971. "Model affect and children's imitative altruism." *Child Development* 42 (6):2061–2065.

Bryan, J. H., J. Redfield, and S. Mader. 1971. "Words and deeds about altruism and the subsequent reinforcement power of the model." *Child Development* 42 (5):1501–1508.

Bryan, J. H., and M. A. Test. 1967. "Models and helping: Naturalistic studies in aiding behavior." *Journal of Personality and Social Psychology* 6:400–407.

Bryan, J. H., and N. H. Walbek. 1970a. "The impact of words and deeds concerning altruism upon children." *Child Development* 41 (3):747–757.

———. 1970b. "Preaching and practicing generosity: Children's actions and reactions." *Child Development* 41 (2):329–353.

Buchan, J. C., S. C. Alberts, J. B. Silk, and J. Altmann. 2003. "True paternal care in a multi-male primate society." *Nature* 425 (6954):179–181.

Burch, E. S. J. 2007. "Traditional native warfare in western Alaska." In *North American Indigenous Warfare and Ritual Violence*, edited by R. J. Chacon and R. G. Mendoza, 11–29. Tucson: University of Arizona Press.

Burkart, J. M., O. Allon, F. Amici, C. Fichtel, C. Finkenwirth, A. Heschl, J. Huber, et al. 2014. "The evolutionary origin of human hyper-cooperation." *Nature Communications* 5. doi:10.1038/ncomms5747.

Burkart, J. M., E. Fehr, C. Efferson, and C. P. van Schaik. 2007. "Other-regarding preferences in a non-human primate: Common marmosets provision food altruistically." *Proceedings of the National Academy of Sciences, USA* 104 (50):19762–19766.

Burkart, J. M., S. B. Hrdy, and C. P. Van Schaik. 2009. "Cooperative breeding and human cognitive evolution." *Evolutionary Anthropology* 18 (5):175–186.

Busnel, R. G., and A. Classe. 1976. *Whistled Languages*. Vol. 13 of Communication and Cybernetics. Berlin: Springer-Verlag.

Buss, D. 1999. *Evolutionary Psychology: The New Science of the Mind*. Boston: Allyn & Bacon.

———. 2007. *Evolutionary Psychology: The New Science of the Mind*. 3rd ed. Boston: Allyn & Bacon.

Buss, D. M., M. G. Haselton, T. K. Shackelford, A. L. Bleske, and J. C. Wakefield. 1998. "Adaptations, exaptations, and spandrels." *American Psychologist* 53 (5):533–548.

Bussey, K., and A. Bandura. 1984. "Influence of gender constancy and social power on sex-linked modeling." *Journal of Personality and Social Psychology* 47 (6):1292–1302.

REFERENCES

Bussey, K., and D. G. Perry. 1982. "Same-sex imitation—The avoidance of cross-sex models or the acceptance of same-sex models." *Sex Roles* 8 (7):773–784.

Buttelmann, D., M. Carpenter, and M. Tomasello. 2009. "Eighteen-month-old infants show false belief understanding in an active helping paradigm." *Cognition* 112 (2):337–342.

Buttelmann, D., N. Zmyj, M. M. Daum, and M. Carpenter. 2012. "Selective imitation of in-group over out-group members in 14-month-old infants." *Child Development* 84(2): 422–428.

Byrne, R. W., and A. Whiten. 1988. *Machiavellian Intelligence: Social Expertise and the Evolution of Intellect in Monkeys, Apes, and Humans*. Oxford: Oxford University Press.

———. 1992. "Cognitive evolution in primates—Evidence from tactical deception." *Man* 27 (3):609–627.

Calvin, W. H. 1993. "The unitary hypothesis: A common neural circuitry for novel manipulations, language, plan-ahead, and throwing?" In *Tools, Language, and Cognition in Human Evolution*, edited by E. R. Gibson and T. Ingold, 230–250. Cambridge: Cambridge University Press.

Camerer, C. F. 1989. "Does the basketball market believe in the 'hot hand'?" *American Economic Review* 79:1257–61.

———. 1995. "Individual decision making." In *The Handbook of Experimental Economics*, edited by J. H. Kagel and A. E. Roth, 587–703. Princeton, NJ: Princeton University Press.

Campbell, B. C. 2011. "Adrenarche and middle childhood." *Human Nature* 22 (3):327–349.

Campbell, D. T. 1965. "Variation and selective retention in socio-cultural evolution." In *Social Change in Developing Areas: A Reinterpretation of Evolutionary Theory*, edited by H. R. Barringer, G. I. Glanksten, and R. W. Mack, 19–49. Cambridge, MA: Schenkman.

Cappelletti, D., W. Guth, and M. Ploner. 2011. "Being of two minds: Ultimatum offers under cognitive constraints." *Journal of Economic Psychology* 32 (6):940–950.

Carney, D. R., A. J. C. Cuddy, and A. J. Yap. 2010. "Power posing: Brief nonverbal displays affect neuroendocrine levels and risk tolerance." *Psychological Science* 21 (10):1363–1368.

Carreiras, M., M. L. Seghier, S. Baquero, A. Estevez, A. Lozano, J. T. Devlin, and C. J. Price. 2009. "An anatomical signature for literacy." *Nature* 461 (7266):983–986. doi:10.1038/nature08461.

Carrier, D. R. 1984. "The energetic paradox of human running and hominid evolution." *Current Anthropology* 25 (4):483–495.

Carrigan, M. A., O. Uryasev, C. B. Frye, B. L. Eckman, C. R. Myers, T. D. Hurley, and S. A. Benner. 2014. "Hominids adapted to metabolize ethanol long before human-directed fermentation." *Proceedings of the National Academy of Sciences, USA*. doi:10.1073/pnas.1404167111.

Caspari, R., and S.-H. Lee. 2004. "Older age becomes common late in human evolution." *Proceedings of the National Academy of Sciences, USA* 101(30):10895–10900. http://www.pnas.org/content/101/30/10895.abstract.

Cassar, A., P. Grosjean, and S. Whitt. 2013. "Legacies of violence: Trust and market development." *Journal of Economic Growth* 18 (3):285–318.

REFERENCES

Castro-Caldas, A., P. C. Miranda, I. Carmo, A. Reis, F. Leote, C. Ribeiro, and E. Ducla-Soares. 1999. "Influence of learning to read and write on the morphology of the corpus callosum." *European Journal of Neurology* 6 (1):23–28. doi:10.1046/j.1468 -1331.1999.610023.x.

Cavalli-Sforza, L. L., and M. Feldman. 1981. *Cultural Transmission and Evolution*. Princeton, NJ: Princeton University Press.

———. 2003. "The application of molecular genetic approaches to the study of human evolution." *Nature Genetics* 33:266–275.

Chalmers, D. K., W. C. Horne, and M. E. Rosenbaum. 1963. "Social agreement and the learning of matching behavior." *Journal of Abnormal & Social Psychology* 66:556–561.

Chandrasekaran, R. 2006. *Imperial Life in the Emerald City: Inside Iraq's Green Zone*. New York: Alfred A. Knopf.

Chapais, B. 2008. *Primeval Kinship: How Pair-Bonding Gave Birth to Human Society*. Cambridge, MA: Harvard University Press.

Chartrand, T. L., and J. A. Bargh. 1999. "The chameleon effect: The perception-behavior link and social interaction." *Journal of Personality and Social Psychology* 76 (6):893–910.

Cheng, J., J. Tracy, T. Foulsham, and A. Kingstone. 2013. "Dual paths to power: Evidence that dominance and prestige are distinct yet viable Avenues to social status." *Journal of Personality and Social Psychology* 104:103–125.

Cheng, J. T., J. L. Tracy, and J. Henrich. 2010. "Pride, personality, and the evolutionary foundations of human social status." *Evolution and Human Behavior* 31 (5):334–347.

Cherry, T. L., P. Frykblom, and J. F. Shogren. 2002. "Hardnose the dictator." *American Economic Review* 92 (4):1218–1221.

Choi, J.-K., and S. Bowles. 2007. "The coevolution of parochial altruism and war." *Science* 318 (5850):636–640.

Chow, V., D. Poulin-Dubois, and J. Lewis. 2008. "To see or not to see: Infants prefer to follow the gaze of a reliable looker." *Developmental Science* 11 (5):761–770. doi:10.1111/j.1467-7687.2008.00726.x.

Christiansen, M. H., and N. Chater. 2008. "Language as shaped by the brain." *Behavioral and Brain Sciences* 31 (5):489–509.

Christiansen, M. H., and S. Kirby. 2003. *Language Evolution*. Studies in the Evolution of Language. Oxford: Oxford University Press.

Chudek, M., S. Heller, S. Birch, and J. Henrich. 2012. "Prestige-biased cultural learning: bystander's differential attention to potential models influences children's learning." *Evolution and Human Behavior* 33 (1):46–56.

Chudek, M., and J. Henrich. 2010. "Culture-gene coevolution, norm-psychology, and the emergence of human prosociality." *Trends in Cognitive Sciences* 15 (5):218–226.

———. n.d. "How exploitation launched human cooperation." Unpublished manuscript.

Chudek, M., J. M. McNamara, S. Birch, P. Bloom, and J. Henrich. n.d. "Developmental and cross-cultural evidence for intuitive dualism." Unpublished manuscript.

Chudek, M., M. Muthukrishna, and J. Henrich. Forthcoming. "Cultural evolution." In *Evolutionary Psychology*, edited by D. Buss. Wiley and Sons.

REFERENCES

Chudek, M., W. Zhao, and J. Henrich. 2013. "Culture-gene coevolution, large-scale cooperation and the shaping of human social psychology." In *Signaling, Commitment, and Emotion*, edited by R. Joyce, K. Sterelny, and B. Calcott, 425–458. Cambridge, MA: MIT Press.

Clancy, B., R. B. Darlington, and B. L. Finlay. 2001. "Translating developmental time across mammalian species." *Neuroscience* 105 (1):7–17.

Cochran, G., and H. Harpending. 2009. *The 10,000 Year Explosion: How Civilization Accelerated Human Evolution*. New York: Basic Books.

Coley, J. D., D. L. Medin, and S. Atran. 1997. "Does rank have its privilege? Inductive inferences within folkbiological taxonomies." *Cognition* 64 (1):73–112.

Collard, M., B. Buchanan, J. Morin, and A. Costopoulos. 2011. "What drives the evolution of hunter-gatherer subsistence technology? A reanalysis of the risk hypothesis with data from the Pacific Northwest." *Philosophical Transactions of the Royal Society B: Biological Sciences* 366 (1567):1129–1138.

Collard, M., B. Buchanan, M. J. O'Brien, and J. Scholnick. 2013. "Risk, mobility or population size? Drivers of technological richness among contact-period western North American hunter-gatherers." *Philosophical Transactions of the Royal Society B: Biological Sciences* 368 (1630). doi:10.1098/rstb.2012.0412.

Collard, M., M. Kemery, and S. Banks. 2005. "Causes of toolkit variation among hunter-gatherers: A test of four competing hypotheses." *Journal of Canadian Archaeology* 29:1–19.

Collard, M., A. Ruttle, B. Buchanan, and M. J. O'Brien. 2012. "Risk of resource failure and toolkit variation in small-scale farmers and herders." *Plos One* 7 (7):e40975.

———. 2013. "Population size and cultural evolution in nonindustrial food-producing societies." *PLoS ONE* 8 (9):e72628. doi:10.1371/journal.pone.0072628.

Colley, S., and R. Jones. 1988. "Rocky Cape revisited—New light on prehistoric Tasmanian fishing." In *The Walking Larder*, edited by J. Clutton-Brock, 336–346. London: Allen & Unwin.

Colloca, L., and F. Benedetti. 2009. "Placebo analgesia induced by social observational learning." *Pain* 144 (1–2):28–34.

Colloca, L., M. Sigaudo, and F. Benedetti. 2008. "The role of learning in nocebo and placebo effects." *Pain* 136 (1–2):211–218.

Coltheart, M.A.X. 2014. "The neuronal recycling hypothesis for reading and the question of reading universals." *Mind & Language* 29 (3):255–269. doi:10.1111/mila.12049.

Conley, T. G., and C. R. Udry. 2010. "Learning about a new technology: Pineapple in Ghana." *American Economic Review* 100 (1):35–69.

Conway, C. M., and M. H. Christiansen. 2001. "Sequential learning in non-human primates." *Trends in Cognitive Sciences* 5 (12):539–546.

Cook, P., and M. Wilson. 2010. "Do young chimpanzees have extraordinary working memory?" *Psychonomic Bulletin & Review* 17 (4):599–600.

Cook, R., G. Bird, G. Lünser, S. Huck, and C. Heyes. 2012. "Automatic imitation in a strategic context: players of rock–paper–scissors imitate opponents' gestures." *Proceedings of the Royal Society B: Biological Sciences* 279 (1729):780–786. doi:10.1098/rspb.2011.1024.

Cookman, S. 2000. *Ice Blink : The Tragic Fate of Sir John Franklin's Lost Polar Expedition*. New York: Wiley.

REFERENCES

Corballis, M. C. 2003. "From Hand to Mouth: The Gestural Origins of Language." In *Language Evolution*, edited by M. H. Christiansen and S. Kirby, 201–218. New York: Oxford University Press.

Corriveau, K., and P. L. Harris. 2009a. "Choosing your informant: weighing familiarity and recent accuracy." *Developmental Science* 12 (3):426–437.

———. 2009b. "Preschoolers continue to trust a more accurate informant 1 week after exposure to accuracy information." *Developmental Science* 12 (1):188–193.

Corriveau, K. H., K. Meints, and P. L. Harris. 2009. "Early tracking of informant accuracy and inaccuracy." *British Journal of Developmental Psychology* 27:331–342.

Cosmides, L., H. C. Barrett, and J. Tooby. 2010. "Adaptive specializations, social exchange, and the evolution of human intelligence." *Proceedings of the National Academy of Sciences, USA* 107:9007–9014.

Cosmides, L., and J. Tooby. 1989. "Evolutionary psychology and the generation of culture: ii. Case study: a computational theory of social exchange." *Ethology & Sociobiology* 10 (1–3):51–97.

Craig, K. D. 1986. "Social modeling influences: Pain in context." In *The Psychology of Pain*, edited by R. A. Sternbach, 67–95. New York: Raven Press.

Craig, K. D., and K. M. Prkachin. 1978. "Social modeling influences on sensory decision-theory and psychophysiological indexes of pain." *Journal of Personality and Social Psychology* 36 (8):805–815.

Crittenden, A. N., and F. W. Marlowe. 2008. "Allomaternal care among the Hadza of Tanzania." *Human Nature* 19 (3):249–262.

Crocker, W. H. 2002. "Canela 'other fathers': Partible paternity and its changing practices." In *Cultures of Multiple Fathers: The Theory and Practice of Partible Paternity in Lowland South America*, edited by S. Beckerman and P. Valentine, 86–104. Gainesville: University Press of Florida.

Crockett, M. J., L. Clark, M. D. Lieberman, G. Tabibnia, and T. W. Robbins. 2010. "Impulsive choice and altruistic punishment are correlated and increase in tandem with serotonin depletion." *Emotion* 10 (6):855–862.

Crockett, M. J., L. Clark, G. Tabibnia, M. D. Lieberman, and T. W. Robbins. 2008. "Serotonin modulates behavioral reactions to unfairness." *Science* 320 (5884): 1739–1739.

Cronin, K. A., K.K.E. Schroeder, E. S. Rothwell, J. B. Silk, and C. T. Snowdon. 2009. "Cooperatively breeding cottontop tamarins (*Saguinus oedipus*) do not donate rewards to their long-term mates." *Journal of Comparative Psychology* 123 (3):231–241.

Csibra, G., and G. Gergely. 2009. "Natural pedagogy." *Trends in Cognitive Sciences* 13 (4):148–153.

Cuatrecasas, P., D. H. Lockwood, and J. R. Caldwell. 1965. "Lactase deficiency in adults—A common occurrence." *Lancet* 1 (7375):14–18

Cummins, D. D. 1996a. "Evidence for the innateness of deontic reasoning." *Mind & Language* 11 (2):160–190.

———. 1996b. "Evidence of deontic reasoning in 3- and 4-year-old children." *Memory & Cognition* 24 (6):823–829.

———. 2013. "Deontic and epistemic reasoning in children revisited: Comment on Dack and Astington." *Journal of Experimental Child Psychology* 116 (3):762–769.

Currie, T. E., and R. Mace. 2009. "Political complexity predicts the spread of ethno-

REFERENCES

linguistic groups." *Proceedings of the National Academy of Sciences, USA* 106 (18):7339–7344. doi:10.1073/pnas.0804698106.

D'Andrade, R. G. 1995. *The Development of Cognitive Anthropology*. Cambridge: Cambridge University Press.

Danenberg, L. O., and H. J. Edenberg. 2005. "The alcohol dehydrogenase 1B (*ADH1B*) and *ADH1C* genes are transcriptionally regulated by DNA methylation and histone deacetylation in hepatoma cells." *Alcoholism—Clinical and Experimental Research* 29 (5):136a.

Darwin, C. 1981. *The Descent of Man and Selection in Relation to Sex*. Princeton, NJ: Princeton University Press. Original published in 1871 by J. Murray, London.

Dawkins, R. 1976. *The Selfish Gene*. Oxford: Oxford University Press.

———. 2006. *The God Delusion*. Boston: Houghton Mifflin.

Deacon, T. W. 1997. *The Symbolic Species: The Co-evolution of Language and the Brain*. New York: Norton.

Dean, L. G., R. L. Kendal, S. J. Schapiro, B. Thierry, and K. N. Laland. 2012. "Identification of the social and cognitive processes underlying human cumulative culture." *Science* 335 (6072):1114–1118.

Deaner, R. O., K. Isler, J. Burkart, and C. van Schaik. 2007. "Overall brain size, and not encephalization quotient, best predicts cognitive ability across non-human primates." *Brain Behavior and Evolution* 70 (2):115–124.

DeBruine, L. 2002. "Facial resemblance enhances trust." *Proceedings of the Royal Society of London Series B: Biological Sciences* 269:1307–1312.

Dediu, D., and D. R. Ladd. 2007. "Linguistic tone is related to the population frequency of the adaptive haplogroups of two brain size genes, *ASPM* and *Microcephalin*." *Proceedings of the National Academy of Sciences, USA* 104 (26):10944–10949.

Dee, T. S. 2005. "A teacher like me: Does race, ethnicity, or gender matter?" *American Economic Review* 95 (2):158–165.

Dehaene, S. 1997. *The Number Sense: How the Mind Creates Mathematics*. New York: Oxford University Press.

———. 2009. *Reading in the Brain: The Science and Evolution of a Human Invention*. New York: Viking.

———. 2014. "*Reading in the Brain* revised and extended: Response to comments." *Mind & Language* 29 (3):320–335. doi:10.1111/mila.12053.

Dehaene, S., F. Pegado, L. W. Braga, P. Ventura, G. Nunes, A. Jobert, G. Dehaene-Lambertz, et al. 2010. "How learning to read changes the cortical networks for vision and language." *Science* 330 (6009):1359–1364.

Delagnes, A., and H. Roche. 2005. "Late Pliocene hominid knapping skills: The case of Lokalalei 2C, West Turkana, Kenya." *Journal of Human Evolution* 48:435–472.

de Quervain, D. J., U. Fischbacher, V. Treyer, M. Schellhammer, U. Schnyder, A. Buck, and E. Fehr. 2004. "The neural basis of altruistic punishment." *Science* 305: 1254–1258.

Derex, M., M.-P. Beugin, B. Godelle, and M. Raymond. 2013. "Experimental evidence for the influence of group size on cultural complexity." *Nature* 503 (7476):389–391. doi:10.1038/nature12774.

d'Errico, F., and L. R. Backwell. 2003. "Possible evidence of bone tool shaping by

REFERENCES

Swartkrans early hominids." *Journal of Archaeological Science* 30 (12):1559–1576. http://www.sciencedirect.com/science/article/pii/S0305440303000529.

d'Errico, F., L. R. Backwell, and L. R. Berger. 2001. "Bone tool use in termite foraging by early hominids and its impact on our understanding of early hominid behaviour." *South African Journal of Science* 97 (3–4):71–75.

Deutscher, G. 2005. *The Unfolding of Language: An Evolutionary Tour of Mankind's Greatest Invention.* New York: Metropolitan Books.

———. 2010. *Through the Language Glass: Why the World Looks Different in Other Languages.* New York: Metropolitan Books/Henry Holt.

de Waal, F.B.M., C. Boesch, V. Horner, and A. Whiten. 2008. "Comparing social skills of children and apes." *Science* 319 (5863):569. doi:10.1126/science.319.5863.569c.

de Waal, F.B.M., K. Leimgruber, and A. R. Greenberg. 2008. "Giving is self-rewarding for monkeys." *Proceedings of the National Academy of Sciences, USA* 105 (36):13685–13689. doi:10.1073/pnas.0807060105.

Diamond, J. 1978. "The Tasmanians: The longest isolation, the simplest technology." *Nature* 273:185–186.

———. 1997. *Guns, Germs, and Steel: The Fates of Human Societies.* New York: W.W. Norton.

Diamond, J., and P. Bellwood. 2003. "Farmers and their languages: The first expansions." *Science* 300 (5619):597–603.

Dove, M. 1993. "Uncertainty, humility and adaptation in the tropical forest: The agricultural augury of the Kantu." *Ethnology* 32 (2):145–167.

Downey, G. 2014. "All forms of writing." *Mind & Language* 29 (3):304–319. doi:10.1111/mila.12052.

Draganski, B., and A. May. 2008. "Training-induced structural changes in the adult human brain." *Behavioural Brain Research* 192 (1):137–142.

Draper, P., and C. Haney. 2005. "Patrilateral bias among a traditionally egalitarian people: Ju/'hoansi naming practice." *Ethnology* 44 (3):243–259.

Dufour, D. L. 1984. "The time and energy-expenditure of indigenous women horticulturalists in the northwest Amazon." *American Journal of Physical Anthropology* 65 (1):37–46.

———. 1985. "Manioc as a dietary staple: Implications for the budgeting of time and energy in the Northwest Amazon." In *Food Energy in Tropical Ecosystem*, edited by D. J. Cattle and K. H. Schwerin, 1–20. New York: Gordon and Breach.

———. 1988a. "Cyanide content of cassava (*Manihot esculenta*, Euphorbiaceae) cultivars used by Tukanoan Indians in northwest Amazonia." *Economic Botany* 42 (2):255–266.

———. 1988b. "Dietary cyanide intake and serum thiocyanate levels in Tukanoan Indians in northwest Amazonia." *American Journal of Physical Anthropology* 75 (2):205.

———. 1994. "Cassava in Amazonia: Lessons in utilization and safety from native peoples." *Acta Horitculturae* 375:175–182.

Dugatkin, L. 1999. *Cheating Monkeys and Citizen Bees.* New York: Free Press.

Dunbar, R.I.M. 1998. "The social brain hypothesis." *Evolutionary Anthropology* 6 (5):178–190.

Duncker, K. 1938. "Experimental modification of children's food preferences through social suggestion." *Journal of Abnormal Psychology* 33:489–507.

REFERENCES

Dunham, Y., A. S. Baron, and M. R. Banaji. 2008. "The development of implicit intergroup cognition." *Trends in Cognitive Sciences* 12 (7):248–253.

Durham, W. H. 1982. "The relationship of genetic and cultural evolution: Models and examples." *Human Ecology* 10 (3):289–323.

———. 1991. *Coevolution: Genes, Culture, and Human Diversity.* Stanford, CA: Stanford University Press.

Durkheim, E. (1915) 1965. *Elementary Forms of Religious Life.* Translated by J. W. Swain. New York: George Allen &Unwin.

Earl, J. W. 1996. "A fatal recipe for Burke and Wills." *Australian Geographic* 43:28–29.

Earl, J. W., and B. V. McCleary. 1994. "Mystery of the poisoned expedition." *Nature* 368 (6473):683–684.

Eckel, C., E. Fatas, and R. Wilson. 2010. "Cooperation and status in organizations." *Journal of Public Economic Theory* 12 (4):737–762.

Eckel, C., and R. Wilson. 2000. "Social learning in a social hierarchy: An experimental study." Unpublished manuscript. http://www.ruf.rice.edu/~rkw/RKW _FOLDER/AAAS2000_ABS.htm.

Edenberg, H. J. 2000. "Regulation of the mammalian alcohol dehydrogenase genes." *Progress in Nucleic Acid Research and Molecular Biology* 64:295–341.

Edenberg, H. J., X. L. Xuei, H. J. Chen, H. J. Tian, L. F. Wetherill, D. M. Dick, L. Almasy, et al. 2006. "Association of alcohol dehydrogenase genes with alcohol dependence: A comprehensive analysis." *Human Molecular Genetics* 15 (9):1539–1549.

Edgerton, R. B. 1992. *Sick Societies: Challenging the Myth of Primitive Harmony.* New York: Free Press.

Efferson, C., R. Lalive, P. J. Richerson, R. McElreath, and M. Lubell. 2008. "Conformists and mavericks: The empirics of frequency-dependent cultural transmission." *Evolution and Human Behavior* 29 (1):56–64. doi:10.1016/j.evolhumbehav.2007 .08.003.

Ehrenreich, B. 2007. *Dancing in the Streets: A History of Collective Joy.* New York: Metropolitan Books.

Eiberg, H., J. Troelsen, M. Nielsen, A. Mikkelsen, J. Mengel-From, K. W. Kjaer, and L. Hansen. 2008. "Blue eye color in humans may be caused by a perfectly associated founder mutation in a regulatory element located within the *HERC2* gene inhibiting *OCA2* expression." *Human Genetics* 123 (2):177–187.

Eibl-Eibesfeldt, I. 2007. *Human Ethology.* New Brunswick, NJ: Aldine Transaction.

Elkin, A. P. 1964. *The Australian Aborigines: How to Understand Them.* Garden City, NY: Anchor Books.

Elliot, R., and R. Vasta. 1970. "The modeling of sharing: Effects associated with vicarious reinforcement, symbolization, age, and generalization." *Journal of Experimental Child Psychology* 10:8–15.

Ember, C. R. 1978. "Myths about hunter-gatherers." *Ethnology* 17 (4):439–448.

———. 2013. "Introduction to 'Coping with environmental risk and uncertainty: Individual and cultural responses.'" *Human Nature* 24 (1):1–4.

Ember, C. R., T. A. Adem, and I. Skoggard. 2013. "Risk, uncertainty, and violence in eastern Africa." *Human Nature* 24 (1):33–58.

Ember, C. R., and M. Ember. 1992. "Resource unpredictability, mistrust, and war—A cross-cultural study." *Journal of Conflict Resolution* 36 (2):242–262.

REFERENCES

———. 2007. "Climate, econiche, and sexuality: Influences on sonority in language." *American Anthropologist* 109 (1):180–185. doi:10.1525/Aa.2007.109.1.180.

Enard, W. 2011. "*FOXP2* and the role of cortico-basal ganglia circuits in speech and language evolution." *Current Opinion in Neurobiology* 21 (3):415–424.

Enard, W., S. Gehre, K. Hammerschmidt, S. M. Holter, T. Blass, M. Somel, M. K. Bruckner, et al. 2009. "A humanized version of *Foxp2* affects cortico-basal ganglia circuits in mice." *Cell* 137 (5):961–971.

Endicott, K. 1988. "Property, power and conflict among the Batek of Malaysia." In *Hunters and Gatherers: Property, Power and Ideology*, edited by T. Ingold, D. Riches, and J. Woodburn, 110–128. Berg: Oxford.

Engelmann, J. M., E. Herrmann, and M. Tomasello. 2012. "Five-year olds, but not chimpanzees, attempt to manage their reputations." *Plos One* 7 (10):e48433. doi:10.1371/journal.pone.0048433.

Engelmann, J. M., H. Over, E. Herrmann, and M. Tomasello. 2013. "Young children care more about their reputation with ingroup members and potential reciprocators." *Developmental Science* 16 (6):952–958.

Ensminger, J., and J. Henrich, eds. 2014. *Experimenting with Social Norms: Fairness and Punishment in Cross-Cultural Perspective*. New York: Russell Sage Press.

Esteban, J., L. Mayoral, and D. Ray. 2012a. "Ethnicity and conflict: An empirical study." *American Economic Review* 102 (4):1310–1342.

———. 2012b. "Ethnicity and conflict: Theory and facts." *Science* 336 (6083):858–865.

Euler, H. A., and B. Weitzel. 1996. "Discriminative grandparental solicitude as reproductive strategy." *Human Nature* 7 (1):39–59.

Evans, N. 2005. "Australian languages reconsidered: A review of Dixon (2002)." *Oceanic Linguistics* 44 (1):216–260.

———. 2012. "An enigma under an enigma: Tracing diversification and dispersal in a continent of hunter-gatherers." Paper presented at the KNAW (Royal Netherlands Academy of Arts and Sciences) conference Patterns of Diversification and Contact: A Global Perspective, Amsterdam, December 11–14.

Evans, N., and P. McConvell. 1998. The enigma of Pama-Nyungan expansion in Australia. In *Archaeology and Language II: Correlating Archaeological and Linguistic Hypotheses*, edited by R. Blench and M. Spriggs, 174–191. London: Routledge.

Everett, D. L. 2005. "Cultural constraints on grammar and cognition in Piraha—Another look at the design features of human language." *Current Anthropology* 46 (4):621–646.

Fairlie, R., F. Hoffmann, and P. Oreopoulos. 2011. "A community college instructor like me: Race and ethnicity interaction in the classroom." Working Paper 17381. National Bureau of Economic Research, Cambridge, MA.

Faisal, A., D. Stout, J. Apel, and B. Bradley. 2010. "The manipulative complexity of Lower Paleolithic stone toolmaking." *Plos One* 5 (11):e13718.

Falk, D. 1990. "Brain evolution in Homo—The radiator theory." *Behavioral and Brain Sciences* 13 (2):333–343.

Fearon, J. D. 2008. "Ethnic mobilization and ethnic violence." In *The Oxford Handbook of Political Economy*, edited by D. A. Wittman and B. R. Weingast, 852–868. Oxford: Oxford University Press.

REFERENCES

Fedzechkina, M., T. F. Jaeger, and E. L. Newport. 2012. "Language learners restructure their input to facilitate efficient communication." *Proceedings of the National Academy of Sciences, USA* 109 (44):17897–17902.

Fehr, E., and C. F. Camerer. 2007. "Social neuroeconomics: The neural circuitry of social preferences." *Trends in Cognitive Sciences* 11 (10):419–427.

Fernandez, R., and A. Fogli. 2006. "Fertility: The role of culture and family experience." *Journal of the European Economic Association* 4 (2–3):552–561.

——. 2009. "Culture: An empirical investigation of beliefs, work, and fertility." *American Economic Journal-Macroeconomics* 1 (1):146–177.

Fessler, D.M.T. 1999. "Toward an understanding of the universality of second order emotions." In *Beyond Nature or Nurture: Biocultural Approaches to the Emotions*, edited by A. Hinton, 75–116. Cambridge: Cambridge University Press.

——. 2002. "Reproductive immunosuppression and diet—An evolutionary perspective on pregnancy sickness and meat consumption." *Current Anthropology* 43 (1):19–61.

——. 2004. "Shame in two cultures." *Journal of Cognition and Culture* 4 (2):207–262.

——. 2006. "A burning desire: Steps toward an evolutionary psychology of fire learning." *Journal of Cognition and Culture* 6 (3–4):429–451.

Fessler, D.M.T., A. P. Arguello, J. M. Mekdara, and R. Macias. 2003. "Disgust sensitivity and meat consumption: A test of an emotivist account of moral vegetarianism." *Appetite* 41 (1):31–41.

Fessler, D.M.T., and C. D. Navarrete. 2003. "Meat is good to taboo: Dietary proscriptions as a product of the interaction of psychological mechanisms and social processes." *Journal of Cognition and Culture* 3 (1):1–40.

——. 2004. "Third-party attitudes toward incest: Evidence for the Westermarck Effect." *Evolution and Human Behavior* 25 (5):277–294.

Finniss, D. G., T. J. Kaptchuk, F. Miller, and F. Benedetti. 2010. "Biological, clinical, and ethical advances of placebo effects." *Lancet* 375 (9715):686–695.

Fischer, P., J. I. Krueger, T. Greitemeyer, C. Vogrincic, A. Kastenmüller, D. Frey, M. Heene, M. Wicher, and M. Kainbacher. 2011. "The bystander-effect: A meta-analytic review on bystander intervention in dangerous and non-dangerous emergencies." *Psychological Bulletin* 137(4): 517–537. http://dx.doi.org/10.1037/a0023304.

Fischer, R., D. Xygalatas, P. Mitkidis, P. Reddish, P. Tok, I. Konvalinka, and J. Bulbulia. 2014. "The fire-walker's high: Affect and physiological responses in an extreme collective ritual." *Plos One* 9 (2):e88355. doi:10.1371/journal.pone.0088355.

Fisher, S. E., and M. Ridley. 2013. "Culture, genes, and the human revolution." *Science* 340 (6135):929–930.

Fiske, A. 1992. "The four elementary forms of sociality: framework for a unified theory of social relations." *Psychological Review* 99 (4):689–723.

Fitch, W. T. 2000. "The evolution of speech: a comparative review." *Trends in Cognitive Sciences* 4 (7):258–267.

Flannery, K. V., and J. Marcus. 2000. "Formative Mexican chiefdoms and the myth of the 'Mother Culture.'" *Journal of Anthropological Archaeology* 19 (1–37).

——. 2012. *The Creation of Inequality: How our Prehistoric Ancestors Set the Stage for Monarchy, Slavery, and Empire.* Cambridge, MA: Harvard University Press.

REFERENCES

Flynn, J. R. 2007. *What Is Intelligence? Beyond the Flynn effect*. Cambridge: Cambridge University Press.

——. 2012. *Are We Getting Smarter? Rising IQ in the Twenty-First Century*. Cambridge: Cambridge University Press.

Foley, C., N. Pettorelli, and L. Foley. 2008. "Severe drought and calf survival in elephants." *Biology Letters* 4 (5):541–544.

Foster, E. A., D. W. Franks, S. Mazzi, S. K. Darden, K. C. Balcomb, J. K. B. Ford, and D. P. Croft. 2012. "Adaptive prolonged postreproductive life span in killer whales." *Science* 337 (6100):1313–1313.

Fought, J. G., R. L. Munroe, C. R. Fought, and E. M. Good. 2004. "Sonority and climate in a world sample of languages: Findings and prospects." *Cross-Cultural Research* 38 (1):27–51. doi:10.1177/1069397103259439.

Fowler, J. H., and N. A. Christakis. 2010. "Cooperative behavior cascades in human social networks." *Proceedings of the National Academy of Sciences, USA* 107 (12):5334–5338.

Frank, M. C., and D. Barner. 2012. "Representing exact number visually using mental abacus." *Journal of Experimental Psychology—General* 141 (1):134–149.

Franklin, A., A. Clifford, E. Williamson, and I. Davies. 2005. "Color term knowledge does not affect categorical perception of color in toddlers." *Journal of Experimental Child Psychology* 90 (114–141).

Frazer, J. G. 1996. *The Golden Bough: A Study in Magic and Religion*. Harmondsworth, UK: Penguin Books.

Fry, A. F., and S. Hale. 1996. "Processing speed, working memory, and fluid intelligence: Evidence for a developmental cascade." *Psychological Science* 7 (4):237–241. doi:10.1111/j.1467-9280.1996.tb00366.x.

Garner, R. 2005. "What's in a name? Persuasion perhaps." *Journal of Consumer Psychology* 15 (2):108–116. http://dx.doi.org/10.1207/s15327663jcp1502_3.

Gaulin, S. J. C., D. H. McBurney, and S. L. Brakeman-Wartell. 1997. "Matrilateral biases in the investment of aunts and uncles: A consequence and measure of paternity uncertainty." *Human Nature* 8 (2):139–151.

Gelman, S. A. 2003. *The Essential Child: Origins of Essentialism in Everyday Thought*. Oxford: Oxford University Press.

Gerbault, P., A. Liebert, Y. Itan, A. Powell, M. Currat, J. Burger, D. M. Swallow, and M. G. Thomas. 2011. "Evolution of lactase persistence: an example of human niche construction." *Philosophical Transactions of the Royal Society B: Biological Sciences* 366 (1566):863–877.

Gerbault, P., C. Moret, M. Currat, and A. Sanchez-Mazas. 2009. "Impact of selection and demography on the diffusion of lactase persistence." *Plos One* 4 (7).e6369.

Gerbault, P., M. Roffet-Salque, R. P. Evershed, and M. G. Thomas. 2013. "How long have adult humans been consuming milk?" *IUBMB Life* 65 (12):983–990.

Gerszten, P. C., and E. M. Gerszten. 1995. "Intentional cranial deformation: A disappearing form of self-mutilation." *Neurosurgery* 37 (3):374–382.

Gil-White, F. 2001. "Are ethnic groups biological 'species' to the human brain? Essentialism in our cognition of some social categories." *Current Anthropology* 42 (4):515–554.

——. 2004. "Ultimatum game with an ethnicity manipulation: Results from Khovdiin Bulgan Cum, Mongolia." In *Foundations of Human Sociality: Economic*

REFERENCES

Experiments and Ethnographic Evidence from Fifteen Small-Scale Societies, edited by J. Henrich, R. Boyd, S. Bowles, C. Camerer, E. Fehr, and H. Gintis, 260–304. New York: Oxford University Press.

Gilberg, R. 1984. "Polar Eskimo." In *Handbook of North American Indians*, edited by D. Damas, 577–594. Washington, DC: Smithsonian Institution Press.

Gillet, J., E. Cartwright, and M. Van Vugt. 2009. "Leadership in a weak-link game." *Economic Inquiry* 51(4): 2028–2043. http://dx.doi.org/10.1111/ecin.12003.

Gilligan, M., J. P. Benjamin, and C. D. Samii. 2011. "Civil war and social capital: Behavioral-game evidence from Nepal." Unpublished manuscript. New York University.

Gilovich, T., D. Griffin , and D. Kahneman, eds. 2002. *Heuristics and Biases: The Psychology of Intuitive Judgment.* New York: Cambridge University Press.

Gilovich, T., R. Vallone, and A. Tversky. 1985. "The hot hand in basketball: On the misperception of random sequences." *Cognitive Psychology* 17 (3):295–314.

Giuliano, P., and A. Alesina. 2010. "The power of the family." *Journal of Economic Growth* 15 (2):93–125.

Gizer, I. R., H. J. Edenberg, D. A. Gilder, K. C. Wilhelmsen, and C. L. Ehlers. 2011. "Association of alcohol dehydrogenase genes with alcohol-related phenotypes in a Native American community sample." *Alcoholism: Clinical and Experimental Research* 35 (11):2008–2018.

Gneezy, A., and D. M. T. Fessler. 2011. "Conflict, sticks and carrots: war increases prosocial punishments and rewards." *Proceedings of the Royal Society B: Biological Sciences.* doi:10.1098/rspb.2011.0805.

Gneezy, U., and A. Rustichini. 2000. "A fine is a price." *Journal of Legal Studies* 29 (1):1–17.

Goldin-Meadow, S., W. C. So, A. Ozyurek, and C. Mylander. 2008. "The natural order of events: How speakers of different languages represent events nonverbally." *Proceedings of the National Academy of Sciences, USA* 105 (27):9163–9168.

Goldman, I. 1979. *The Cubeo.* Urbana: University of Illinois Press.

Goldstein, J., J. Davidoff, and D. Roberson. 2009. "Knowing color terms enhances recognition: Further evidence from English and Himba." *Journal of Experimental Child Psychology* 102 (2):219–238.

Goldstein, R., J. Almenberg, A. Dreber, J. W. Emerson, A. Herschkowitsch, and J. Katz. 2008. "Do more expensive wines taste better? Evidence from a large sample of blind tastings." *Journal of Wine Economics* 3 (01):1–9. doi:10.1017/S1931436 100000523.

Goodwin, R. 2008. *Crossing the Continent, 1527–1540: The Story of the First African-American Explorer of the American South.* New York: Harper.

Gordon, P. 2005. "Numerical cognition without words: Evidence from Amazonia." *Science* 306:496–499.

Goren-Inbar, N. 2011. "Culture and cognition in the Acheulian industry: A case study from Gesher Benot Ya'aqov. *Philosophical Transactions of the Royal Society B: Biological Sciences* 366 (1567):1038–1049.

Goren-Inbar, N., N. Alperson, M. E. Kislev, O. Simchoni, Y. Melamed, A. Ben-Nun, and E. Werker. 2004. "Evidence of hominin control of fire at Gesher Benot Ya'aqov, Israel." *Science* 304 (5671):725–727.

Goren-Inbar, N., G. Sharon, Y. Melamed, and M. Kislev. 2002. "Nuts, nut cracking,

REFERENCES

and pitted stones at Gesher Benot Ya'aqov, Israel." *Proceedings of the National Academy of Sciences, USA* 99 (4):2455–2460.

Gott, B. 2002. "Fire-making in Tasmania: Absence of evidence is not evidence." *Current Anthropology* 43 (4):650–655.

Gottfried, A. E., and P. A. Katz. 1977. "Influence of belief, race, and sex similarities between child observers and models on attitudes and observational learning." *Child Development* 48 (4):1395–1400. http://www.jstor.org/stable/1128498.

Goubert, L., J. W. S. Vlaeyen, G. Crombez, and K. D. Craig. 2011. "Learning about pain from others: An observational learning account." *Journal of Pain* 12 (2):167–174.

Gowdy, J., R. Iorgulescu, and S. Onyeiwu. 2003. "Fairness and retaliation in a rural Nigerian village." *Journal of Economic Behavior & Organization* 52:469–479.

Grant, F., and M. A. Hogg. 2012. "Self-uncertainty, social identity prominence and group identification." *Journal of Experimental Social Psychology* 48 (2):538–542.

Greene, J. D., L. E. Nystrom, A. D. Engell, J. M. Darley, and J. D. Cohen. 2004. "The neural bases of cognitive conflict and control in moral judgment." *Neuron* 44 (2):389–400.

Greenfield, N., and J. T. Kuznicki. 1975. "Implied competence, task complexity, and imitative behavior." *Journal of Social Psychology* 95:251–261.

Gregor, T. 1977. *Mehinaku: The Drama of Daily Life in a Brazilian Indian Village.* Chicago: University of Chicago Press.

Gregory, S. W., K. Dagan, and S. Webster. 1997. "Evaluating the relation of vocal accommodation in conversation partners' fundamental frequencies to perceptions of communication quality." *Journal of Nonverbal Behavior* 21 (1):23–43.

Gregory, S. W., and S. Webster. 1996. "A nonverbal signal in voices of interview partners effectively predicts communication accommodation and social status perceptions." *Journal of Personality and Social Psychology* 70 (6):1231–1240.

Gregory, S. W., S. Webster, and G. Huang. 1993. "Voice pitch and amplitude convergence as a metric of quality in dyadic interviews." *Language & Communication* 13 (3):195–217.

Greif, M. L., D. G. K. Nelson, F. C. Keil, and F. Gutierrez. 2006. "What do children want to know about animals and artifacts? Domain-specific requests for information." *Psychological Science* 17 (6):455–459.

Grosjean, P. 2011. "A history of violence: The culture of honor as a determinant of homocide in the US South." Unpublished manuscript. http://papers.ssrn.com/sol3/papers.cfm?abstract_id=1917113.

Grossmann, I., J. Y. Na, M. E. W. Varnum, D. C. Park, S. Kitayama, and R. E. Nisbett. 2010. "Reasoning about social conflicts improves into old age." *Proceedings of the National Academy of Sciences, USA* 107 (16):7246–7250.

Gruber, T., M. N. Muller, V. Reynolds, R. Wrangham, and K. Zuberbuhler. 2011. "Community-specific evaluation of tool affordances in wild chimpanzees." *Scientific Reports* 1.

Gruber, T., M. N. Muller, P. Strimling, R. Wrangham, and K. Zuberbuhler. 2009. "Wild chimpanzees rely on cultural knowledge to solve an experimental honey acquisition task." *Current Biology* 19 (21):1806–1810.

Grusec, J. E. 1971. "Power and the internalization of self-denial." *Child Development* 42 (1):93–105.

REFERENCES

Guess, H. A. 2002. *The Science of the Placebo: Toward an Interdisciplinary Research Agenda*. London: BMJ Books.

Guiso, L., P. Sapienza, and L. Zingales. 2006. "Does culture affect economic outcomes?" *Journal of Economic Perspectives* 20 (2):23–48.

———. 2009. "Cultural biases in economic exchange?" *Quarterly Journal of Economics* 124 (3):1095–1131.

Gurven, M. 2004a. "To give and to give not: The behavioral ecology of human food transfers." *Behavioral and Brain Sciences* 27 (4):543–559.

———. 2004b. "Tolerated reciprocity, reciprocal scrounging, and unrelated kin: Making sense of multiple models." *Behavioral and Brain Sciences* 27 (4):572–579.

Gurven, M., and H. Kaplan. 2007. "Longevity among hunter-gatherers: A cross-cultural examination." *Population and Development Review* 33 (2):321–365.

Gurven, M., J. Stieglitz, P. L. Hooper, C. Gomes, and H. Kaplan. 2012. "From the womb to the tomb: The role of transfers in shaping the evolved human life history." *Experimental Gerontology* 47 (10):807–813.

Guth, W., M. V. Levati, M. Sutter, and E. Van der Heijden. 2007. "Leading by example with and without exclusion power in voluntary contribution experiments." *Journal of Public Economics* 91 (5–6):1023–1042.

Haidt, J. 2012. *The Righteous Mind: Why Good People Are Divided by Politics and Religion*. New York: Pantheon Books.

Hämäläinen, P. 2008. *The Comanche Empire*. Lamar Series in Western History. New Haven, CT: Yale University Press.

Hamlin, J. K. 2013a. "Failed attempts to help and harm: Intention versus outcome in preverbal infants' social evaluations." *Cognition* 128 (3):451–474. http://www.sciencedirect.com/science/article/pii/S0010027713000796.

———. 2013b. "Moral judgment and action in preverbal infants and toddlers: Evidence for an innate moral core." *Current Directions in Psychological Science* 22 (3):186–193.

Hamlin, J. K., E. V. Hallinan, and A. L. Woodward. 2008. "Do as I do: 7-month-old infants selectively reproduce others' goals." *Developmental Science* 11 (4):487–494.

Hamlin, J. K., N. Mahajan, Z. Liberman, and K. Wynn. 2013. "Not like me = Bad: infants prefer those who harm dissimilar others." *Psychological Science* 24 (4):589–594.

Hamlin, J. K., and K. Wynn. 2011. "Young infants prefer prosocial to antisocial others." *Cognitive Development* 26 (1):30–39.

Hamlin, J. K., K. Wynn, and P. Bloom. 2007. "Social evaluation by preverbal infants." *Nature* 450 (7169):557–559.

Hamlin, J. K., K. Wynn, P. Bloom, and N. Mahajan. 2011. "How infants and toddlers react to antisocial others." *Proceedings of the National Academy of Sciences, USA* 108 (50):19931–19936.

Harbaugh, W. T., U. Mayr, and D. R. Burghart. 2007. "Neural responses to taxation and voluntary giving reveal motives for charitable donations." *Science* 316 (5831):1622–1625.

Hardacre, E. 1880. "Eighteen years alone." *Schribner's Monthly* 20 (5):657–664.

Hare, B., J. Call, B. Agnetta, and M. Tomasello. 2000. "Chimpanzees know what conspecifics do and do not see." *Animal Behaviour* 59:771–785.

REFERENCES

Hare, B., and M. Tomasello. 2004. "Chimpanzees are more skillful in competitive than in cooperative tasks." *Animal Behaviour* 68:571–581.

Harris, M. B. 1970. "Reciprocity and generosity: Some determinants of sharing in children." *Child Development* 41:313–328.

———. 1971. "Models, norms and sharing." *Psychological Reports* 29:147–153.

Harris, P. L., and K. H. Corriveau. 2011. "Young children's selective trust in informants." *Philosophical Transactions of the Royal Society B: Biological Sciences* 366 (1567):1179–1187.

Harris, P. L., and M. Nunez. 1996. "Understanding of permission rules by preschool children." *Child Development* 67 (4):1572–1591.

Harris, P. L., M. Nunez, and C. Brett. 2001. "Let's swap: Early understanding of social exchange by British and Nepali children." *Memory & Cognition* 29 (5):757–764.

Harris, P. L., E. S. Pasquini, S. Duke, J. J. Asscher, and F. Pons. 2006. "Germs and angels: The role of testimony in young children's ontology." *Developmental Science* 9 (1):76–96.

Hay, J., and L. Bauer. 2007. "Phoneme inventory size and population size." *Language* 83 (2):388–400.

Hayek, F. A. v., and W. W. Bartley. 1988. *The Fatal Conceit: The Errors of Socialism*. London: Routledge.

Hayes, M. G., J. B. Coltrain, and D. H. O'Rourke. 2003. "Mitochondrial analyses of Dorset, Thule, Sadlermiut, and Aleut skeletal samples from the prehistoric North American arctic." In *Mummies in a New Millennium: Proceedings of the 4th World Congress on Mummy Studies*, edited by N. Lynnerup, C. Andreasen, and J. Berglund, 125–128. Copenhagen: Danish Polar Center.

Hedden, T., S. Ketay, A. Aron, H. R. Markus, and J. D. E. Gabrieli. 2008. "Cultural influences on neural substrates of attentional control." *Psychological Science* 19:12–17.

Heine, B., and T. Kuteva. 2002a. On the evolution of grammatical forms. In *The Transition to Language*, edited by A. Wray, 376–397. New York: Oxford University Press.

———. 2002b. *World Lexicon of Grammaticalization*. New York: Cambridge University Press.

———. 2007. *The Genesis of Grammar: A Reconstruction* Studies in the Evolution of Language. Oxford: Oxford University Press.

Heine, S. J. 2008. *Cultural Psychology*. New York: W. W. Norton.

Heine, S. J., T. Proulx, and K. D. Vohs. 2006. "The meaning maintenance model: On the coherence of social motivations." *Personality and Social Psychology Review* 10 (2):88–110.

Heinrich, B. 2002. *Why We Run: A Natural History*. New York: Ecco.

Heinz, H. 1994. *Social Organization of the !Ko Bushmen*. Cologne: Rudiger Koppe.

Henrich, J. 2000. "Does culture matter in economic behavior: Ultimatum game bargaining among the Machiguenga." *American Economic Review* 90 (4):973–980.

———. 2002. "Decision-making, cultural transmission and adaptation in economic anthropology." In *Theory in Economic Anthropology*, edited by J. Ensminger, 251–295. Walnut Creek, CA: AltaMira Press.

———. 2004a. "Cultural group selection, coevolutionary processes and large-scale cooperation." *Journal of Economic Behavior & Organization* 53:3–35.

REFERENCES

————. 2004b. "Demography and cultural evolution: Why adaptive cultural processes produced maladaptive losses in Tasmania." *American Antiquity* 69 (2):197–214.

————. 2004c. "Inequity aversion in Capuchins?" *Nature* 428:139.

————. 2006. "Understanding cultural evolutionary models: A reply to Read's critique." *American Antiquity* 71 (4):771–782.

————. 2008. "A cultural species." In *Explaining Culture Scientifically*, edited by M. Brown, 184–210. Seattle: University of Washington Press.

————. 2009a. "The evolution of costly displays, cooperation, and religion: Credibility enhancing displays and their implications for cultural evolution." *Evolution and Human Behavior* 30:244–260.

————. 2009b. "The evolution of innovation-enhancing institutions." In *Innovation in Cultural Systems: Contributions in Evolution Anthropology*, edited by S. J. Shennan and M. J. O'Brien, 99–120. Cambridge, MA: MIT Press.

————. 2014. "Rice, psychology and innovation." *Science* 344:593.

Henrich, J., W. Albers, R. Boyd, K. McCabe, G. Gigerenzer, H. P. Young, and A. Ockenfels. 2001. "Is culture important in bounded rationality?" In *Bounded Rationality: The Adaptive Toolbox*, edited by G. Gigerenzer and R. Selten, 343–359. Cambridge, MA: MIT Press.

Henrich, J., and R. Boyd. 2001. "Why people punish defectors: Weak conformist transmission can stabilize costly enforcement of norms in cooperative dilemmas." *Journal of Theoretical Biology* 208:79–89.

Henrich, J., R. Boyd, S. Bowles, C. Camerer, E. Fehr, and H. Gintis, eds. 2004. *Foundations of Human Sociality: Economic Experiments and Ethnographic Evidence from Fifteen Small-Scale Societies*. Oxford: Oxford University Press.

Henrich, J., R. Boyd, S. Bowles, C. Camerer, E. Fehr, H. Gintis, and R. McElreath. 2004. "Overview and Synthesis." In *Foundations of Human Sociality: Economic Experiments and Ethnographic Evidence from Fifteen Small-Scale Societies*, edited by J. Henrich, R. Boyd, Samuel Bowles, C. Camerer, E. Fehr, and H. Gintis, 9–51. Oxford: Oxford University Press.

Henrich, J., R. Boyd, S. Bowles, C. Camerer, E. Fehr, H. Gintis, R. McElreath, M. Alvard, A. Barr, J. Ensminger, N. S. Henrich, K. Hill, F. Gil-White, M. Gurven, F. W. Marlowe, J. Q. Patton, and D. Tracer. 2005. "'Economic man' in cross-cultural perspective: Behavioral experiments in 15 small-scale societies." *Behavioral and Brain Sciences* 28 (6):795–855.

Henrich, J., R. Boyd, S. Bowles, H. Gintis, C. Camerer, E. Fehr, and R. McElreath. 2001. "In search of Homo economicus: Experiments in 15 small-scale societies." *American Economic Review* 91:73–78.

Henrich, J., R. Boyd, and P. J. Richerson. 2012. "The puzzle of monogamous marriage." *Philosophical Transactions of the Royal Society B: Biological Sciences* 367: 657–669.

Henrich, J., and J. Broesch. 2011. "On the nature of cultural transmission networks: Evidence from Fijian villages for adaptive learning biases." *Philosophical Transactions of the Royal Society B: Biological Sciences* 366:1139–1148.

Henrich, J., J. Ensminger, R. McElreath, A. Barr, C. Barrett, A. Bolyanatz, J. C. Cardenas, et al. 2010. "Market, religion, community size and the evolution of fairness and punishment." *Science* 327:1480–1484.

REFERENCES

Henrich, J., and F. Gil-White. 2001. "The evolution of prestige: Freely conferred deference as a mechanism for enhancing the benefits of cultural transmission." *Evolution and Human Behavior* 22 (3):165–196.

Henrich, J., S. J. Heine, and A. Norenzayan. 2010a. "Beyond WEIRD: Towards a broad-based behavioral science." *Behavioral and Brain Sciences* 33 (2/3):51–75.

———. 2010b. "The weirdest people in the world?" *Behavioral and Brain Sciences* 33 (2/3):1–23.

Henrich, J., and N. Henrich. 2010. "The evolution of cultural adaptations: Fijian taboos during pregnancy and lactation protect against marine toxins." *Proceedings of the Royal Society B: Biological Sciences* 366:1139–1148.

———. 2014. "Fairness without Punishment: Behavioral Experiments in the Yasawa Island, Fiji." In *Experimenting with Social Norms: Fairness and Punishment in Cross-Cultural Perspective*, edited by J. Ensminger and J. Henrich, 225–258. New York: Russell Sage Press.

Henrich, J., and R. McElreath. 2003. "The evolution of cultural evolution." *Evolutionary Anthropology* 12 (3):123–135.

Henrich, J., R. McElreath, J. Ensminger, A. Barr, C. Barrett, A. Bolyanatz, J. C. Cardenas, et al. 2006. "Costly punishment across human societies." *Science* 312:1767–1770.

Henrich, J., and J. B. Silk. 2013. "Interpretative problems with chimpanzee ultimatum game." White paper. Social Science Research Network (SSRN). http://www.pnas.org/content/110/33/E3049.full?ijkey=ca7e16c2e252064447c5e7447884aab7c8cb598e&keytype2=tf_ipsecsha.

Henrich, J., and N. Smith. 2004. "Comparative experimental evidence from Machiguenga, Mapuche, and American populations." In *Foundations of Human Sociality: Economic Experiments and Ethnographic Evidence from Fifteen Small-Scale Societies*, edited by J. Henrich, R. Boyd, S. Bowles, H. Gintis, E. Fehr, and C. Camerer, 125–167. Oxford: Oxford University Press.

Henrich, J., and C. Tennie. Forthcoming. "Cultural evolution in chimpanzees and humans." In *Chimpanzees and Human Evolution*, edited by M. Muller, R. Wrangham, and D. Pilbream. Cambridge, MA: Harvard University Press.

Henrich, N., and J. Henrich. 2007. *Why Humans Cooperate: A Cultural and Evolutionary Explanation* Oxford: Oxford University Press.

Herbranson, W. T., and J. Schroeder. 2010. "Are birds smarter than mathematicians? Pigeons (*Columba livia*) perform optimally on a version of the Monty Hall Dilemma." *Journal of Comparative Psychology* 124 (1):1–13.

Herrmann, E., J. Call, M. V. Hernandez-Lloreda, B. Hare, and M. Tomasello. 2007. "Humans have evolved specialized skills of social cognition: The cultural intelligence hypothesis." *Science* 317 (5843):1360–1366.

Herrmann, E., M. V. Hernandez-Lloreda, J. Call, B. Hare, and M. Tomasello. 2010. "The structure of individual differences in the cognitive abilities of children and chimpanzees." *Psychological Science* 21 (1):102–110.

Herrmann, P. A., C. H. Legare, P. L. Harris, and H. Whitehouse. 2013. "Stick to the script: The effect of witnessing multiple actors on children's imitation." *Cognition* 129 (3):536–543. http://dx.doi.org/10.1016/j.cognition.2013.08.010.

Hewlett, B., and S. Winn. 2012. "Allomaternal gatherers. nursing among hunter-gatherers." *American Journal of Physical Anthropology* 147:165–165.

REFERENCES

Heyes, C. 2012a. "Grist and mills: On the cultural origins of cultural learning." *Philosophical Transactions of the Royal Society B: Biological Sciences* 367 (1599):2181–2191.

———. 2012b. "What's social about social learning?" *Journal of Comparative Psychology* 126 (2):193–202.

Hill, K. 2002. "Altruistic cooperation during foraging by the Ache and the evolved human predisposition to cooperate." *Human Nature* 13 (1):105–128.

Hill, K., and A. M. Hurtado. 1996. *Ache Life History*. New York: Aldine de Gruyter.

———. 2009. "Cooperative breeding in South American hunter-gatherers." *Proceedings of the Royal Society B: Biological Sciences* 276 (1674):3863–3870.

Hill, K., and K. Kintigh. 2009. "Can anthropologists distinguish good and poor hunters? Implications for hunting hypotheses, sharing conventions, and cultural transmission." *Current Anthropology* 50 (3):369–377.

Hill, K., R. S. Walker, M. Božičević, J. Eder, T. Headland, B. Hewlett, A. M. Hurtado, F. Marlowe, P. Wiessner, and B. Wood. 2011. "Co-residence patterns in hunter-gatherer societies show unique human social structure." *Science* 331 (6022):1286–1289. doi:10.1126/science.1199071.

Hill, K., B. Wood, J. Baggio, A. M. Hurtado, and R. Boyd. 2014{YES AU: OK?}. "Hunter-gatherer Inter-band Interaction Rates: Implications for Cumulative Culture." *PLoS ONE* 9 (7):e102806.

Hilmert, C. J., J. A. Kulik, and N. J. S. Christenfeld. 2006. "Positive and negative opinion modeling: The influence of another's similarity and dissimilarity." *Journal of Personality and Social Psychology* 90 (3):440–452.

Hirschfeld, L. A. 1996. *Race in the Making: Cognition, Culture, and the Child's Construction of Human Kinds*. Cambridge, MA: MIT Press.

Hoenig, S. B. 1953. *The Great Sanhedrin: A Study of the Origin, Development, Composition, and Functions of the Bet Din ha-Gadol during the Second Jewish Commonwealth*. Philadelphia,: Dropsie College for Hebrew and Cognate Learning.

Hoffmann, F., and P. Oreopoulos. 2009. "A professor like me: The influence of instructor gender on college achievement." *Journal of Human Resources* 44 (2): 479–494.

Hogg, M. A., and J. Adelman. 2013. "Uncertainty-identity theory: Extreme groups, radical behavior, and authoritarian leadership." *Journal of Social Issues* 69 (3):436–454.

Hölldobler, B., and E. O. Wilson. 1990. *The Ants*. Cambridge, MA: Belknap Press of Harvard University Press.

Holmberg, A. R. 1950. *Nomads of the Long Bow*. Smithsonian Institution Institute of Social Anthropology Publ. No. 10. Washington DC: United States Government Printing Office.

Hoppitt, W., and K. N. Laland. 2013. *Social Learning: An Introduction to Mechanisms, Methods, and Models*. Princeton, NJ: Princeton University Press.

Horner, V., D. Proctor, K. E. Bonnie, A. Whiten, and F.B.M. de Waal. 2010. "Prestige affects cultural learning in chimpanzees." *PLoS ONE* 5 (5):e10625.

Horner, V., and A. Whiten. 2005. "Causal knowledge and imitation/emulation switching in chimpanzees (*Pan trogiodytes*) and children (*Homo sapiens*)." *Animal Cognition* 8 (3):164–181.

Horwitz, R. I., C. M. Viscoli, L. Berkman, R. M. Donaldson, S. M. Horwitz, C. J. Mur-

REFERENCES

ray, D. F. Ransohoff, and J. Sindelar. 1990. "Treatment adherence and risk of death after a myocardial infarction." *Lancet* 336 (8714):542–545.

House, B., J. Henrich, B. Sarnecka, and J. B. Silk. 2013. "The development of contingent reciprocity in children." *Evolution and Human Behavior* 34 (2):86–93.

House, B. R., J. B. Silk, J. Henrich, H. C. Barrett, B. A. Scelza, A. H. Boyette, B. S. Hewlett, R. McElreath, and S. Laurence. 2013. "Ontogeny of prosocial behavior across diverse societies." *Proceedings of the National Academy of Sciences, USA* 110 (36):14586–14591. doi:10.1073/pnas.1221217110.

Hove, M. J., and J. L. Risen. 2009. "It's all in the timing: Interpersonal synchrony increases affiliation." *Social Cognition* 27 (6):949–960.

Hrdy, S. B. 2009. *Mothers and Others: The Evolutionary Origins of Mutual Understanding*. Cambridge, MA: Belknap Press of Harvard University Press.

Hua, C. 2001. *A Society without Fathers or Husbands*. New York: Zone Books.

Hudson, T. 1981. "Recently discovered accounts concerning the "lone woman" of San Nicolas Island." *Journal of California and Great Basin Anthropology* 3 (2): 187–199.

Humphrey, N. 1976. "The social function of intellect." In *Growing Points in Ethology*, edited by P.P.G. Bateson, and R. A. Hinde, 303–317. Cambridge: Cambridge University Press.

———. 2012. "This chimp will kick your ass at memory games—But how the hell does he do it?" *Trends in Cognitive Sciences* 16 (7):353–355. doi:10.1016/j.tics.2012.05.002.

Hyde, T. M., S. V. Mathew, T. Ali, B. K. Lipska, A. J. Law, O. E. Metitiri, D. R. Weinberger, and J. E. Kleinman. 2009. "Cation chloride cotransporters: Expression patterns in development and schizophrenia." *Schizophrenia Bulletin* 35:150–151.

Ichikawa, M. 1987. "Food restrictions of the Mbuti Pygmies, Eastern Zaire." *African Study Monographs* Suppl. (6):97–121.

Ingram, C.J.E., C. A. Mulcare, Y. Itan, M. G. Thomas, and D. M. Swallow. 2009. "Lactose digestion and the evolutionary genetics of lactase persistence." *Human Genetics* 124 (6):579–591.

Ingram, C.J.E., T. O. Raga, A. Tarekegn, S. L. Browning, M. F. Elamin, E. Bekele, M. G. Thomas, et al. 2009. "Multiple rare variants as a cause of a common phenotype: Several different lactase persistence associated alleles in a single ethnic group." *Journal of Molecular Evolution* 69 (6):579–588.

Inoue, S., and T. Matsuzawa. 2007. "Working memory of numerals in chimpanzees." *Current Biology* 17 (23):R1004–R1005.

Isler, K., and C. P. Van Schaik. 2009. "Why are there so few smart mammals (but so many smart birds)?" *Biology Letters* 5 (1):125–129.

———. 2012. "Allomaternal care, life history and brain size evolution in mammals." *Journal of Human Evolution* 63 (1):52–63.

Isler, K., J. T. Van Woerden, A. F. Navarrete, and C. P. Van Schaik. 2012. "The "gray ceiling": Why apes are not as large-brained as humans." *American Journal of Physical Anthropology* 147:173–173.

Itan, Y., B. L. Jones, C. J. E. Ingram, D. M. Swallow, and M. G. Thomas. 2010. "A worldwide correlation of lactase persistence phenotype and genotypes." *BMC Evolutionary Biology* 10. doi:10.1186/1471-2148-10-36.

REFERENCES

Jablonski, N. G., and G. Chaplin. 2000. "The evolution of human skin coloration." *Journal of Human Evolution* 39 (1):57–106.

————. 2010. "Human skin pigmentation as an adaptation to UV radiation." *Proceedings of the National Academy of Sciences, USA* 107:8962–8968.

Jackson, F.L.C., and R. T. Jackson. 1990. "The role of cassava in African famine prevention." In *African Food Systems in Crisis*, Part 2, *Contending with Change*, edited by R. Huss-Ashmore, 207–225. Amsterdam: Gordon and Breach.

Jaeggi, A. V., L. P. Dunkel, M. A. Van Noordwijk, S. A. Wich, A. A. L. Sura, and C. P. Van Schaik. 2010. "Social learning of diet and foraging skills by wild immature Bornean orangutans: Implications for culture." *American Journal of Primatology* 72 (1):62–71.

James, P. A., G. Mitchell, M. Bogwitz, and G. J. Lindeman. 2013. "The Angelina Jolie effect." *Medical Journal of Australia* 199 (10):646–646.

Jaswal, V. K., and L. S. Malone. 2007. "Turning believers into skeptics: 3-year-olds' sensitivity to cues to speaker credibility." *Journal of Cognition and Development* 8 (3):263–283.

Jaswal, V. K., and L. A. Neely. 2006. "Adults don't always know best: Preschoolers use past reliability over age when learning new words." *Psychological Science* 17 (9):757–758.

Jensen, K., J. Call, and M. Tomasello. 2007a. "Chimpanzees are rational maximizers in an ultimatum game." *Science* 318 (5847):107–109.

————. 2007b. "Chimpanzees are vengeful but not spiteful." *Proceedings of the National Academy of Sciences, USA* 104 (32):13046–13050.

————. 2013. "Chimpanzee responders still behave like rational maximizers." *Proceedings of the National Academy of Sciences, USA*. doi:10.1073/pnas.1303627110.

Jensen, K., B. Hare, J. Call, and M. Tomasello. 2006. "What's in it for me? Self-regard precludes altruism and spite in chimpanzees." *Proceedings of the Royal Society B: Biological Sciences* 273 (1589):1013–1021.

Jerardino, A., and C. W. Marean. 2010. "Shellfish gathering, marine paleoecology and modern human behavior: Perspectives from cave PP13B, Pinnacle Point, South Africa." *Journal of Human Evolution* 59 (3–4):412–424.

Jobling, J. W., and W. Petersen. 1916. "The epidemiology of pellagra in Nashville Tennessee." *Journal of Infectious Diseases* 18 (5):501–567.

Johnson, A., and T. Earle. 2000. *The Evolution of Human Societies*. 2nd ed. Stanford, CA: Stanford University Press.

Johnson, R. T., J. A. Burk, and L. A. Kirkpatrick. 2007. "Dominance and prestige as differential predictors of aggression and testosterone levels in men." *Evolution and Human Behavior* 28 (5):345–351. doi:10.1016/j.evolhumbehav.2007.04.003.

Jonas, K. 1992. "Modeling and suicide: A test of the Werther effect." *British Journal of Social Psychology* 31:295–306.

Jones, B. C., L. M. DeBruine, A. C. Little, R. P. Burriss, and D. R. Feinberg. 2007. "Social transmission of face preferences among humans." *Proceedings of the Royal Society B: Biological Sciences* 274 (1611):899–903.

Jones, R. 1974. "Tasmanian tribes." In *Aboriginal Tribes of Australia*, edited by N. B. Tindale, 319–354. San Francisco: UCLA Press.

————. 1976. "Tasmania: Aquatic machines and off-shore islands." In *Problems in*

Economic and Social Archaeology, edited by G. Sieveking, I. H. Longworth, and K. E. Wilson, 235–263. London: Duckworth.

———. 1977a. "Man as an element of a continental fauna: The case of the sundering of the Bassian bridge." In *Sunda and Sahul: Prehistoric Studies in Southeast Asia, Melanesia and Australia*, edited by J. Allen, J. Golson, and Rhys Jones, 317–386. London: Academic Press.

———. 1977b. "The Tasmanian paradox." In *Stone Tools as Cultural Markers: Change, Evolution and Complexity*, edited by V. S. Wright, 189–204. Atlantic Highlands, NJ: Humanities Press.

———. 1977c. "Why did the Tasmanians stop eating fish?" In *Explorations in Ethnoarchaeology*, edited by R. Gould, 11–47. Santa Fe: University of New Mexico Press.

———. 1990. "From Kakadu to Kutikina: The southern continent at 18,000 years ago." In *Low Latitudes*, Vol. 2 of *The World at 18,000 B. P.*, edited by C. Gamble and O. Soffer, 264–295. London: Unwin Hyman.

———. 1995. "Tasmanian archaeology: Establishing the sequence." *Annual Review of Anthropology* 24:423–46.

Kagel, J. C., D. McDonald, and R. C. Battalio. 1990. "Tests of "fanning out" of indifference curves: Results from animal and human experiments." *American Economic Review* 80:912–21.

Kahneman, D. 2011. *Thinking, Fast and Slow*. New York: Farrar, Straus and Giroux.

Kahneman, D., P. Slovic, and A. Tversky. 1982. *Judgment under Uncertainty: Heuristics and Biases*. Cambridge: Cambridge University Press.

Kail, R. V. 2007. "Longitudinal evidence that increases in processing speed and working memory enhance children's reasoning." *Psychological Science* 18 (4):312–313. doi:10.1111/j.1467–9280.2007.01895.x.

Kalmar, I. 1985. "Are there really no primitive languages?" In *Literacy, Language and Learning: The Nature and Consequences of Reading and Writing*, edited by D. R. Olson, N. Torrance, and A. Hildyard, 148–166. Cambridge: Cambridge University Press.

Kanovsky, M. 2007. "Essentialism and folksociology: Ethnicity again." *Journal of Cognition and Culture* 7:241–281.

Kaplan, H., M. Gurven, J. Winking, P. L. Hooper, and J. Stieglitz. 2010. "Learning, menopause, and the human adaptive complex." *Reproductive Aging* 1204:30–42.

Kaplan, H., K. Hill, J. Lancaster, and A. M. Hurtado. 2000. "A theory of human life history evolution: Diet, intelligence, and longevity." *Evolutionary Anthropology* 9 (4):156–185.

Katz, S. H., M. L. Hediger, and L. A. Valleroy. 1974. "Traditional maize processing techniques in the New World: Traditional alkali processing enhances the nutritional quality of maize." *Science* 184 (17 May):765–773.

Kay, P. 2005. "Color categories are not arbitrary." *Cross-Cultural Research* 39 (1):39–55.

Kay, P., and T. Regier. 2006. "Language, thought and color: recent developments." *Trends in Cognitive Sciences* 10 (2):51–54.

Kayser, M., F. Liu, A.C.J.W. Janssens, F. Rivadeneira, O. Lao, K. van Duijn, M. Vermeulen, et al. 2008. "Three genome-wide association studies and a linkage analysis

REFERENCES

identify *HERC2* as a human iris color gene." *American Journal of Human Genetics* 82 (2):411–423.

Keeley, L. 1997. *War before Civilization*. Oxford: Oxford University Press.

Kelly, R. C. 1985. *The Nuer Conquest*. Ann Arbor: University of Michigan Press.

Kelman, H. C. 1958. "Compliance, identification, and internalization: Three processes of attitude change." *Journal of Conflict Resolution* 2:51–60.

Kendon, A. 1988. *Sign Languages of Aboriginal Australia: Cultural, Semiotic, and Communicative Perspectives*. Cambridge: Cambridge University Press.

Kenward, B. 2012. "Over-imitating preschoolers believe unnecessary actions are normative and enforce their performance by a third party." *Journal of Experimental Child Psychology* 112 (2):195–207.

Kessler, R. C., G. Downey, and H. Stipp. 1988. "Clustering of teenage suicides after television news stories about suicide: A reconsideration." *American Journal of Psychiatry* 145:1379–83.

Kessler, R. C., and H. Stipp. 1984. "The impact of fictional television suicide stories on U.S. fatalities: A replication." *American Journal of Sociology* 90 (1):151–167.

Khaldûn, I. 2005. *The Muqaddimah: An Introduction to History*. Princeton, NJ: Princeton University Press.

Kim, G., and K. Kwak. 2011. "Uncertainty matters: Impact of stimulus ambiguity on infant social referencing." *Infant and Child Development* 20 (5):449–463. doi:10 .1002/icd.708.

Kimbrough, E., and A. Vostroknutov. 2013. "Norms make preferences social." Unpublished manuscript. Simon Fraser University.

Kinzler, K. D., K. H. Corriveau, and P. L. Harris. 2011. "Children's selective trust in native-accented speakers." *Developmental Science* 14 (1):106–111.

Kinzler, K. D., and J. B. Dautel. 2012. "Children's essentialist reasoning about language and race." *Developmental Science* 15 (1):131–138.

Kinzler, K. D., E. Dupoux, and E. S. Spelke. 2007. "The native language of social cognition." *Proceedings of the National Academy of Sciences, USA* 104 (30):12577–12580.

Kinzler, K. D., K. Shutts, J. Dejesus, and E. S. Spelke. 2009. "Accent trumps race in guiding children's social preferences." *Social Cognition* 27 (4):623–634.

Kinzler, K. D., K. Shutts, and E. S. Spelke. 2012. "Language-based social preferences among children in South Africa." *Language Learning and Development* 8 (215–232).

Kirby, S. 1999. *Function, Selection, and Innateness: The Emergence of Language Universals*. Oxford: Oxford University Press.

Kirby, S., M. H. Christiansen, and N. Chater. 2013. "Syntax as an adaptation to the learner." In Biological Foundations and Origin of Syntax, edited by D. Bickerton and E. Szathmáry, 325–344. Cambridge, MA: MIT Press.

Kirby, S., H. Cornish, and K. Smith. 2008. "Cumulative cultural evolution in the laboratory: An experimental approach to the origins of structure in human language." *Proceedings of the National Academy of Sciences, USA* 105 (31):10681–10686.

Kirschner, S., and M. Tomasello. 2009. "Joint drumming: Social context facilitates synchronization in preschool children." *Journal of Experimental Child Psychology* 102 (3):299–314.

REFERENCES

————. 2010. "Joint music making promotes prosocial behavior in 4-year-old children." *Evolution and Human Behavior* 31 (5):354–364.

Klein, R. G. 2009. *The Human Career: Human Biological and Cultural Origins.* 3rd ed. Chicago: University of Chicago Press.

Kline, M. A., and R. Boyd. 2010. "Population size predicts technological complexity in Oceania." *Proceedings of the Royal Society B: Biological Sciences* 277 (1693): 2559–2564.

Klucharev, V., K. Hytonen, M. Rijpkema, A. Smidts, and G. Fernandez. 2009. "Reinforcement learning signal predicts social conformity." *Neuron* 61 (1):140–151.

Knauft, B. M. 1985. *Good Company and Violence: Sorcery and Social Action in a Lowland New Guinea Society.* Berkeley: University of California Press.

Kobayashi, Y., and K. Aoki. 2012. "Innovativeness, population size and cumulative cultural evolution." *Theoretical Population Biology* 82 (1):38–47.

Koenig, M. A., and P. L. Harris. 2005. "Preschoolers mistrust ignorant and inaccurate speakers." *Child Development* 76 (6):1261–1277.

Kong, J., R. L. Gollub, G. Polich, I. Kirsch, P. LaViolette, M. Vangel, B. Rosen, and T. J. Kaptchuk. 2008. "A functional magnetic resonance imaging study on the neural mechanisms of hyperalgesic nocebo effect." *Journal of Neuroscience* 28 (49): 13354–13362.

Konvalinka, I., D. Xygalatas, J. Bulbulia, U. Schjodt, E. M. Jegindo, S. Wallot, G. Van Orden, and A. Roepstorff. 2011. "Synchronized arousal between performers and related spectators in a fire-walking ritual." *Proceedings of the National Academy of Sciences, USA* 108 (20):8514–8519.

Krackle, W. H. 1978. *Force and Persuasion: Leadership in an Amazonian Society.* Chicago: University of Chicago Press.

Krakauer, J. 1997. *Into Thin Air: A Personal Account of the Mount Everest Disaster.* New York: Villard.

Kramer, K. L. 2010. "Cooperative breeding and its significance to the demographic success of humans." *Annual Review of Anthropology* 39:417–436.

Kroeber, A. L. 1925. *Handbook of the Indians of California.* Bulletin of the Smithsonian Institution. Bureau of American Ethnology. Washington, DC: U.S. Government Printing Office.

————. 1958. "Sign language inquiry." *International Journal of American Linguistics* 24 (1):1–19. doi:10.2307/1264168.

Kroll, Y., and H. Levy. 1992. "Further tests of the Separation Theorem and the Capital Asset Pricing Model." *American Economic Review* 82 (3):664–670.

Kuhl, P. K. 2000. "A new view of language acquisition." *Proceedings of the National Academy of Sciences, USA* 97 (22):11850–11857. doi:10.1073/pnas.97.22.11850.

Kumru, C. S., and L. Vesterlund. 2010. "The effect of status on charitable giving." *Journal of Public Economic Theory* 12 (4):709–735.

Kwok, V., Z. D. Niu, P. Kay, K. Zhou, L. Mo, Z. Jin, K. F. So, and L. H. Tan. 2011. "Learning new color names produces rapid increase in gray matter in the intact adult human cortex." *Proceedings of the National Academy of Sciences, USA* 108 (16):6686–6688.

Lachmann, M., and C. T. Bergstrom. 2004. "The disadvantage of combinatorial communication." *Proceedings of the Royal Society of London Series B: Biological Sciences* 271 (1555):2337–2343.

REFERENCES

Laland, K. N. 2004. "Social learning strategies." *Learning & Behavior* 32 (1):4–14.

Laland, K. N., N. Atton, and M. M. Webster. 2011. "From fish to fashion: Experimental and theoretical insights into the evolution of culture." *Philosophical Transactions of the Royal Society B: Biological Sciences* 366 (1567):958–968.

Laland, K. N., J. Odling-Smee, and S. Myles. 2010. "How culture shaped the human genome: Bringing genetics and the human sciences together." *Nature Reviews Genetics* 11 (2):137–148.

Lambert, A. D. 2009. *The Gates of Hell : Sir John Franklin's Tragic Quest for the North West Passage.* New Haven, CT: Yale University Press.

Lambert, P. M. 1997. Patterns of violence in prehistoric hunter-gatherer societies of coastal southern California. In *Troubled Times: Violence and Warfare in the Past,* edited by D. L. Martin and D. W. Frayer, 77–109. Amsterdam: Gordon and Breach.

Langergraber, K. 2012. "Cooperation among kin." In *The Evolution of Primate Societies,* edited by J. C. Mitani, J. Call, P. M. Kappeler, R. A. Palombit, and J. B. Silk, 491–513. Chicago: University of Chicago Press.

Langergraber, K., J. Mitani, and L. Vigilant. 2009. "Kinship and social bonds in female chimpanzees *(Pan troglodytes).*" *American Journal of Primatology* 71 (10): 840–851.

Langergraber, K. E., J. C. Mitani, and L. Vigilant. 2007. "Wild male chimpanzees preferentially affiliate and cooperate with maternal but not paternal siblings." *American Journal of Physical Anthropology*:150–150.

Lappan, S. 2008. "Male care of infants in a siamang *(Symphalangus syndactylus)* population including socially monogamous and polyandrous groups." *Behavioral Ecology and Sociobiology* 62 (8):1307–1317.

Lawless, R. 1975. "Effects of population growth and environment changes on divination practices in northern Luzon." *Journal of Anthropological Research* 31 (1): 18–33.

Leach, H. M. 2003. "Human domestication reconsidered." *Current Anthropology* 44 (3):349–368.

Ledyard, J. O. 1995. "Public goods: A survey of experimental research." In *The Handbook of Experimental Economics,* edited by John H. Kagel and Alvin E. Roth, 111–194. Princeton, NJ: Princeton University Press.

Lee, R. B. 1979. *The !Kung San: Men, Women, and Work in a Foraging Society.* Cambridge: Cambridge University Press.

———. 1986. "!Kung Kin terms: The name relationship and the process of discovery." In *The Past and Future of !Kung Ethnography: Essays in Honor of Lorna Marshall,* edited by M. Biesele, R. Gordon, and R. B. Lee, 77–102. Hamburg: Helmut Buske.

Lee, R. B., and R. H. Daly. 1999. *The Cambridge Encyclopedia of Hunters and Gatherers.* Cambridge: Cambridge University Press.

Lee, S. H., and M. H. Wolpoff. 2002. "Pattern of brain size increase in Pleistocene Homo." *Journal of Human Evolution* 42 (3):A19–A20.

Lehmann, L., K. Aoki, and M. W. Feldman. 2011. "On the number of independent cultural traits carried by individuals and populations." *Philosophical Transactions of the Royal Society B: Biological Sciences* 366 (1563):424–435.

Leonard, W. R., M. L. Robertson, J. J. Snodgrass, and C. W. Kuzawa. 2003. "Metabolic

REFERENCES

correlates of hominid brain evolution." *Comparative Biochemistry and Physiology A: Molecular & Integrative Physiology* 136 (1):5–15.

Leonard, W. R., J. J. Snodgrass, and M. L. Robertson. 2007. "Effects of brain evolution on human nutrition and metabolism." *Annual Review of Nutrition* 27:311–327.

Leonardi, M., P. Gerbault, M. G. Thomas, and J. Burger. 2012. "The evolution of lactase persistence in Europe. A synthesis of archaeological and genetic evidence." *International Dairy Journal* 22 (2):88–97.

Lesorogol, C., and J. Ensminger. 2013. "Double-blind dictator games in Africa and the U.S.: Differential experimenter effects." In *Experimenting with Social Norms: Fairness and Punishment in Cross-Cultural Perspective*, edited by J. Ensminger and J. Henrich, 149–157. New York: Russell Sage Press.

Li, H., S. Gu, Y. Han, Z. Xu, A. J. Pakstis, L. Jin, J. R. Kidd, and K. K. Kidd. 2011. "Diversification of the *ADH1B* gene during expansion of modern humans." *Annals of Human Genetics* 75:497–507.

Liebenberg, L. 1990. *The Art of Tracking: The Origin of Science*. Cape Town, South Africa: David Philip Publishers.

———. "Persistence hunting by modern hunter-gatherers." *Current Anthropology* 47 (6):1017–1025.

Lieberman, D. 2013. *The Story of the Human Body: Evolution, Health, and Disease*. New York: Random House.

Lieberman, D. E., D. M. Bramble, D. A. Raichlen, and J. J. Shea. 2009. "The evolutionary question posed by human running capabilities." In *The First Humans: Origin and Early Evolution of the Genus Homo*, edited by F. E. Grine, J. G. Fleagel, and R. E. Leakey, 77–92. New York: Springer.

Lieberman, D., D. M. T. Fessler, and A. Smith. 2011. "The relationship between familial resemblance and sexual attraction: An update on Westermarck, Freud, and the incest taboo." *Personality and Social Psychology Bulletin* 37 (9):1229–1232.

Lieberman, D., J. Tooby, and L. Cosmides. 2003. "Does morality have a biological basis? An empirical test of the factors governing moral sentiments relating to incest." *Proceedings of the Royal Society B: Biological Sciences* 270 (1517):819–826.

Lieberman, D. E., M. Venkadesan, W. A. Werbel, A. I. Daoud, S. D'Andrea, I. S. Davis, R. O. Mang'Eni, and Y. Pitsiladis. 2010. "Foot strike patterns and collision forces in habitually barefoot versus shod runners." *Nature* 463 (7280):531–U149.

Lind, J., and P. Lindenfors. 2010. "The number of cultural traits is correlated with female group size but not with male group size in chimpanzee communities." *PLoS ONE* 5 (3):e9241. doi:10.1371/journal.pone.0009241.

Lindblom, B. 1986. "Phonetic universals in vowel systems." In *Experimental Phonology*, edited by J. J. Ohala and J. J. Jaeger, 13–44.Waltham, MA: Academic Press.

Little, A. C., R. P. Burriss, B. C. Jones, L. M. DeBruine, and C. A. Caldwell. 2008. "Social influence in human face preference: men and women are influenced more for long-term than short-term attractiveness decisions." *Evolution and Human Behavior* 29 (2):140–146.

Little, A. C., B. C. Jones, L. M. DeBruine, and C. A. Caldwell. 2011. "Social learning and human mate preferences: a potential mechanism for generating and maintaining between-population diversity in attraction." *Philosophical Transactions of the Royal Society B: Biological Sciences* 366 (1563):366–375.

Lombard, M. 2011. "Quartz-tipped arrows older than 60 ka: Further use-trace evi-

REFERENCES

dence from Sibudu, KwaZulu-Natal, South Africa." *Journal of Archaeological Science* 38 (8):1918–1930.

Lomer, M. C. E., G. C. Parkes, and J. D. Sanderson. 2008. "Review article: Lactose intolerance in clinical practice—myths and realities." *Alimentary Pharmacology & Therapeutics* 27 (2):93–103.

Lopez, A., S. Atran, J. D. Coley, D. L. Medin, and E. E. Smith. 1997. "The tree of life: Universal and cultural features of folkbiological taxonomies and inductions." *Cognitive Psychology* 32 (3):251–295.

Lorenzen, E. D., D. Nogues-Bravo, L. Orlando, J. Weinstock, J. Binladen, K. A. Marske, A. Ugan, et al. 2011. "Species-specific responses of Late Quaternary megafauna to climate and humans." *Nature* 479 (7373):359–365.

Lothrop, S. K. 1928. *The Indians of Tierra del Fuego*. New York: Museum of the American Indian and Heye Foundation.

Lovejoy, C. O. 2009. "Reexamining human origins in light of *Ardipithecus ramidus*." *Science* 326 (5949).

Lozoff, B. 1983. "Birth and bonding in non-industrial societies." *Developmental Medicine and Child Neurology* 25 (5):595–600.

Luczak, S. E., S. J. Glatt, and T. L. Wall. 2006. "Meta-analyses of *ALDH2* and *ADH1B* with alcohol dependence in Asians." *Psychological Bulletin* 132 (4):607–621.

Lumsden, C., and E. O. Wilson. 1981. *Genes, Mind and Culture*. Cambridge, MA: Harvard University Press.

Lupyan, G., and R. Dale. 2010. "Language structure is partly determined by social structure." *Plos One* 5 (1):e8559. doi:10.1371/journal.pone.0008559.

Lyons, D. E., A. G. Young, and F. C. Keil. 2007. "The hidden structure of overimitation." *Proceedings of the National Academy of Sciences* 104 (50):19751–19756.

Maguire, E. A., K. Woollett, and H. J. Spiers. 2006. "London taxi drivers and bus drivers: A structural MRI and neuropsychological analysis." *Hippocampus* 16 (12): 1091–1101.

Mallery, G. 2001 (1881). *Sign Language among North American Indians*. Kindle ed. New York: Dover Publications.

Mann, C. C. 2012. *1493: Uncovering the New World Columbus Created*. New York: Vintage Books.

Marlowe, F. W. 2004. "What explains Hadza food sharing?" In *Research in Economic Anthropology: Aspects of Human Behavioral Ecology*, edited by M. Alvard, 67–86. Greenwich, CT: JAI Press.

Marlowe, F. 2010. *The Hadza: Hunter-Gatherers of Tanzania*. Berkeley: University of California Press.

Marshall, L. 1976. *The !Kung of Nyae Nyae*. Cambridge, MA: Harvard University Press.

Martin, C. F., R. Bhui, P. Bossaerts, T. Matsuzawa, and C. F. Camerer. 2014. "Experienced chimpanzees are more strategic than humans in competitive games." *Scientific Reports* 4:5182. doi:10.1038/srep05182.

Martin, C. L., L. Eisenbud, and H. Rose. 1995. "Children's gender-based reasoning about toys." *Child Development* 66 (5):1453–1471.

Martin, C. L., and J. K. Little. 1990. "The relation of gender understanding to children's sex-type preferences and gender stereotypes." *Child Development* 61 (5):1427–1439.

REFERENCES

Martinez, I., M. Rosa, R. Quam, P. Jarabo, C. Lorenzo, A. Bonmati, A. Gomez-Olivencia, A. Gracia, and J. L. Arsuaga. 2013. "Communicative capacities in Middle Pleistocene humans from the Sierra de Atapuerca in Spain." *Quaternary International* 295:94–101.

Mascaro, O., and G. Csibra. 2012. "Representation of stable social dominance relations by human infants." *Proceedings of the National Academy of Sciences, USA* 109 (18):6862–6867.

Mathew, S. n.d. "Second-order free rider elicit moral punitive sentiments in a small-scale society." Unpublished manuscript. Arizona State Univ.

Mathew, S., and R. Boyd. 2011. "Punishment sustains large-scale cooperation in prestate warfare." *Proceedings of the National Academy of Sciences, USA* 108 (28): 11375–11380.

Mathew, S., R. Boyd, and M. van Veelen. 2013. "Human cooperation among kin and close associates may require enforcement of norms by third parties." In *Cultural Evolution*, edited by P. J. Richerson and M. H. Christiansen, 45–60. Cambridge, MA: MIT Press.

Mausner, B. 1954. "The effect of prior reinforcement on the interaction of observe pairs." *Journal of Abnormal Social Psychology* 49:65–68.

Mausner, B., and B. L. Bloch. 1957. "A study of the additivity of variables affecting social interaction." *Journal of Abnormal Social Psychology* 54:250–256.

Maxwell, M. S. 1984. "Pre-Dorset and Dorset prehistory of Canada." In *Arctic*, Vol. 5 of *Handbook of North American Indians*, edited by D. Damas, 359–368. Washington, DC: Smithsonian Institution Press.

Maynard Smith, J., and E. R. Szathmáry. 1999. *The Origins of Life: From the Birth of Life to the Origin of Language*. Oxford: Oxford University Press.

McAuliffe, K., and H. Whitehead. 2005. "Eusociality, menopause and information in matrilineal whales." *Trends in Ecology and Evolution* 20 (12):650.

McBrearty, S., and A. Brooks. 2000. "The revolution that wasn't: A new interpretation of the origin of modern human behavior." *Journal of Human Evolution* 39:453–563.

Mccleary, B. V., and B. F. Chick. 1977. "Purification and properties of a thiaminase I enzyme from nardoo (*Marsilea drummondii*)." *Phytochemistry* 16 (2):207–213.

McCollum, M. A., C. C. Sherwood, C. J. Vinyard, C. O. Lovejoy, and F. Schachat. 2006. "Of muscle-bound crania and human brain evolution: The story behind the *MYH16* headlines." *Journal of Human Evolution* 50 (2):232–236.

McComb, K., C. Moss, S. M. Durant, L. Baker, and S. Sayialel. 2001. "Matriarchs as repositories of social knowledge in African elephants." *Science* 292 (5516): 491–494.

McComb, K., G. Shannon, S. M. Durant, K. Sayialel, R. Slotow, J. Poole, and C. Moss. 2011. "Leadership in elephants: The adaptive value of age." *Proceedings of the Royal Society B: Biological Sciences* 278 (1722):3270–3276.

McComb, K., G. Shannon, K. N. Sayialel, and C. Moss. 2014. "Elephants can determine ethnicity, gender, and age from acoustic cues in human voices." *Proceedings of the National Academy of Sciences, USA* 111 (14):5433–5438.

McConvell, P. 1985. "The origin of subsections in northern Australia." *Oceania* 56 (1):1–33. doi:10.2307/40330845.

REFERENCES

————. 1996. "Backtracking to Babel: The chronology of Pama-Nyungan expansion in Australia." *Archaeology in Oceania* 31 (3):125–144. doi:10.2307/40387040.

McDonough, C. M., A. Tellezgiron, M. Gomez, and L. W. Rooney. 1987. "Effect of cooking time and alkali content on the structure of corn and sorghum nixtamal." *Cereal Foods World* 32 (9):660–661.

McDougall, C. 2009. *Born to Run: A Hidden Tribe, Superathletes, and the Greatest Race the World Has Never Seen.* New York: Alfred A. Knopf.

McElreath, R., A. V. Bell, C. Efferson, M. Lubell, P. J. Richerson, and T. Waring. 2008. "Beyond existence and aiming outside the laboratory: estimating frequency-dependent and pay-off-biased social learning strategies." *Philosophical Transactions of the Royal Society B: Biological Sciences* 363 (1509):3515–3528. doi:10.1098/rstb.2008.0131.

McElreath, R., R. Boyd, and P. J. Richerson. 2003. "Shared norms and the evolution of ethnic markers." *Current Anthropology* 44 (1):122–129.

McElreath, R., M. Lubell, P. J. Richerson, T. M. Waring, W. Baum, E. Edsten, C. Efferson, and B. Paciotti. 2005. "Applying evolutionary models to the laboratory study of social learning." *Evolution and Human Behavior* 26 (6):483–508. doi:10.1016/j.evolhumbehav.2005.04.003.

McGhee, R. 1984. "Thule prehistory of Canada." In *Arctic*, Vol. 5 of *Handbook of North American Indians*, edited by D. Damas. Washington, DC: Smithsonian Institution Press.

McGovern, P. E., J. H. Zhang, J. G. Tang, Z. Q. Zhang, G. R. Hall, R. A. Moreau, A. Nunez, et al. 2004. "Fermented beverages of pre- and proto-historic China." *Proceedings of the National Academy of Sciences, USA* 101 (51):17593–17598.

McGuigan, N. 2012. "The role of transmission biases in the cultural diffusion of irrelevant actions." *Journal of Comparative Psychology* 126 (2):150–160.

————. 2013. "The influence of model status on the tendency of young children to over-imitate." *Journal of Experimental Child Psychology* 116 (4):962–969. http://dx.doi.org/10.1016/j.jecp.2013.05.004.

McGuigan, N., D. Gladstone, and L. Cook. 2012. "Is the cultural transmission of irrelevant tool actions in adult humans (*Homo sapiens*) best explained as the result of an evolved conformist bias?" *Plos One* 7 (12):e50863. doi:10.1371/journal.pone.0050863.

McGuigan, N., J. Makinson, and A. Whiten. 2011. "From over-imitation to super-copying: Adults imitate causally irrelevant aspects of tool use with higher fidelity than young children." *British Journal of Psychology* 102:1–18.

McGuigan, N., A. Whiten, E. Flynn, and V. Horner. 2007. "Imitation of causally opaque versus causally transparent tool use by 3-and 5-year-old children." *Cognitive Development* 22 (3):353–364.

McNeill, W. H. 1995. *Keeping Together in Time: Dance and Drill in Human History.* Cambridge, MA: Harvard University Press.

McPherron, S. P., Z. Alemseged, C. W. Marean, J. G. Wynn, D. Reed, D. Geraads, R. Bobe, and H. A. Bearat. 2010. "Evidence for stone-tool-assisted consumption of animal tissues before 3.39 million years ago at Dikika, Ethiopia." *Nature* 466 (7308):857–860.

Medin, D. L., and S. Atran. 1999. *Folkbiology.* Cambridge, MA: MIT Press.

————. 2004. "The native mind: Biological categorization and reasoning in development and across cultures." *Psychological Review* 111 (4):960–983.

Mellars, P., and J. C. French. 2011. "Tenfold population increase in Western Europe at the Neandertal-to-modern human transition." *Science* 333 (6042):623–627.

Meltzoff, A. N., A. Waismeyer, and A. Gopnik. 2012. "Learning about causes from people: Observational causal learning in 24-month-old infants." *Developmental Psychology* 48 (5):1215–1228. doi:10.1037/a0027440.

Mesoudi, A. 2011a. "An experimental comparison of human social learning strategies: Payoff-biased social learning is adaptive but underused." *Evolution and Human Behavior* 32 (5):334–342.

————. 2011b. "Variable cultural acquisition costs constrain cumulative cultural evolution." *Plos One* 6 (3):e18239. http://dx.doi.org/10.1371%2Fjournal.pone .0018239.

Mesoudi, A., L. Chang, K. Murray, and H. J. Lu. 2014. "Higher frequency of social learning in China than in the West shows cultural variation in the dynamics of cultural evolution." *Proceedings of the Royal Society B: Biological Sciences* 282 (1798). doi:10.1098/rspb.2014.2209.

Mesoudi, A., and M. O Brien. 2008. "The cultural transmission of Great Basin projectile-point technology: An experimental simulation." *American Antiquity* 73 (1):3–28.

Meulman, E. J. M., C. M. Sanz, E. Visalberghi, and C. P. van Schaik. 2012. "The role of terrestriality in promoting primate technology." *Evolutionary Anthropology* 21 (2):58–68.

Meyer, J. 2004. "Bioacoustics of human whistled languages: An alternative approach to the cognitive processes of language." *Anais da Academia Brasileira de Ciencias* 76 (2):405–412.

Meyers, J. L., D. Shmulewitz, E. Aharonovich, R. Waxman, A. Frisch, A. Weizman, B. Spivak, H. J. Edenberg, J. Gelernter, and D. S. Hasin. 2013. "Alcohol-metabolizing genes and alcohol phenotypes in an Israeli household sample." *Alcoholism: Clinical and Experimental Research* 37 (11):1872–1881.

Midlarsky, E., and J. H. Bryan. 1972. "Affect expressions and children's imitative altruism." *Journal of Experimental Child Psychology* 6:195–203.

Miller, D. J., T. Duka, C. D. Stimpson, S. J. Schapiro, W. B. Baze, M. J. McArthur, A. J. Fobbs, et al. 2012. "Prolonged myelination in human neocortical evolution." *Proceedings of the National Academy of Sciences, USA* 109 (41):16480–16485.

Miller, N. E., and J. Dollard. 1941. *Social Learning and Imitation.* New Haven, CT: Yale University Press.

Mischel, W., and R. M. Liebert. 1966. "Effects of discrepancies between observed and imposed reward criteria on their acquisition and transmission." *Journal of Personality and Social Psychology* 3:45–53.

Mitani, J. C., D. P. Watts, and S. J. Amsler. 2010. "Lethal intergroup aggression leads to territorial expansion in wild chimpanzees." *Current Biology* 20 (12):R507–R508. doi:10.1016/j.cub.2010.04.021.

Mithun, M. 1984. "How to avoid subordination." *Berkeley Linguistics Society* 10:493–523.

Moerman, D. 2002. "Explanatory mechanisms for placebo effects: Cultural influences and the meaning response." In *The Science of the Placebo: Toward an Interdisciplinary Research Agenda*, edited by H. A. Guess, A. Kleinman, J. W. Kusek, and L. W. Engel, 77–107. London: BMJ Books.

———. 2000. "Cultural variations in the placebo effect: Ulcers, anxiety, and blood pressure." *Medical Anthropology Quarterly* 14 (1):51–72.

Mokyr, J. 1990. *The Lever of Riches*. New York: Oxford University Press.

Moll, J., F. Krueger, R. Zahn, M. Pardini, R. de Oliveira-Souzat, and J. Grafman. 2006. "Human fronto-mesolimbic networks guide decisions about charitable donation." *Proceedings of the National Academy of Sciences, USA* 103 (42):15623–15628.

Moore, O. K. 1957. "Divination—A new perspective." *American Anthropologist* 59: 69–74.

Moran, S., D. McCloy, and R. Wright. 2012. "Revisiting population size vs. phoneme inventory size." *Language* 88 (4):877–893.

Morgan, R. 1979. "An account of the discovery of a whale-bone house on San Nicolas Island." *Journal of California and Great Basin Anthropology* 1 (1):171–177.

Morgan, T.J.H., and K. Laland. 2012. "The biological bases of conformity." *Frontiers in Neuroscience* 6 (87):1–7.

Morgan, T.J.H., L. E. Rendell, M. Ehn, W. Hoppitt, and K. N. Laland. 2012. "The evolutionary basis of human social learning." *Proceedings of the Royal Society B: Biological Sciences* 279 (1729):653–662.

Morgan, T.J.H., N. T. Uomini, L. E. Rendell, L. Chouinard-Thuly, S. E. Street, H. M. Lewis, C. P. Cross, et al. 2015. "Experimental evidence for the co-evolution of hominin tool-making teaching and language." *Nature Communications* 6. http://dx.doi.org/10.1038/ncomms7029.

Morris, I. 2014. *War, What Is It Good For? The Role of Conflict in Civilisation, from Primates to Robots*. London: Profile Books.

Morse, J. M., C. Jehle, and D. Gamble. 1990. "Initiating breastfeeding: A world survey of the timing of postpartum breastfeeding." *International Journal of Nursing Studies* 27 (3):303–313. http://dx.doi.org/10.1016/0020–7489(90)90045-K.

Mowat, F. 1960. *Ordeal by Ice, His The Top of the World*, Vol. 1. [Toronto]: McClelland & Stewart.

Moya, C., R. Boyd, and J. Henrich. Forthcoming. "Reasoning about cultural and genetic transmission: Developmental and cross-cultural evidence from Peru, Fiji, and the US on how people make inferences about trait and identity transmission." *Topics in Cognitive Science*.

Muller, M., R. Wrangham, and D. Pilbeam, eds. Forthcoming. *Chimpanzees and Human Evolution*. Cambridge, MA: Harvard University Press.

Munroe, R. L., J. G. Fought, and R. K. S. Macaulay. 2009. "Warm climates and sonority classes not simply more vowels and fewer consonants." *Cross-Cultural Research* 43 (2):123–133. doi:10.1177/1069397109331485.

Muthukrishna, M., S. J. Heine, W. Toyakawa, T. Hamamura, T. Kameda, and J. Henrich. n.d. "Overconfidence is universal? Depends what you mean." Unpublished manuscript. http://coevolution.psych.ubc.ca/pdfs/OverconfidenceManuscript 2014.pdf.

Muthukrishna, M., T. Morgan, and J. Henrich. Forthcoming. "The when and who of social learning and conformist transmission." *Evolution and Human Behavior*.

REFERENCES

Muthukrishna, M., B. W. Shulman, V. Vasilescu, and J. Henrich. 2014. "Sociality influences cultural complexity." *Proceedings of the Royal Society B: Biological Sciences* 281 (1774). doi:10.1098/rspb.2013.2511.

Mutinda, H., J. H. Poole, and C. J. Moss. 2011. "Decision making and leadership in using the ecosystem." In *The Amboseli Elephants: A Long-Term Perspective on a Long-Lived Mammal*, edited by C. J. Moss, H. Croze, and P. C. Lee, 246–259. Chicago: University of Chicago Press.

Myers, F. 1988. "Burning the truck and holding the country: Property, time and the negotiation of identity among the Pintupi Aborigines." In *Hunters and Gatherers: Property, Power and Ideology*, edited by T. Ingold, D. Riches, and J. Woodburn, 15–43. Oxford: Berg.

Naber, M., M. V. Pashkam, and K. Nakayama. 2013. "Unintended imitation affects success in a competitive game." *Proceedings of the National Academy of Sciences, USA* 110 (50):20046–20050.

Nakahashi, W., J. Y. Wakano, and J. Henrich. 2012. "Adaptive social learning strategies in temporally and spatially varying environments: How temporal vs. spatial variation, number of cultural traits, and costs of learning influence the evolution of conformist-biased transmission, payoff-biased transmission, and individual learning." *Human Nature* 23 (4):386–418.

Neff, B. D. 2003. "Decisions about parental care in response to perceived paternity." *Nature* 422 (6933):716–719.

Nettle, D. 2007. "Language and genes: A new perspective on the origins of human cultural diversity." *Proceedings of the National Academy of Sciences, USA* 104 (26):10755–10756. doi:10.1073/pnas.0704517104.

———. 2012. "Social scale and structural complexity in human languages." *Philosophical Transactions of the Royal Society B: Biological Sciences* 367 (1597):1829–1836.

Newman, R. W. 1970. "Why man is such a sweaty and thirsty naked animal—A speculative review." *Human Biology* 42 (1):12–27.

Newmeyer, F. J. 2002. "Uniformitarian assumptions and language evolution research." In *The Transition to Language*, edited by A. Wray, 359–375. Oxford: Oxford University Press.

Nhassico, D., H. Muquingue, J. Cliff, A. Cumbana, and J. H. Bradbury. 2008. "Rising African cassava production, diseases due to high cyanide intake and control measures." *Journal of the Science of Food and Agriculture* 88 (12):2043–2049.

Nielsen, M. 2012. "Imitation, pretend play, and childhood: Essential elements in the evolution of human culture?" *Journal of Comparative Psychology* 126 (2):170–181.

Nielsen, M., and K. Tomaselli. 2010. "Overimitation in Kalahari Bushman children and the origins of human cultural cognition." *Psychological Science* 21 (5):729–736.

Nisbett, R. E. 2003. *The Geography of Thought: How Asians and Westerners Think Differently ... and Why*. New York: Free Press.

Nisbett, R. E., and D. Cohen. 1996. *Culture of Honor*. Boulder, CO: Westview Press.

Nixon, L. A., and M. D. Robinson. 1999. "The educational attainment of young women: Role model effects of female high school faculty." *Demography* 36 (2):185–194.

REFERENCES

Noell, A. M., and D. K. Himber. 1979. *The History of Noell's Ark Gorilla Show: The Funniest Show on Earth, Which Features the "World's Only Athletic Apes."* Tarpon Springs, FL: Noell's Ark Publisher.

Norenzayan, A. 2013. *Big Gods: How Religion Transformed Cooperation and Conflict.* Princeton, NJ: Princeton University Press.

Norenzayan, A., A. F. Shariff, W. M. Gervais, A. Willard, R. McNamara, E. Slingerland, and J. Henrich. Forthcoming. "The cultural evolution of prosocial religions." *Behavioral and Brain Sciences.*

Nunez, M., and P. L. Harris. 1998. "Psychological and deontic concepts: Separate domains or intimate connection?" *Mind & Language* 13 (2):153–170.

O'Brien, M. J., and K. N. Laland. 2012. "Genes, culture, and agriculture: An example of human niche construction." *Current Anthropology* 53 (4):434–470.

O'Connor, S., R. Ono, and C. Clarkson. 2011. "Pelagic fishing at 42,000 years before the present and the maritime skills of modern humans." *Science* 334 (6059):1117–1121. doi:10.1126/science.1207703.

Offerman, T., J. Potters, and J. Sonnemans. 2002. "Imitation and belief learning in an oligopoly experiment." *Review of Economic Studies* 69 (4):973–998.

Offerman, T., and J. Sonnemans. 1998. "Learning by experience and learning by imitating others." *Journal of Economic Behavior and Organization* 34 (4):559–575.

Oota, H., W. Settheetham-Ishida, D. Tiwawech, T. Ishida, and M. Stoneking. 2001. "Human mtDNA and Y-chromosome variation is correlated with matrilocal versus patrilocal residence." *Nature Genetics* 29 (1):20–21.

Over, H., and M. Carpenter. 2012. "Putting the social into social learning: Explaining both selectivity and fidelity in children's copying behavior." *Journal of Comparative Psychology* 126 (2):182–192.

———. 2013. "The social side of imitation." *Child Development Perspectives* 7 (1): 6–11.

Paciotti, B., and C. Hadley. 2003. "The ultimatum game in southwestern Tanzania." *Current Anthropology* 44 (3):427–432.

Padmaja, G. 1995. "Cyanide detoxification in cassava for food and feed uses." *Critical Reviews in Food Science and Nutrition* 35 (4):299–339.

Pagel, M. D. 2012. *Wired for Culture: Origins of the Human Social Mind.* New York: W. W. Norton.

Paige, D. M., T. M. Bayless, and G. G. Graham. 1972. "Milk programs—Helpful or harmful to Negro children?" *American Journal of Public Health and the Nations Health* 62 (11):1486-1488. doi:10.2105/Ajph.62.11.1486.

Paine, R. 1971. "Animals as capital—Comparisons among northern nomadic herders and hunters." *Anthropological Quarterly* 44 (3):157–172.

Paladino, M. P., M. Mazzurega, F. Pavani, and T. W. Schubert. 2010. "Synchronous multisensory stimulation blurs self-other boundaries." *Psychological Science* 21 (9):1202–1207.

Panchanathan, K., and R. Boyd. 2004. "Indirect reciprocity can stabilize cooperation without the second-order free rider problem." *Nature* 432:499–502.

Panger, M. A., A. S. Brooks, B. G. Richmond, and B. Wood. 2002. "Older than the Oldowan? Rethinking the emergence of hominin tool use." *Evolutionary Anthropology* 11 (6):235–245.

Pashos, A. 2000. "Does paternal uncertainty explain discriminative grandparental

REFERENCES

solicitude? A cross-cultural study in Greece and Germany." *Evolution and Human Behavior* 21 (2):97–109.

Pawley, A. 1987. "Encoding events in Kalam and English: Different logics for reporting experience." In *Coherence and Grounding in Discourse*, edited by R. Tomlin, 329–360. Amsterdam: John Benjamins.

Pearce, E., C. Stringer, and R. I. M. Dunbar. 2013. "New insights into differences in brain organization between Neanderthals and anatomically modern humans." *Proceedings of the Royal Society B: Biological Sciences* 280 (1758). doi:10.1098/rspb.2013.0168.

Peng, Y., H. Shi, X. B. Qi, C. J. Xiao, H. Zhong, R. L. Z. Ma, and B. Su. 2010. "The *ADH1B Arg47His* polymorphism in East Asian populations and expansion of rice domestication in history." *BMC Evolutionary Biology* 10:15. doi:10.1186/1471-2148-10-15.

Perreault, C. 2012. "The pace of cultural evolution." *PLoS ONE* 7 (9):e45150. doi:10.1371/journal.pone.0045150.

Perreault, C., P. J. Brantingham, S. L. Kuhn, S. Wurz, and X. Gao. 2013. "Measuring the complexity of lithic technology." *Current Anthropology* 54 (S8):S397–S406. doi:10.1086/673264.

Perreault, C., C. Moya, and R. Boyd. 2012. "A Bayesian approach to the evolution of social learning." *Evolution and Human Behavior* 33 (5):449–459.

Perry, D. G., and K. Bussey. 1979. "Social-learning theory of sex differences—Imitation is alive and well." *Journal of Personality and Social Psychology* 37 (10):1699–1712.

Perry, G. H., N. J. Dominy, K. G. Claw, A. S. Lee, H. Fiegler, R. Redon, J. Werner, et al. 2007. "Diet and the evolution of human amylase gene copy number variation." *Nature Genetics* 39 (10):1256–1260.

Perry, G. H., B. C. Verrelli, and A. C. Stone. 2005. "Comparative analyses reveal a complex history of molecular evolution for human *MYH16*." *Molecular Biology and Evolution* 22 (3):379–382.

Peterson, S., F. Legue, T. Tylleskar, E. Kpizingui, and H. Rosling. 1995. "Improved cassava-processing can help reduce iodine deficiency disorders in the Central African Republic." *Nutrition Research* 15 (6):803–812.

Peterson, S., H. Rosling, T. Tylleskar, M. Gebremedhin, and A. Taube. 1995. "Endemic goiter in Guinea." *Lancet* 345 (8948):513–514.

Phillips, D. P., T. E. Ruth, and L. M. Wagner. 1993. "Psychology and survival." *Lancet* 342 (8880):1142–1145.

Phoenix, D. 2003. "Burke and Wills: Melbourne to the Gulf—A brief history of the Victorian Exploring Expedition of 1860–1." http://www.burkeandwills.net.au/downloads/.

Pietraszewski, D., L. Cosmides, and J. Tooby. 2014. "The content of our cooperation, not the color of our skin: An alliance detection system regulates categorization by coalition and race, but not sex." *Plos One* 9 (2):e88534. doi:10.1371/journal.pone.0088534.

Pietraszewski, D., and A. Schwartz. 2014a. "Evidence that accent is a dedicated dimension of social categorization, not a byproduct of coalitional categorization." *Evolution and Human Behavior* 35 (1):51–57.

———. 2014b. "Evidence that accent is a dimension of social categorization, not a byproduct of perceptual salience, familiarity, or ease-of-processing." *Evolution and Human Behavior* 35 (1):43–50.

Pike, T. W., and K. N. Laland. 2010. "Conformist learning in nine-spined stickle-backs' foraging decisions." *Biology Letters* 6 (4):466–468.

Pingle, M. 1995. "Imitation vs. rationality: An experimental perspective on decision-making." *Journal of Socio-Economics* 24:281–315.

Pingle, M., and R. H. Day. 1996. "Modes of economizing behavior: Experimental evidence." *Journal of Economic Behavior & Organization* 29:191–209.

Pinker, S. 1997. *How the Mind Works*. New York: W. W. Norton.

———. 2002. *The Blank Slate: The Modern Denial of Human Nature*. New York: Viking.

———. 2010. "The cognitive niche: Coevolution of intelligence, sociality, and language." *Proceedings of the National Academy of Sciences* 107 (Suppl. 2):8993–8999. doi:10.1073/pnas.0914630107.

———. 2011. *The Better Angels of Our Nature: Why Violence Has Declined*. New York: Viking.

Pinker, S., and P. Bloom. 1990. "Natural language and natural selection." *Behavioral and Brain Sciences* 13 (4):707–726.

Pitchford, N. J., and K. T. Mullen. 2002. "Is the acquisition of basic-colour terms in young children constrained?" *Perception* 31 (11):1349–1370.

Place, S. S., P. M. Todd, L. Penke, and J. B. Asendorpf. 2010. "Humans show mate copying after observing real mate choices." *Evolution and Human Behavior* 31 (5):320–325.

Plassmann, H., J. O'Doherty, B. Shiv, and A. Rangel. 2008. "Marketing actions can modulate neural representations of experienced pleasantness." *Proceedings of the National Academy of Sciences, USA* 105 (3):1050–1054.

Plummer, T. 2004. "Flaked stones and old bones: Biological and cultural evolution at the dawn of technology." *Yearbook of Physical Anthropology* 47:118–164.

Potters, J., M. Sefton, and L. Vesterlund. 2005. "After you—Endogenous sequencing in voluntary contribution games." *Journal of Public Economics* 89 (8):1399–1419.

———. 2007. "Leading-by-example and signaling in voluntary contribution games: An experimental study." *Economic Theory* 33 (1):169–182.

Poulin-Dubois, D., I. Brooker, and A. Polonia. 2011. "Infants prefer to imitate a reliable person." *Infant Behavior and Development* 34 (2):303–309. http://dx.doi.org/10.1016/j.infbeh.2011.01.006.

Powell, A., S. Shennan, and M. G. Thomas. 2009. "Late Pleistocene demography and the appearance of modern human behavior." *Science* 324 (5932):1298–1301.

Pradhan, G. R., C. Tennie, and C. P. van Schaik. 2012. "Social organization and the evolution of cumulative technology in apes and hominins." *Journal of Human Evolution* 63 (1):180–190.

Presbie, R. J., and P. F. Coiteux. 1971. "Learning to be generous or stingy: Imitation of sharing behavior as a function of model generosity and vicarious reinforcement." *Child Development* 42 (4):1033–1038.

Price, D. D., D. G. Finniss, and F. Benedetti. 2008. "A comprehensive review of the placebo effect: Recent advances and current thought." *Annual Review of Psychology* 59:565–590.

Price, T. D., and J. A. Brown, eds. 1988. *Prehistoric Hunter Gathers: The Emergence of Cultural Complexity*. New York: Academic Press.

Proctor, D., R. A. Williamson, F.B.M. de Waal, and S. F. Brosnan. 2013. "Chimpanzees play the ultimatum game." *Proceedings of the National Academy of Sciences, USA* 110 (6):2070–2075.

Pulliam, H. R., and C. Dunford. 1980. *Programmed to Learn: An Essay on the Evolution of Culture*. New York: Columbia University Press.

Puurtinen, M., and T. Mappes. 2009. "Between-group competition and human cooperation." *Proceedings of the Royal Society B: Biological Sciences* 276 (1655):355–360.

Rabinovich, R., S. Gaudzinski-Windheuser, and N. Goren-Inbar. 2008. "Systematic butchering of fallow deer (Dama) at the early middle Pleistocene Acheulian site of Gesher Benot Ya'aqov (Israel)." *Journal of Human Evolution* 54 (1):134–149.

Radcliffe-Brown, A. R. 1964. *The Andaman Islanders*. New York: Free Press.

Rakoczy, H., N. Brosche, F. Warneken, and M. Tomasello. 2009. "Young children's understanding of the context-relativity of normative rules in conventional games." *British Journal of Developmental Psychology* 27:445–456.

Rakoczy, H., F. Wameken, and M. Tomasello. 2008. "The sources of normativity: Young children's awareness of the normative structure of games." *Developmental Psychology* 44 (3):875–881.

Rand, D. G., J. D. Greene, and M. A. Nowak. 2012. "Spontaneous giving and calculated greed." *Nature* 489 (7416):427–430.

———. 2013. "Intuition and cooperation reconsidered. Reply." *Nature* 498 (7452): E2–E3.

Rand, D. G., A. Peysakhovich, G. T. Kraft-Todd, G. E. Newman, O. Wurzbacher, M. A. Nowak, and J. D. Greene. 2014. "Social heuristics shape intuitive cooperation." *Nature Communications* 5:3677. doi:10.1038/ncomms4677.

Rasmussen, K., G. Herring, and H. Moltke. 1908. *The People of the Polar North: A Record*. London: K. Paul, Trench, Trübner & Co.

Reader, S. M., Y. Hager, and K. N. Laland. 2011. "The evolution of primate general and cultural intelligence." *Philosophical Transactions of the Royal Society B: Biological Sciences* 366 (1567):1017–1027.

Reader, S. M., and K. N. Laland. 2002. "Social intelligence, innovation, and enhanced brain size in primates." *Proceedings of the National Academy of Sciences, USA* 99 (7):4436–4441.

Real, L. A. 1991. "Animal choice behavior and the evolution of cognitive architecture." *Science* 253:980–86.

Reali, F., and M. H. Christiansen. 2009. "Sequential learning and the interaction between biological and linguistic adaptation in language evolution." *Interaction Studies* 10 (1):5–30.

Rendell, L., L. Fogarty, W. J. E. Hoppitt, T. J. H. Morgan, M. M. Webster, and K. N. Laland. 2011. "Cognitive culture: Theoretical and empirical insights into social learning strategies." *Trends in Cognitive Sciences* 15 (2):68–76.

Rendell, L., and H. Whitehead. 2001. "Culture in whales and dolphins." *Behavioral and Brain Sciences* 24 (2):309–382

Reyes-Garcia, V., J. Broesch, L. Calvet-Mir, N. Fuentes-Pelaez, T. W. McDade, S. Parsa, S. Tanner, et al. 2009. "Cultural transmission of ethnobotanical knowledge and

REFERENCES

skills: An empirical analysis from an Amerindian society." *Evolution and Human Behavior* 30 (4):274–285.

Reyes-Garcia, V., J. L. Molina, J. Broesch, L. Calvet, T. Huanca, J. Saus, S. Tanner, W. R. Leonard, and T. W. McDade. 2008. "Do the aged and knowledgeable men enjoy more prestige? A test of predictions from the prestige-bias model of cultural transmission." *Evolution and Human Behavior* 29 (4):275–281.

Rice, M. E., and J. E. Grusec. 1975. "Saying and doing: Effects of observer performance." *Journal of Personality and Social Psychology* 32:584–593.

Richerson, P. J., R. Baldini, A. Bell, K. Demps, K. Frost, V. Hillis, S. Mathew, et al. Forthcoming. "Cultural group selection plays an essential role in explaining human cooperation: A sketch of the evidence." *Behavioral & Brain Sciences.*

Richerson, P. J., and R. Boyd. 1998. "The evolution of ultrasociality." In *Indoctrinability, Ideology and Warfare,* edited by I. Eibl-Eibesfeldt and F. K. Salter, 71–96. New York: Berghahn Books.

————. 2000a. "Climate, culture and the evolution of cognition." In *The Evolution of Cognition,* edited by C. M. Heyes, 329–345. Cambridge, MA: MIT Press.

————. 2000b. "Built for speed: Pliestocene climate variation and the origins of human culture" In *Perspectives in Ethology,* edited by F. Tonneau and N. Thompson, 1–45. New York: Springer.

————. 2005. *Not by Genes Alone: How Culture Transformed Human Evolution.* Chicago: University of Chicago Press.

Richerson, P. J., R. Boyd, and R. L. Bettinger. 2001. "Was agriculture impossible during the Pleistocene but mandatory during the Holocene? A climate change hypothesis." *American Antiquity* 66 (3):387–411.

Richerson, P. J., R. Boyd, and J. Henrich. 2010. "Gene-culture coevolution in the age of genomics." *Proceedings of the National Academy of Sciences* 107 (Suppl. 2):8985–8992. doi:10.1073/pnas.0914631107.

Richerson, P. J., and J. Henrich. 2012. "Tribal social instincts and the cultural evolution of institutions to solve collective action problems." *Cliodynamics* 3 (1): 38–80.

Ridley, M. 2010. *The Rational Optimist: How Prosperity Evolves.* New York: Harper.

Rilling, J. K., A. G. Sanfey, L. E. Nystrom, J. D. Cohen, D. A. Gutman, T. R. Zeh, G. Pagnoni, G. S. Berns, and C. D. Kilts. 2004. "Imaging the social brain with fMRI and interactive games." *International Journal of Neuropsychopharmacology* 7:S477–S478.

Rivers, W.H.R. 1931. "The disappearance of useful arts." In *Source Book in Anthropology,* edited by A. L. Kroeber and T. Waterman, 524–534. New York: Harcourt Brace.

Roach, N. T., and D. E. Lieberman. 2012. "Derived anatomy of the shoulder and wrist enable throwing ability in Homo." *American Journal of Physical Anthropology* 147:250–250.

Roach, N. T., and D. E. Lieberman. 2013. "The biomechanics of power generation during human high-speed throwing." *American Journal of Physical Anthropology* 150:233–233.

Roach, N. T., M. Venkadesan, M. J. Rainbow, and D. E. Lieberman. 2013. "Elastic energy storage in the shoulder and the evolution of high-speed throwing in Homo." *Nature* 498 (7455):483–486.

REFERENCES

Rodahl, K., and T. Moore. 1943. "The vitamin A content and toxicity of bear and seal liver." *Biochemical Journal* 37:166–168.

Roe, D. A. 1973. *A Plague of Corn: The Social History of Pellagra*. Ithaca, NY: Cornell University Press.

Rogers, E. M. 1995a. *Diffusion of Innovations*. New York: Free Press.

Rosekrans, M. A. 1967. "Imitation in children as a function of perceived similarity to a social model and vicarious reinforcement." *Journal of Personality and Social Psychology* 7 (3):307–315.

Rosenbaum, M., and R. R. Blake. 1955. "The effect of stimulus and background factors on the volunteering response." *Journal Abnormal Social Psychology* 50:193–196.

Rosenbaum, M. E., and I. F. Tucker. 1962. "The competence of the model and the learning of imitation and nonimitation." *Journal of Experimental Psychology* 63 (2):183–190.

Rosenthal, T. L., and B. J. Zimmerman. 1978. *Social Learning and Cognition*. New York: Academic Press.

Roth, G., and U. Dicke. 2005. "Evolution of the brain and intelligence." *Trends in Cognitive Sciences* 9 (5):250–257.

Rozin, P., L. Ebert, and J. Schull. 1982. "Some like it hot—A temporal analysis of hedonic responses to chili pepper." *Appetite* 3 (1):13–22.

Rozin, P., L. Gruss, and G. Berk. 1979. "Reversal of innate aversions—Attempts to induce a preference for chili peppers in rats." *Journal of Comparative and Physiological Psychology* 93 (6):1001–1014.

Rozin, P., and K. Kennel. 1983. "Acquired preferences for piquant foods by chimpanzees." *Appetite* 4 (2):69–77.

Rozin, P., M. Mark, and D. Schiller. 1981. "The role of desensitization to capsaicin in chili pepper ingestion and preference." *Chemical Senses* 6 (1):23–31.

Rozin, P., and D. Schiller. 1980. "The nature and acquisition of a preference for chili pepper by humans." *Motivation and Emotion* 4 (1):77–101. doi:10.1007/bf0099 5932.

Rubinstein, D. H. 1983. "Epidemic Suicide among Micronesian Adolescents." *Social Science Medicine* 17 (10):657–665.

Rushton, J. P. 1975. "Generosity in children: Immediate and long term effects of modeling, preaching, and moral judgement." *Journal of Personality and Social Psychology* 31:459–466.

Rushton, J. P., and A. C. Campbell. 1977. "Modeling, vicarious reinforcement and extraversion on blood donating in adults. Immediate and long-term effects." *European Journal of Social Psychology* 7 (3):297–306.

Ryalls, B. O., R. E. Gul, and K. R. Ryalls. 2000. "Infant imitation of peer and adult models: Evidence for a peer model advantage." *Merrill-Palmer Quarterly Journal of Developmental Psychology* 46 (1):188–202.

Saaksvuori, L., T. Mappes, and M. Puurtinen. 2011. "Costly punishment prevails in intergroup conflict." *Proceedings of the Royal Society B: Biological Sciences* 278 (1723):3428–3436.

Sabbagh, M. A., and D. A. Baldwin. 2001. "Learning words from knowledgeable versus ignorant speakers: Links between preschoolers' theory of mind and semantic development." *Child Development* 72 (4):1054–1070.

Sahlins, M. 1961. "The segmentary lineage: An organization of predatory expansion." *American Anthropologist* 63 (2):322–345.

Salali, G. D., M. Juda, and J. Henrich. 2015. "Transmission and development of costly punishment in children." *Evolution and Human Behavior* 36 (2): 86–94.

Sandgathe, D., H. Dibble, P. Goldberg, S. J. P. McPherron, A. Turq, L. Niven, and J. Hodgkins. 2011a. "On the role of fire in Neanderthal Adaptations in Western Europe: Evidence from Pech de l'Aze and Roc de Marsal, France." *PaleoAnthropology* 2011:216–242.

———. 2011b. "Timing of the appearance of habitual fire use." *Proceedings of the National Academy of Sciences, USA* 108 (29):E298–E298.

Sanfey, A. G., J. K. Rilling, J. A. Aronson, L. E. Nystrom, and J. D. Cohen. 2003. "The neural basis of economic decision-making in the ultimatum game." *Science* 300:1755–1758.

Sanz, C. M., and D. B. Morgan. 2007. "Chimpanzee tool technology in the Goualougo Triangle, Republic of Congo." *Journal of Human Evolution* 52 (4):420–433.

———. 2011. "Elemental variation in the termite fishing of wild chimpanzees (*Pan troglodytes*)." *Biology Letters* 7 (4):634–637. doi:10.1098/rsbl.2011.0088.

Scally, A., J. Y. Dutheil, L. W. Hillier, G. E. Jordan, I. Goodhead, J. Herrero, A. Hobolth, et al. 2012. "Insights into hominid evolution from the gorilla genome sequence." *Nature* 483 (7388):169–175.

Schachter, S., and R. Hall. 1952. "Group-derived restraints and audience persuasion." *Human Relations* 5:397–406.

Schapera, I. 1930. *The Khoisan Peoples of South Africa*. London: Routledge.

Schick, K. D., N. Toth, G. Garufi, E. S. Savage-Rumbaugh, D. Rumbaugh, and R. Sevcik. 1999. "Continuing investigations into the stone tool-making and tool-using capabilities of a bonobo (*Pan paniscus*)." *Journal of Archaeological Science* 26:821–832.

Schmelz, M., J. Call, and M. Tomasello. 2011. "Chimpanzees know that others make inferences." *Proceedings of the National Academy of Sciences, USA* 108 (7):3077–3079.

———. 2013. "Chimpanzees predict that a competitor's preference will match their own." *Biology Letters* 9 (1).

Schmidt, M. F. H., H. Rakoczy, and M. Tomasello. 2012. "Young children enforce social norms selectively depending on the violator's group affiliation." *Cognition* 124 (3):325–333.

Schmidt, M. F. H., and M. Tomasello. 2012. "Young children enforce social norms." *Current Directions in Psychological Science* 21 (4):232–236.

Schnall, E., and M. J. Greenberg. 2012. "Groupthink and the Sanhedrin: An analysis of the ancient court of Israel through the lens of modern social psychology." *Journal of Management History* 18 (3):285–294.

Scholz, M. N., K. D'Août, M. F. Bobbert, and P. Aerts. 2006. "Vertical jumping performance of bonobo (*Pan paniscus*) suggests superior muscle properties." *Proceedings of the Royal Society B: Biological Sciences* 273 (1598):2177–2184. doi:10.1098/rspb.2006.3568.

Scofield, J., and D. A. Behrend. 2008. "Learning words from reliable and unreliable

REFERENCES

speakers." *Cognitive Development* 23 (2):278–290. doi:10.1016/j.cogdev.2008.01 .003.

Scott, D. J., C. S. Stohler, C. M. Egnatuk, H. Wang, R. A. Koeppe, and J. K. Zubieta. 2008. "Placebo and nocebo effects are defined by opposite opioid and dopaminergic responses." *Archives of General Psychiatry* 65 (2):220–231.

Scott, R. M., R. Baillargeon, H. J. Song, and A. M. Leslie. 2010. "Attributing false beliefs about non-obvious properties at 18 months." *Cognitive Psychology* 61 (4):366–395.

Sear, R., and R. Mace. 2008. "Who keeps children alive? A review of the effects of kin on child survival." *Evolution and Human Behavior* 29 (1):1–18.

Selten, R., and J. Apesteguia. 2005. "Experimentally observed imitation and cooperation in price competition on the circle." *Games and Economic Behavior* 51 (1): 171–192.

Sepher, J. 1983. *Incest; The Biosocial View*. New York: Academic Press.

Sharon, G., N. Alperson-Afil, and N. Goren-Inbar. 2011. "Cultural conservatism and variability in the Acheulian sequence of Gesher Benot Ya'aqov." *Journal of Human Evolution* 60 (4):387–397.

Shea, J. J. 2006. "The origins of lithic projectile point technology: evidence from Africa, the Levant, and Europe." *Journal of Archaeological Science* 33 (6):823–846.

Shea, J. J., and M. Sisk. 2010. "Complex projectile technology and *Homo sapiens* dispersal into Western Europe." *PaleoAnthropology*:100–122.

Shennan, S. 2001. "Demography and cultural innovation: A model and its implications for the emergence of modern human culture." *Cambridge Archaeology Journal* 11 (1):5–16.

Sherman, P. W., and J. Billing. 1999. "Darwinian gastronomy: Why we use spices." *BioScience* 49 (6):453–463.

Sherman, P. W., and S. M. Flaxman. 2001. "Protecting ourselves from food." *American Scientist* 89 (2):142–151.

Sherman, P. W., and G. A. Hash. 2001. "Why vegetable recipes are not very spicy." *Evolution and Human Behavior* 22 (3):147–163.

Shimelmitz, R., S. L. Kuhn, A. J. Jelinek, A. Ronen, A. E. Clark, and M. Weinstein-Evron. 2014. "'Fire at will': The emergence of habitual fire use 350,000 years ago." *Journal of Human Evolution* 77:196–203. http://dx.doi.org/10.1016/j.jhevol.2014 .07.005.

Shutts, K., M. R. Banaji, and E. S. Spelke. 2010. "Social categories guide young children's preferences for novel objects." *Developmental Science* 13 (4):599–610.

Shutts, K., K. D. Kinzler, and J. M. DeJesus. 2013. "Understanding infants' and children's social learning about foods: Previous research and new prospects." *Developmental Psychology* 49 (3):419–425.

Shutts, K., K. D. Kinzler, C. B. Mckee, and E. S. Spelke. 2009. "Social information guides infants' selection of foods." *Journal of Cognition and Development* 10 (1–2):1–17.

Silberberg, A., and D. Kearns. 2009. "Memory for the order of briefly presented numerals in humans as a function of practice." *Animal Cognition* 12 (2):405–407.

Silk, J. B. 2002. "Practice random acts of aggression and senseless acts of intimida-

REFERENCES

tion: The logic of status contests in social groups." *Evolutionary Anthropology* 11 (6):221–225.

Silk, J. B., S. F. Brosnan, J. Vonk, J. Henrich, D. J. Povinelli, A. S. Richardson, S. P. Lambeth, J. Mascaro, and S. J. Shapiro. 2005. "Chimpanzees are indifferent to the welfare of unrelated group members." *Nature* 437:1357–1359.

Silk, J. B., and B. R. House. 2011. "Evolutionary foundations of human prosocial sentiments." *Proceedings of the National Academy of Sciences, USA* 108:10910–10917.

Silverman, P., and R. J. Maxwell. 1978. "How do I respect thee? Let me count the ways: Deference towards elderly men and women." *Behavior Science Research* 13 (2):91–108.

Simmons, L. W. 1945. *The Role of the Aged in Primitive Society*. New Haven, CT: Yale University Press.

Simon, H. 1990. "A mechanism for social selection and successful altruism." *Science* 250:1665–1668.

Simoons, F. J. 1970. "Primary adult lactose intolerance and the milking habit: A problem in biologic and cultural interrelations: II. A culture historical hypothesis." *American Journal of Digestive Diseases* 15 (8):695–710.

Slingerland, E., J. Henrich, and A. Norenzayan. 2013. The evolution of prosocial religions. In *Cultural Evolution: Society, Technology, Language and Religion*, edited by P. J. Richerson and M. H. Christiansen, 335–348. Cambridge: MIT Press.

Sloane, S., R. Baillargeon, and D. Premack. 2012. "Do infants have a sense of fairness?" *Psychological Science* 23 (2):196–204.

Smaldino, P. E., J. C. Schank, and R. McElreath. 2013. "Increased costs of cooperation help cooperators in the long run." *American Naturalist* 181 (4):451–463.

Smil, V. 2002. "Biofixation and nitrogen in the biosphere and in global food production." In *Nitrogen Fixation: Global Perspectives*, edited by T. M. Finan, M. R. O'Brian, D. B. Layzell, and J. K. Vessey, 7–10. Wallingford, UK: CAB International.

Smil, V. 2011. "." *Population and Development Review* 37 (4):613–636

Smith, E. A., and B. Winterhalder. 1992. *Evolutionary Ecology and Human Behavior*. New York: Aldine de Gruyter.

Smith, J. R., M. A. Hogg, R. Martin, and D. J. Terry. 2007. "Uncertainty and the influence of group norms in the attitude-behaviour relationship." *British Journal of Social Psychology* 46:769–792.

Smith, K., and S. Kirby. 2008. "Cultural evolution: Implications for understanding the human language faculty and its evolution." *Philosophical Transactions of the Royal Society B: Biological Sciences* 363 (1509):3591–3603.

Snyder, J. K., L. A. Kirkpatrick, and H. C. Barrett. 2008. "The dominance dilemma: Do women really prefer dominant mates?" *Personal Relationships* 15 (4):425–444. doi:10.1111/j.1475-6811.2008.00208.x.

Soler, R. 2010. "Costly signaling, ritual and cooperation: Findings from Candomblé, an Afro-Brazilian religion." Unpublished manuscript.

Soltis, J., R. Boyd, and P. J. Richerson. 1995. "Can group-functional behaviors evolve by cultural group selection? An empirical test." *Current Anthropology* 36 (3):473–494.

Sosis, R., H. Kress, and J. Boster. 2007. "Scars for war: Evaluating signaling explana-

REFERENCES

tions for cross-cultural variance in ritual costs." *Evolution and Human Behavior* 28:234–247.

Spencer, B., and F. J. Gillen. 1968. *The Native Tribes of Central Australia*. New York: Dover Publications.

Spencer, C., and E. Redmond. 2001. "Multilevel selection and political evolution in the Valley of Oaxaca." *Journal of Anthropological Archaeology* 20:195–229.

Spencer, R. F. 1984. "North Alaska Coast Eskimo." In *Arctic*, Vol. 5 of *Handbook of North American Indians*, edited by D. Damas. Washington, DC: Smithsonian Institution Press.

Sperber, D. 1996. *Explaining Culture: A Naturalistic Approach*. Oxford: Blackwell.

Sperber, D., F. Clement, C. Heintz, O. Mascaro, H. Mercier, G. Origgi, and D. Wilson. 2010. "Epistemic vigilance." *Mind & Language* 25 (4):359–393.

Stack, S. 1987. "Celebrities and suicide: A taxonomy and analysis, 1948–1983." *American Sociological Review* 52 (3):401–412.

———. 1990. "Divorce, suicide, and the mass media: An analysis of differential identification, 1948–1980." *Journal of Marriage & the Family* 52 (2):553–560.

———. 1992. "Social correlates of suicide by age: Media impacts." In *Life Span Perspectives of Suicide: Time-Lines in the Suicide Process*, edited by A. Leenaars, 187–213. New York: Plenum Press.

———. 1996. "The effect of the media on suicide: Evidence from Japan, 1955–1985." *Suicide & Life-Threatening Behavior* 26 (2):132–142.

Stanovich, K. E. 2013. "Why humans are (sometimes) less rational than other animals: Cognitive complexity and the axioms of rational choice." *Thinking & Reasoning* 19 (1):1–26.

Stedman, H. H., B. W. Kozyak, A. Nelson, D. M. Thesier, J. B. Shrager, C. R. Bridges, N. Minugh-Purvis, and M. A. Mitchell. 2003. "Inactivating mutation in the MYH 16 'superfast' myosin gene abruptly reduced the size of the jaw closing muscles in a recent human ancestor." *Molecular Therapy* 7 (5):S106–S106.

Stedman, H. H., B. W. Kozyak, A. Nelson, D. M. Thesier, L. T. Su, D. W. Low, C. R. Bridges, J. B. Shrager, N. Minugh-Purvis, and M. A. Mitchell. 2004. "Myosin gene mutation correlates with anatomical changes in the human lineage." *Nature* 428 (6981):415–418.

Stenberg, G. 2009. "Selectivity in infant social referencing." *Infancy* 14 (4):457–473.

Sterelny, K. 2012a. *The Evolved Apprentice: How Evolution Made Humans Unique*. The Jean Nicod Lectures. Cambridge, MA: The MIT Press.

———. 2012b. "Language, gesture, skill: The co-evolutionary foundations of language." *Philosophical Transactions of the Royal Society B: Biological Sciences* 367 (1599):2141–2151.

Stern, T. 1957. "Drum and whistle languages—An analysis of speech surrogates." *American Anthropologist* 59 (3):487–506.

Stewart, K. M. 1994. "Early hominid utilization of fish resources and implications for seasonality and behavior." *Journal of Human Evolution* 27 (1–3):229–245.

Stout, D. 2002. "Skill and cognition in stone tool production—An ethnographic case study from Irian Jaya." *Current Anthropology* 43 (5):693–722.

———. 2011. "Stone toolmaking and the evolution of human culture and cognition." *Philosophical Transactions of the Royal Society B: Biological Sciences* 366 (1567):1050–1059.

REFERENCES

Stout, D., and T. Chaminade. 2007. "The evolutionary neuroscience of tool making." *Neuropsychologia* 45 (5):1091–1100.

———. 2012. "Stone tools, language and the brain in human evolution." *Philosophical Transactions of the Royal Society B: Biological Sciences* 367 (1585):75–87.

Stout, D., S. Semaw, M. J. Rogers, and D. Cauche. 2010. "Technological variation in the earliest Oldowan from Gona, Afar, Ethiopia." *Journal of Human Evolution* 58 (6):474–491.

Stout, D., N. Toth, K. Schick, and T. Chaminade. 2008. "Neural correlates of Early Stone Age toolmaking: Technology, language and cognition in human evolution." *Philosophical Transactions of the Royal Society B: Biological Sciences* 363 (1499):1939–1949.

Striedter, G. F. 2004. *Principles of Brain Evolution.* Sunderland, MA: Sinauer Associates.

Stringer, C. 2012. *Lone Survivors: How We Came to Be the Only Humans on Earth.* New York: Henry Holt and Company.

Sturm, R. A., D. L. Duffy, Z. Z. Zhao, F.P.N. Leite, M. S. Stark, N. K. Hayward, N. G. Martin, and G. W. Montgomery. 2008. "A single SNP in an evolutionary conserved region within intron 86 of the *HERC2* gene determines human blue-brown eye color." *American Journal of Human Genetics* 82 (2):424–431.

Sturtevant, W. C. 1978. *Arctic,* Vol. 5 of *Handbook of North American Indians.* Washington, DC: Smithsonian Institution Press.

Surovell, T. A. 2008. "Extinction of big game." In *Encyclopedia of Archaeology,* edited by D. Pearsall, 1365–1374 . New York: Academic Press.

Sutton, M. Q. 1986. "Warfare and expansion: An ethnohistoric perspective on the Numic spread." *Journal of California and Great Basin Anthropology* 8 (1):65–82.

———. 1993. "The Numic expansion in Great Basin Oral Tradition." *Journal of California and Great Basin Anthropology* 15 (1):111–128.

Suwa, G., B. Asfaw, R. T. Kono, D. Kubo, C. O. Lovejoy, and T. D. White. 2009. "The *Ardipithecus ramidus* skull and its implications for hominid origins." *Science* 326 (5949).

Szwed, M., F. Vinckier, L. Cohen, and S. Dehaene. 2012. "Towards a universal neurobiological architecture for learning to read." *Behavioral and Brain Sciences* 35 (5):308–309.

Tabibnia, G., A. B. Satpute, and M. D. Lieberman. 2008. "The sunny side of fairness—Preference for fairness activates reward circuitry (and disregarding unfairness activates self-control circuitry)." *Psychological Science* 19 (4):339–347.

Talhelm, T., X. Zhang, S. Oishi, C. Shimin, D. Duan, X. Lan, and S. Kitayama. 2014. "Large-scale psychological differences within China explained by rice versus wheat agriculture." *Science* 344 (6184):603–608.

Testart, A. 1988. "Some major problems in the social anthropology of hunter-gatherers." *Current Anthropology* 29 (1):1–31.

Thompson, J., and A. Nelson. 2011. "Middle childhood and modern human origins." *Human Nature* 22 (3):249–280. doi:10.1007/s12110–011–9119–3.

Thomsen, L., W. E. Frankenhuis, M. Ingold-Smith, and S. Carey. 2011. "Big and mighty: Preverbal infants mentally represent social dominance." *Science* 331 (6016):477–80. doi:10.1126/science.1199198.

Tolstrup, J. S., B. G. Nordestgaard, S. Rasmussen, A. Tybjaerg-Hansen, and M. Gron-

REFERENCES

baek. 2008. "Alcoholism and alcohol drinking habits predicted from alcohol dehydrogenase genes." *Pharmacogenomics Journal* 8 (3):220–227.

Tomasello, M. 1999. *The Cultural Origins of Human Cognition*. Cambridge, MA: Harvard University Press.

———. 2000a. "Culture and cognitive development." *Current Directions in Psychological Science* 9 (2):37–40.

———. 2000b. "Primate cognition: Introduction to the issue." *Cognitive Science* 24 (3):351–361.

———. 2010. *Origins of Human Communication*. Cambridge. MA: MIT Press.

Tomasello, M., R. Strosberg, and N. Akhtar. 1996. "Eighteen-month-old children learn words in non-ostensive contexts." *Journal of Child Language* 23 (1):157–176.

Tomblin, J. B., E. Mainela-Arnold, and X. Zhang. 2007. "Procedural learning in adolescents with and without specific language impairment." *Language Learning and Development* 3 (4):269–293. doi:10.1080/15475440701377477.

Tomkins, W. 1936. *Universal Sign Language of the Plains Indians of North America*. San Diego: Frye & Smith.

Tooby, J., and L. Cosmides. 1992. "The psychological foundations of culture." In *The Adapted Mind*, edited by J. Barkow, L.Cosmides, and J. Tooby, 19–136. New York: Oxford University Press.

Toth, N., and K. Schick. 2009. "The Oldowan: The tool making of early hominins and chimpanzees compared." *Annual Review of Anthropology* 38:289–305. doi:10.1146/annurev-anthro-091908–164521.

Tracy, J. L., and D. Matsumoto. 2008. "The spontaneous expression of pride and shame: Evidence for biologically innate nonverbal displays." *Proceedings of the National Academy of Sciences, USA* 105 (33):11655–11660. doi:10.1073/pnas.0802 686105.

Tracy, J. L., and R. W. Robins. 2008. "The nonverbal expression of pride: Evidence for cross-cultural recognition." *Journal of Personality and Social Psychology* 94 (3): 516–530.

Tracy, J. L., R. W. Robins, and K. H. Lagattuta. 2005. "Can children recognize pride?" *Emotion* 5 (3):251–257.

Tracy, J. L., A. F. Shariff, W. Zhao, and J. Henrich. 2013. "Cross-cultural evidence that the pride expression is a universal automatic status signal." *Journal of Experimental Psychology-General* 142:163–180.

Tubbs, R. S., E. G. Salter, and W. J. Oakes. 2006. "Artificial deformation of the human skull: A review." *Clinical Anatomy* 19 (4):372–377.

Turchin, P. 2005. *War and Peace and War: The Life Cycles of Imperial Nations*. New York: Pearson Education.

———. 2010. "Warfare and the evolution of social complexity: A multilevel-selection approach." *Structure and Dynamics* 4 (3). http://escholarship.org/uc/item/7j11945r.

Tuzin, D. 1976. *The Ilahita Arapesh*. Berkeley: University of California Press.

———. 2001. *Social Complexity in the Making: A Case Study among the Arapesh of New Guinea*. London: Routledge.

Tylleskar, T., M. Banea, N. Bikangi, R. D. Cooke, N. H. Poulter, and H. Rosling. 1992. "Cassava cyanogens and konzo, an upper motoneuron disease found in Africa." *Lancet* 339 (8787):208–211.

REFERENCES

Tylleskar, T., M. Banea, N. Bikangi, L. Fresco, L. A. Persson, and H. Rosling. 1991. "Epidemiologic evidence from Zaire for a dietary etiology of konzo, an upper motor-neuron disease." *Bulletin of the World Health Organization* 69 (5):581–589.

Tylleskar, T., W. P. Howlett, H. T. Rwiza, S. M. Aquilonius, E. Stalberg, B. Linden, A. Mandahl, H. C. Larsen, G. R. Brubaker, and H. Rosling. 1993. "Konzo—A distinct disease entity with selective upper motor-neuron damage." *Journal of Neurology Neurosurgery and Psychiatry* 56 (6):638–643.

Valdesolo, P., and D. DeSteno. 2011. "Synchrony and the social tuning of compassion." *Emotion* 11 (2):262–266.

Valdesolo, P., J. Ouyang, and D. DeSteno. 2010. "The rhythm of joint action: Synchrony promotes cooperative ability." *Journal of Experimental Social Psychology* 46 (4):693–695.

van Schaik, C. P., M. Ancrenaz, B. Gwendolyn, B. Galdikas, C. D. Knott, I. Singeton, A. Suzuki, S. S. Utami, and M. Merrill. 2003. "Orangutan cultures and the evolution of material culture." *Science* 299:102–105.

van Schaik, C. P., and J. M. Burkart. 2011. "Social learning and evolution: the cultural intelligence hypothesis." *Philosophical Transactions of the Royal Society B: Biological Sciences* 366 (1567):1008–1016. doi:10.1098/rstb.2010.0304.

van Schaik, C. P., K. Isler, and J. M. Burkart. 2012. "Explaining brain size variation: from social to cultural brain." *Trends in Cognitive Sciences* 16 (5):277–284.

van Schaik, C. P., and G. R. Pradhan. 2003. "A model for tool-use traditions in primates: implications for the coevolution of culture and cognition." *Journal of Human Evolution* 44:645–664.

VanderBorght, M., and V. K. Jaswal. 2009. "Who knows best? Preschoolers sometimes prefer child informants over adult informants." *Infant and Child Development* 18 (1):61–71.

van't Wout, M., R. S. Kahn, A. G. Sanfey, and A. Aleman. 2006. "Affective state and decision-making in the ultimatum game." *Experimental Brain Research* 169 (4): 564–568.

Ventura, P., T. Fernandes, L. Cohen, J. Morais, R. Kolinsky, and S. Dehaene. 2013. "Literacy acquisition reduces the influence of automatic holistic processing of faces and houses." *Neuroscience Letters* 554:105–109.

Vitousek, P. M., H. A. Mooney, J. Lubchenco, and J. M. Melillo. 1997. "Human domination of Earth's ecosystems." *Science* 277 (5325):494–499. doi:10.1126/science .277.5325.494.

von Rueden, C., M. Gurven, and H. Kaplan. 2008. "The multiple dimensions of male social status in an Amazonian society." *Evolution and Human Behavior* 29 (6): 402–415.

von Rueden, C., M. Gurven, and H. Kaplan. 2011. "Why do men seek status? Fitness payoffs to dominance and prestige." *Proceedings of the Royal Society B: Biological Sciences* 278 (1715):2223–2232.

Vonk, J., S. F. Brosnan, J. B. Silk, J. Henrich, A. S. Richardson, S. P. Lambeth, S. J. Schapiro, and D. J. Povinelli. 2008. "Chimpanzees do not take advantage of very low cost opportunities to deliver food to unrelated group members." *Animal Behaviour* 75:1757–1770.

Voors, M. J., E. E. M. Nillesen, P. Verwimp, E. H. Bulte, R. Lensink, and D. P. Van

REFERENCES

Soest. 2012. "Violent conflict and behavior: A field experiment in Burundi." *American Economic Review* 102 (2):941–964.

Wade, N. 2009. *The Faith Instinct: How Religion Evolved and Why It Endures.* New York: Penguin Press.

———. 2014. *A Troublesome Inheritance: Genes, Race, and Human History.* New York: Penguin Press.

Wadley, L. 2010. "Compound-adhesive manufacture as a behavioral proxy for complex cognition in the Middle Stone Age." *Current Anthropology* 51:S111-S119.

Wadley, L., T. Hodgskiss, and M. Grant. 2009. "Implications for complex cognition from the hafting of tools with compound adhesives in the Middle Stone Age, South Africa." *Proceedings of the National Academy of Sciences, USA* 106 (24): 9590–9594.

Wakano, J. Y., and K. Aoki. 2006. "A mixed strategy model for the emergence and intensification of social learning in a periodically changing natural environment." *Theoretical Population Biology* 70 (4):486–497.

Wakano, J. Y., K. Aoki, and M. W. Feldman. 2004. "Evolution of social learning: A mathematical analysis." *Theoretical Population Biology* 66 (3):249–258.

Walden, T. A., and G. Kim. 2005. "Infants' social looking toward mothers and strangers." *International Journal of Behavioral Development* 29 (5):356–360.

Walker, R. S., M. V. Flinn, and K. R. Hill. 2010. "Evolutionary history of partible paternity in lowland South America." *Proceedings of the National Academy of Sciences, USA* 107 (45):19195–19200. doi:10.1073/pnas.1002598107.

Ward, C. V., M. W. Tocheri, J. M. Plavcan, F. H. Brown, and F. K. Manthi. 2013. "Earliest evidence of distinctive modern human-like hand morphology from West Turkana, Kenya." *American Journal of Physical Anthropology* 150:284–284.

———. 2014. "Early Pleistocene third metacarpal from Kenya and the evolution of modern human-like hand morphology." *Proceedings of the National Academy of Sciences, USA* 111 (1):121–124.

Wasserman, I. M., S. Stack, and J. L. Reeves. 1994. "Suicide and the media: *The New York Times*'s presentation of front-page suicide stories between 1910 and 1920." *Journal of Communication* 44 (2):64–83.

Watts, D. J. 2011. *Everything Is Obvious*: How Common Sense Fails Us*. New York: Crown Business.

Webb, W. P. 1959. *The Great Plains.* Waltham, MA: Blaisdell.

Webster, M. A., and P. Kay. 2005. "Variations in color naming within and across populations." *Behavioral and Brain Sciences* 28 (4):512–513.

Wertz, A. E., and K. Wynn. 2014a. "Selective social learning of plant edibility in 6- and 18-month-old infants." *Psychological Science* 25 (4):874–882.

———. 2014b. "Thyme to touch: Infants possess strategies that protect them from dangers posed by plants." *Cognition* 130 (1):44–49.

White, T. D., B. Asfaw, Y. Beyene, Y. Haile-Selassie, C. O. Lovejoy, G. Suwa, and G. WoldeGabriel. 2009. "*Ardipithecus ramidus* and the paleobiology of early hominids." *Science* 326 (5949):75–86.

Whitehouse, H. 1996. "Rites of terror: Emotion, metaphor and memory in Melanesian initiation cults." *Journal of the Royal Anthropological Institute* 2 (4):703–715.

———. 2004. *Modes of Religiosity: A Cognitive Theory of Religious Transmission.* Lanham, MD: Altamira Press.

REFERENCES

Whitehouse, H., and J. A. Lanman. 2014. "The ties that bind us: Ritual, fusion, and identification." *Current Anthropology* 55 (6):674–695. doi:10.1086/678698.

Whitehouse, H., B. McQuinn, M. Buhrmester, and W. B. Swann. 2014. "Brothers in arms: Libyan revolutionaries bond like family." *Proceedings of the National Academy of Sciences, USA* 111 (50):17783–17785. doi:10.1073/pnas.1416284111.

Whiten, A., and C. P. van Schaik. 2007. "The evolution of animal 'cultures' and social intelligence." *Philosophical Transactions of the Royal Society B: Biological Sciences* 362 (1480):603–620.

Whiting, M. G. 1963. "Toxicity of cycads." *Economic Botany* 17:271–302.

Wichmann, S., T. Rama, and E. W. Holman. 2011. "Phonological diversity, word length, and population sizes across languages: The ASJP evidence." *Linguistic Typology* 15:177–197.

Wiessner, P. 1982. "Risk, reciprocity and social influences on !Kung San economics." In *Politics and History in Band Societies*, edited by E. Leacock and R. B. Lee, 61–84. New York: Cambridge University Press.

———. 1998. "On network analysis: The potential for understanding (and misunderstanding) !Kung Hxaro." *Current Anthropology* 39 (4):514–517.

———. 2002. "Hunting, healing, and hxaro exchange—A long-term perspective on !Kung (Ju/'hoansi) large-game hunting." *Evolution and Human Behavior* 23 (6): 407–436.

———. 2005. "Norm enforcement among the Ju/'hoansi Bushmen—A case of strong reciprocity?" *Human Nature* 16 (2):115–145.

Wiessner, P., and A. Tumu. 1998. *Historical Vines*. Smithsonian Series in Ethnographic Inquiry, edited by William Merrill and Ivan Karp. Washington DC: Smithsonian Institution Press.

Wiley, A. S. 2004. "'Drink milk for fitness': The cultural politics of human biological variation and milk consumption in the United States." *American Anthropologist* 106 (3):506–517.

Wilkins, J., and M. Chazan. 2012. "Blade production similar to 500 thousand years ago at Kathu Pan 1, South Africa: Support for a multiple origins hypothesis for early Middle Pleistocene blade technologies." *Journal of Archaeological Science* 39 (6):1883–1900.

Wilkins, J., B. J. Schoville, K. S. Brown, and M. Chazan. 2012. "Evidence for early hafted hunting technology." *Science* 338 (6109):942–946.

Willard, A. K., J. Henrich, and A. Norenzayan. n.d. "The role of memory, belief, and familiarity in the transmission of counterintuitive content." Unpublished manuscript. http://coevolution.psych.ubc.ca/pdfs/AKWillard_CogSci_Total.pdf.

Williams, T. I. 1987. *The History of Invention*. New York: Facts on File.

Wills, W. J., W. Wills, and G. Farmer. 1863. *A Successful Exploration through the Interior of Australia, from Melbourne to the Gulf of Carpentaria*. London: R. Bentley.

Wilson, D. S. 2005. "Evolution for everyone: How to increase acceptance of, interest in, and knowledge about evolution." *Plos Biology* 3 (12):2058–2065.

Wilson, D. S., and E. O. Wilson. 2007. "Rethinking the theoretical foundation of sociobiology." *Quarterly Review of Biology* 82 (4):327–348.

Wilson, E. O. 2012. *The Social Conquest of Earth*. New York: Liveright.

Wilson, M. L., C. Boesch, T. Furuichi, I. C. Gilby, C. Hashimoto, G. Hohmann, N. Itoh, et al. 2012. "Rates of lethal aggression in chimpanzees depend on the number of adult males rather than measures of human disturbance." *American Journal of Physical Anthropology* 147:305–305.

Wilson, M. L., and R. W. Wrangham. 2003. "Intergroup relations in chimpanzees." *Annual Review of Anthropology* 32:363–392.

Wilson, W., and D. L. Dufour. 2002. "Why 'bitter' cassava? Productivity of 'bitter' and 'sweet' cassava in a Tukanoan Indian settlement in the northwestern Amazon." *Economic Botany* 56 (1):49–57.

Wiltermuth, S. S., and C. Heath. 2009. "Synchrony and cooperation." *Psychological Science* 20 (1):1–5.

Wolf, A. P. 1995. *Sexual Attraction and Childhood Association: A Chinese Brief for Edward Westermarck*. Stanford, CA: Stanford University Press.

Wolf, T. M. 1973. "Effects of live modeled sex-inappropriate play behavior in a naturalistic setting." *Developmental Psychology* 9 (1):120–123.

———. 1975. "Influence of age and sex of model on sex-inappropriate play." *Psychological Reports* 36 (1):99–105.

Wolff, P., D. L. Medin, and C. Pankratz. 1999. "Evolution and devolution of folkbiological knowledge." *Cognition* 73 (2):177–204.

Woodburn, J. 1982. "Egalitarian societies." *Man* 17 (3):431–451.

———. "'Sharing is not a form of exchange': An analysis of property-sharing in immediate-return hunter-gatherer societies." In *Property Relations*, edited by C. M. Hann, 48–237. Cambridge: Cambridge University Press.

Woodman, D. C. 1991. *Unravelling the Franklin Mystery Inuit Testimony*. Montreal: McGill-Queen's University Press.

Woollett, K., and E. A. Maguire. 2009. "Navigational expertise may compromise anterograde associative memory." *Neuropsychologia* 47 (4):1088–1095.

———. 2011. "Acquiring 'the knowledge' of London's layout drives structural brain changes." *Current Biology* 21 (24):2109–2114.

Woollett, K., H. J. Spiers, and E. A. Maguire. 2009. "Talent in the taxi: A model system for exploring expertise." *Philosophical Transactions of the Royal Society B: Biological Sciences* 364 (1522):1407–1416.

Woolley, A. W., C. F. Chabris, A. Pentland, N. Hashmi, and T. W. Malone. 2010. "Evidence for a collective intelligence factor in the performance of human groups." *Science* 330 (6004):686–688.

World Bank Group. 2015. *Mind, Society and Behavior*. World Development Report 2015. Washington DC: World Bank.

Wrangham, R. W. 2009. *Catching Fire: How Cooking Made Us Human*. New York: Basic Books.

Wrangham, R., and R. Carmody. 2010. "Human adaptation to the control of fire." *Evolutionary Anthropology* 19 (5):187–199.

Wrangham, R., and N. Conklin-Brittain. 2003. "Cooking as a biological trait." *Comparative Biochemistry and Physiology A: Molecular & Integrative Physiology* 136 (1):35–46.

Wrangham, R., Z. Machanda, and R. McCarthy. 2005. "Cooking, time-budgets, and the sexual division of labor." *American Journal of Physical Anthropology*:226–227.

Wrangham, R. W., and L. Glowacki. 2012. "Intergroup aggression in chimpanzees and war in nomadic hunter-gatherers evaluating the chimpanzee model." *Human Nature* 23 (1):5–29.

REFERENCES

Wray, A., and G. W. Grace. 2007. "The consequences of talking to strangers: Evolutionary corollaries of socio-cultural influences on linguistic form." *Lingua* 117 (3):543–578.

Xu, J., M. Dowman, and T. L. Griffiths. 2013. "Cultural transmission results in convergence towards colour term universals." *Proceedings of the Royal Society B: Biological Sciences* 280 (1758).

Xygalatas, D., P. Mitkidis, R. Fischer, P. Reddish, J. Skewes, A. W. Geertz, A. Roepstorff, and J. Bulbulia. 2013. "Extreme rituals promote prosociality." *Psychological Science* 24 (8):1602–1605.

Yellen, J. E., A. S. Brooks, E. Cornelissen, M. J. Mehlman, and K. Stewart. 1995. "A Middle Stone Age worked bone industry from Katanda, Upper Semliki Valley, Zaire." *Science* 268:553–556.

Young, D., and R. L. Bettinger. 1992. "The Numic spread: A computer simulation." *American Antiquity* 57 (1):85–99.

Zahn, R., R. de Oliveira-Souza, I. Bramati, G. Garrido, and J. Moll. 2009. "Subgenual cingulate activity reflects individual differences in empathic concern." *Neuroscience Letters* 457 (2):107–110.

Zaki, J., and J. P. Mitchell. Forthcoming. "Intuitive prosociality." *Current Directions in Psychological Science*.

Zaki, J., J. Schirmer, and J. P. Mitchell. 2011. "Social influence modulates the neural computation of value." *Psychological Science* 22 (7):894–900.

Zesch, S. 2004. *The Captured: A True Story of Indian Abduction on the Texas Frontier.* New York: St. Martin's Press.

Zink, K. D., D. E. Lieberman, and P. W. Lucas. 2014. "Food material properties and early hominin processing techniques." *Journal of Human Evolution* 77 (0):155–166. http://dx.doi.org/10.1016/j.jhevol.2014.06.012.

Zmyj, N., D. Buttelmann, M. Carpenter, and M. M. Daum. 2010. "The reliability of a model influences 14-month-olds' imitation." *Journal of Experimental Child Psychology* 106 (4):208–220. http://dx.doi.org/10.1016/j.jecp.2010.03.002.

ILLUSTRATION CREDITS

Figure 2.1. Source: Museum Victoria.

Figure 2.2. Drawn from data in Herrmann et al., 2007.

Figure 2.3. Reprinted from *Current Biology*, 17/23. Sana Inoue and Tetsuro Matsuzawa. "Working memory of numerals in chimpanzees." R10004-R1005. Copyright 2007, with permission from Elsevier.

Figure 2.4. Redrawn from "Chimpanzee choice rates in competitive games match equilibrium game theory predictions." Christopher Flynn Martin, Rahul Bhui, Peter Bossaerts, Tetsuro Matsuzawa, Colin Camerer. *Scienctific Reports*, Jun, 2014. Published by Nature Publishing Group.

Figure 3.1. Museum of Cultural History, University of Oslo / Photographers: Anette Slettnes and Nina Wallin Hansen.

Figure 3.2. Painting by Scott Melbourne. In William John Wills, *Successful Exploration through the Interior of Australia from Melbourne to the Gulf of Carpentaria. From the Journals and Letters of William John Wills*. Edited by his father, William Wills. London: Richard Bently, 1863.

Figure 6.1. Reprinted from *The American Journal of Human Genetics*, 84.1, Borinskaya et al., "Distribution of the Alcohol Dehydrogenase ADH18*47His Allele in Eurasia." 89–92. Copyright © 2009 The American Society of Human Genetics, with permission from Elsevier Ltd. and the American Society of Human Genetics.

Figure 6.2. Gerbault et al., "Evolution of lactase persistence: An example of human niche construction." *Philosophical Transactions B*, 2011, 366, 1566, by permission of the Royal Society.

Figure 7.1. Redrawn from Dufour, 1994.

Figure 8.1. Copyright 2004 Studio Southwest, Bob Willingham.

Figure 9.1. Redrawn from Hill et al., 2011.

Figure 10.1. Redrawn from Evans, 2005.

Figure 11.1. "Young Children Enforce Social Norms." Marco F. H. Schmidt, Michael Tomasello, *Current Directions in Psychological Science* (21:4). Copyright © 2012 by the Association for Psychological Science. Reprinted by Permission of SAGE Publications.

Figure 11.2. Redrawn from Rand, Greene, and Nowak, 2012.

Figure 11.3. Adapted by permission from Macmillan Publishers, Ltd.: *Nature* ("Spontaneous giving and calculated greed." David G. Rand, Joshua D. Greene, Martin A. Nowak. September 19, 2012.). Copyright 2012.

Figure 11.4. Paul Kane. *Caw-Wacham*. About 1848, oil on canvas. The Montreal Museum of Fine Arts, purchase, William Gilman Cheney, bequest. Photo: The Montreal Museum of Fine Arts, Christine Guest.

Figure 12.2. Adapted from Kline 2010, #9767.

Figure 12.3. Adapted from Muthukrishna et al, 2014.

Figure 13.1. William Tomkins. *Indian Sign Language*. Dover Publications, 1931. Reprinted with kind permission of Dover Publications.

Figure 14.1. Sylvanus Griswold Morley. "An Introduction to the Study of the Maya Hieroglyphs." *Smithsonian Institution Bureau of American Ethnology, Bulletin 57.* Washington DC: Washington Government Printing Office, 1915.

Figure 14.3. Adapted from Hedden et al, 2008.

Figure 14.4. Redrawn from Phillips, Ruth, and Wagner, 1993.

Figure 15.1. Reproduced by kind permission of the KGA Research Project. Originally published in *PNAS*, 2013, Beyene et al. Figure S1.

INDEX

abacus, 7, 230, 267; mental, 278, 321
Abelam group, 175, 355n.17
aboriginal American societies, 4, 31
aboriginal Australian societies, 28; cultural adaptation of, 28–30; sign languages of, 236–37; tool kits of, 220–
   21. *See also* Tasmanian
   hunter-gatherers
accumulated knowledge: in aging killer
   whales, 135–36; in elephant matriarchs, 136–37; longevity and,
   133–35
Aché bands: communal rituals among,
   159–62; different relationships in,
   155; meat sharing in, 156–59; multiple fathers in, 151; social life features in, 162–64; sociality and cooperation among, 155–64; weak
   marriage norms of, 149–50. *See also*
   Yandruwandrha
adaptive cultural information, 3–5; of
   high-prestige individuals, 42–43;
   loss of with shrinking population,
   218–24
ADH genes, alcohol-inhibiting variant,
   86–88
ADH1B gene: alcohol-inhibiting properties of, 86–88; global distribution
   of, 87
adolescence, emergence of, 63–64
affinal ties, 145, 147, 156, 159, 215, 224;

ceremonial gatherings and, 178; in
   hunter-gatherer bands, 163
Africa, human expansion out of,
   183–84
age, political leadership and, 350n.30
age bias, 46–48
age cues, children's use of, 338n.22
age-related wisdom, 131–37; genetic
   consequences and implications of,
   60
age status, 118; in small-scale societies,
   338n.23
agriculture: in culture-gene coevolution, 92–93; development of, 84–85;
   fermented beverages and, 88; rice,
   86–88, 106
alloparenting, 307–11, 313, 371n.15
altruism, 359n.19; kin-based, 205–7;
   prestige and, 128–31. *See also* generosity; reciprocity
Amundsen, Roald, 26
AMY1 gene, in chimpanzees and humans, 93
Andaman Islanders, 118
animals: cognitive system for learning
   about, 78–81, 341–42n.33; domestication of, 90–91, 92
ants, specialized speciation of, 10
apes: pair-bonding in, 304–6, 351n.11;
   social learning in, 14–15. *See also*
   chimpanzees

431